GENETICALLY
ENGINEERED
VACCINES

ADVANCES IN EXPERIMENTAL MEDICINE AND BIOLOGY

Recent Volumes in this Series

GENETICALLY ENGINEERED VACCINES

Edited by

Joseph E. Ciardi

National Institute of Dental Research
Bethesda, Maryland

Jerry R. McGhee

University of Alabama at Birmingham
Birmingham, Alabama

and

Jerry M. Keith

National Institute of Dental Research
Bethesda, Maryland

SPRINGER SCIENCE+BUSINESS MEDIA, LLC

Library of Congress Cataloging-in-Publication Data

Genetically engineered vaccines / edited by Joseph E. Ciardi, Jerry R.
  McGhee, and Jerry M. Keith.
      p.   cm. -- (Advances in experimental medicine and biology ; v.
  327)
      "Proceedings of a workshop sponsored by the National Institute of
  Dental Research ..., held November 6-8, 1991, in Bethesda,
  Maryland"--T.p. verso.
      Includes bibliographical references and indexes.
      ISBN 978-1-4613-6507-5   ISBN 978-1-4615-3410-5 (eBook)
      DOI 10.1007/978-1-4615-3410-5
      1. Synthetic vaccines--Congresses. 2. Genetic engineering-
  -Congresses. 3. Recombinant microorganisms--Congresses. 4. Mouth-
  -Diseases--Vaccination--Congresses.   I. Ciardi, J. E. II. McGhee,
  Jerry R. III. Keith, Jerry M. IV. National Institute of Dental
  Research (U.S.) V. Series.
      [DNLM: 1. Biological Availability--congresses. 2. Immunity-
  -physiology--congresses. 3. Mouth Diseases--prevention & control-
  -congresses. 4. Mucuous Membrane--immunology--congresses.
  5. Recombination, Genetic--congresses. 6. Vaccines--congresses.
  W1 AD559 v.327 / WU 166 G328 1991]
  QR189.2.G43 1992
  615'.372--dc20
  DNLM/DLC
  for Library of Congress   92-48784
                                                                 CIP

Proceedings of a workshop sponsored by the National Institute of
Dental Research on Genetically Engineered Vaccines: Prospects for
Oral Disease Prevention, held November 6-8, 1991, in Bethesda, Maryland

ISBN 978-1-4613-6507-5

Published 1992 by Springer Science+Business Media New York
Originally published by Plenum Press in 1992
Softcover reprint of the hardcover 1st edition 1992

PREFACE

The National Institute of Dental Research sponsored a workshop on "Genetically Engineered Vaccines: Prospects for Oral Disease Prevention," held at the National Institutes of Health (NIH) on November 6-8, 1991. The purpose of the workshop was to convene molecular biologists and immunologists to address the state of the science in vaccine development and to explore the potential of developing vaccines for prevention of oral diseases. The goal was to elicit new research initiatives and recommendations for vaccine development with emphasis on the prevention of oral diseases and diseases affecting the orofacial tissues.

The workshop was attended by more than 100 persons who heard 30 presentations, and the speakers provided the papers for this volume. The workshop focused on the following topics: oral diseases and host immune responses, update on vaccines and vaccine development, vaccines and the mucosal immune system, optimizing mucosal and systemic immune responses, delivery systems and immune analysis, target antigen selection and vaccine development, immunological correlates of protection and future directions/recommendations. Three key areas were identified: Optimizing the Mucosal Immune Response, Antigen Delivery Systems, and Target Antigens and Immunological Correlates of Protection. The summary and recommendations from these deliberations is included at the end of this volume.

The workshop was dynamic and stimulating, yielding an excellent overview of the field of vaccine development and the prospects for developing vaccines for preventing oral diseases. The participants, who included dental and nondental scientists and who represented two different fields of research (molecular biology and immunology), were pleased to have the opportunity to exchange research ideas and results. The excitement of their discussions conveyed the energy driving research today as new opportunities and challenges combine to broaden the scope of possibilities for preventing disease. It is now conceivable that techniques and methodologies being developed in other areas of vaccine research can be applied to oral diseases as well, and that this research will have large benefit for many populations worldwide. At the same time, vaccine research on oral disease(s) promises to shed light on common body processes and immune response.

Many contributed to the success of this workshop. We express our sincere gratitude and thanks to: Douglas Brown, Claudia Gentry-Weeks, Matthew Kinnard, David Klein, Francis Macrina, Frank Robey, and James Rooney who all served with us on the planning committee; Linda Richardson for expert editorial assistance; and Debra Clisby, Michael Simmons, Sheila Witherspoon, and Pennie Yates for invaluable help in the preparation of manuscripts.

Joseph E. Ciardi
Jerry R. McGhee
Jerry M. Keith

CONTENTS

INTRODUCTORY ADDRESS

KEYNOTE PRESENTATION

SESSION I: ORAL DISEASES AND HOST IMMUNE RESPONSES

SESSION II: UPDATE ON VACCINES AND VACCINE DEVELOPMENT

SESSION V: DELIVERY SYSTEMS AND IMMUNE ANALYSIS

A. PASSIVE IMMUNITY

B. ACTIVE IMMUNITY

SESSION VI: TARGET ANTIGEN SELECTION AND VACCINE DEVELOPMENT

THE NEW BIOLOGY AND VACCINE RESEARCH

Harald Löe

National Institute of Dental Research
National Institutes of Health
Bethesda, MD 20892

I am here to welcome you to the workshop on <u>Genetically Engineered Vaccines:</u> <u>Prospects for Oral Disease Prevention</u>, which I expect will be one of the most exciting and promising workshops to be held at the NIH. Most people--lay and learned--seem to agree that the concept of a vaccine has a fundamental appeal: here is a way of building the body's own defenses against disease-causing microorganisms so that if they invade, they can do no harm. It is an ingenious concept; indeed, it was the work of genius.

The genius, of course, was Louis Pasteur. And the critical experiment took place exactly 110 years ago at Pouille-le-Fort in France on May 5, May 31, and June 2, 1881. Those were the dates on which Pasteur first vaccinated 24 sheep against anthrax and left 24 others untreated, then inoculated all 48 animals with a lethal dose of virulent anthrax bacilli, and, finally, displayed 24 dead or dying unvaccinated sheep on the grass while, nearby, the 24 vaccinated sheep grazed in perfect health.

Microorganisms and their role in disease had been discovered. In the context of all of human history, Pasteur's finding was a brilliant flash in the darkness of ignorance and helplessness. It also marked the advent of a scientific approach to the treatment of human disease.

In the decades that followed, other great scientists would make other brilliant discoveries: Koch with the tubercle bacillus; von Behring with tetanus and diphtheria antitoxins; Beijerinck with the discovery of a new type of organism--the mosaic virus of plants. And, as research in bacteriology and virology grew, so did the study of the host response: the new science of immunology was born, pioneered by Ehrlich, Metchnikoff, and others.

Finally, lest we forget, there was the dental scientist W. D. Miller who worked in Berlin next door to the laboratory of the great Herr Professor Koch himself. It was there that Miller developed the bacterio-chemical theory of dental caries. However, he died too soon to see his work vindicated.

All that history is barely a century old--a century which for long has been celebrated for its conquest of infectious diseases, either through vaccines or drugs.

However, the events of the past decade have shattered our complacency. AIDS happened; also, we are experiencing an increase in tuberculosis; in measles; we know that

Genetically Engineered Vaccines, Edited by
J.E. Ciardi *et al.*, Plenum Press, New York, 1992

immunization programs are not reaching all children; and some vaccines--for pertussis, for influenza--have gotten a bad name or a bad press.

In the developing world, there never was cause for complacency. Not only are measles, polio, and cholera commonplace, but diseases like malaria are on the rise again.

Meanwhile, the two most costly and endemic oral infectious diseases--caries and periodontal diseases--have yet to capitalize on a vaccine approach.

A CHALLENGING AGENDA

So the agenda for vaccine research and development continues to be long and challenging. Fortunately, what makes this workshop different from all others in the past is the New Biology. Today, not only can we generate candidate antigens of great purity, we can combine them with adjuvants and insert them into carefully crafted vectors for use as oral vaccines. All this would not be possible if at the same time we were not seeing great progress in understanding the complexities of the immune system, and of mucosal immunity in particular.

In the course of the next couple of days we will have the opportunity to hear from world experts in these many fields--individuals who are presently designing and testing candidate vaccines for caries and periodontal diseases, as well as for oral herpes and other oral viral and bacterial diseases, HIV, rotavirus, and pertussis. Others will explore the host response, the use of monoclonal antibodies, and other aspects of immune defense.

I believe that this workshop and its program illustrate how far oral health research has come into the forefront of biomedical research. We have gone beyond an exclusive focus on diseases of teeth and gums to a broadened program of research that explores the whole range of conditions affecting the face, the mouth, and jaws and their interactions with other body systems.

It follows that whatever is said with regard to oral disease, oral vaccines, and oral immunity has implications for disease elsewhere in the body as well. Indeed, as the prospect of developing vaccines for oral diseases grows closer to reality, we may very well want to add other immunogens to produce polyvalent vaccines that would provide protection against a number of oral and systemic diseases. The concept--like the original concept of a vaccine itself--is appealing and exciting.

As I said at the outset, I am delighted to welcome you on behalf of NIDR; I am pleased that this workshop came about and I want publicly to thank Joseph Ciardi and his committee for their work in planning the meeting and assembling such a distinguished group of speakers and moderators. I thank you all for coming and I wish you well in your deliberations. I shall look forward to hearing your conclusions and recommendations, which, I can assure you, will be most seriously considered as we refine our research plans for the coming years.

MUCOSAL IMMUNITY TO VACCINES: CURRENT CONCEPTS FOR VACCINE DEVELOPMENT AND IMMUNE RESPONSE ANALYSIS

Jerry R. McGhee and Hiroshi Kiyono

The Immunobiology Vaccine Center, The Mucosal Immunization Research Group, Departments of Microbiology and Oral Biology, The University of Alabama at Birmingham, Medical Center, Birmingham, AL 35294

INTRODUCTION

Considerable experimental evidence supports the existence of a separate mucosal immune system. Further, the importance of mucosal immunity has recently been given more emphasis in the areas of immunology, infectious diseases and vaccine development, since it is realized that most infectious diseases are initiated through invasion of mucosal tissues. However, mucosal immunology has often been ignored because of the difficulties in the isolation and characterization of lymphoid cells and measurement of secretory antibodies from various mucosa-associated tissues. In order to solve the mystery of the mucosal immune system, a large number of mucosal immunology laboratories have undertaken the challenge of this important area and have applied concepts of modern immunology and molecular biology to this important area of science. In this review, we have briefly summarized some of the current dogma of mucosal immunology which should be considered for the development of mucosal vaccines.

WHY IS THE MUCOSAL IMMUNE SYSTEM IMPORTANT FOR THE DEVELOPMENT OF EFFECTIVE VACCINES?

The seeming complexity of the mucosal immune system, which spans a large surface area (over 400 m^2) and which consists of sites where antigens are encountered, processed and which initially trigger B and T cells, and separate areas where immune cells actually function, has made it difficult for assessment of vaccines. Thus, all vaccines except one (Sabin OPV) are today administered by systemic routes, and these vaccines result in effective cell-mediated and antibody responses in systemic tissues, but are essentially ineffective for induction of mucosal immunity in humans who have not previously suffered a mucosal infection by the causative organism. Furthermore, attempts to induce mucosal immunity by vaccination through thought to be appropriate routes, e.g., oral immunization, are sometimes successful but often are not. If would therefore be useful to attempt to simplify our approaches to mucosal immunization and place emphasis on some of the relevant lymphoid cells and molecules involved in mucosal immunity.

Estimates of immunoglobulin (Ig) isotype synthesis have clearly shown that IgA predominates in humans, and represents greater than 60% of all antibody isotypes produced.[1,2] Most IgA produced in mammals is derived from plasma cells in mucosal effector sites, for example lamina propria (LP) regions of the gastrointestinal (GI) and upper respiratory tracts (URT) and in ascinar regions of glands.[3] Approximately 80% of

Genetically Engineered Vaccines, Edited by
J.E. Ciardi *et al.*, Plenum Press, New York, 1992

mucosal (S-IgA) antibodies are derived from the GI tract (Fig. 1). This amount of S-IgA is required, since if one could flatten out the entire surface area of the upper and lower intestine, it would cover greater than 80 percent of a basketball court (Fig. 1)! This clearly suggests that a large commitment has been made by the host for T and B cell immunity in the intestine. Thus, many studies in humans and especially in mice have focused on the GI tract, and much of our understanding of the mucosal immune system has been derived from these studies. Further, it has been shown that oral immunization induces antigen-specific S-IgA responses in the oral cavity as well.[4] It would be well advised to ask if this has relevance to other sites commonly infected by pathogens and thus targets of vaccines, e.g., the URT and lower lungs and the genitourinary tract. The answer at this point would be that it may be an over-simplification to assume that GI tract (oral) immunization will provide the protection required for adequate immunity in the lungs and reproductive tract (see below). In addition, it is also important to consider other routes of mucosal immunization (e.g., vaccine targeting to the nasal-associated lymphoreticular tissues) in addition to oral immunization for the induction of mucosal immunity in the URT and the oral cavity. Although most investigators now agree that the concept of the common mucosal immune system (CMIS) is important for the induction of antigen-specific immune responses in different mucosa-associated tissues, we now must also realize that a better understanding of how the CMIS is compartmentalized will be essential in order for us to use the mucosal immune system for the development of effective vaccines.

EVIDENCE FOR COMPARTMENTS WITHIN THE COMMON MUCOSAL IMMUNE SYSTEM

The distribution of IgA1 and IgA2 plasma cells in various human mucosal effector sites provides additional clues that separate mucosal inductive sites, e.g., bronchus-

Predominance Of S-IgA Antibodies In Human Mucosa - Associated Tissues

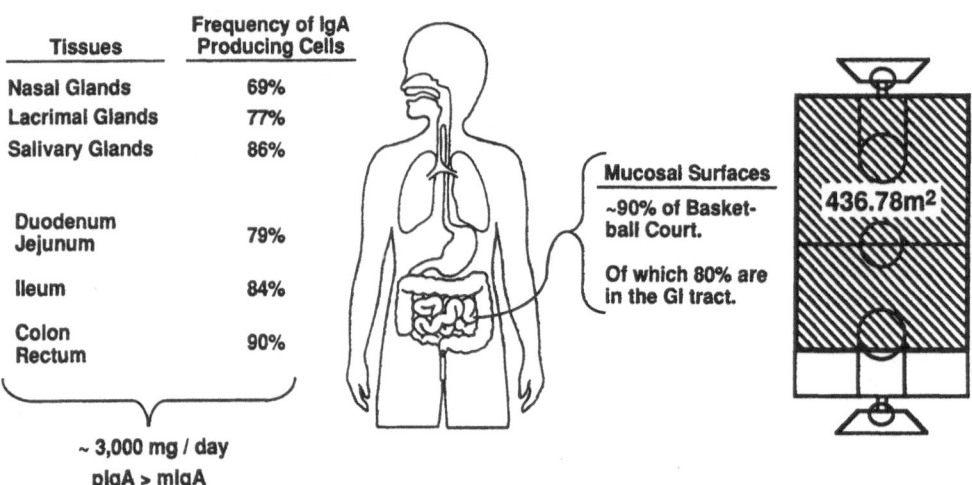

Figure 1. Relative frequencies of IgA producing plasma cells in various mucosal effector tissues. The majority of S-IgA antibodies are produced in the GI tract and the mucosal surfaces in this region would cover over 80% of a basketball court.

associated and nasal-associated versus gut-associated lymphoreticular tissues (BALT and NALT vs GALT) may differ in their role for provision of the necessary B cell precursors for S-IgA responses. For example, the nasal mucosa, lacrymal glands and lamina propria of the URT contain mainly IgA1 plasma cells; IgA2[+] cells are infrequent.[5] In contrast, higher numbers of IgA2 plasma cells are seen in the lamina propria of the lower intestinal tract. If one assumes that the tonsils are a major representation of NALT/BALT in humans, and similarly that the appendix is somewhat representative of GALT, then a pattern emerges. For example, pokeweed mitogen triggered B cell cultures derived from tonsils produce all major Ig isotypes, including IgA.[6] Interestingly, greater than 95% of the IgA produced is of the IgA1 subclass, supporting the notion that tonsils are a possible site for precursor IgA1[+] B cells.[7] Likewise, human appendix, which may be representative of GALT, contains both IgA1[+] and IgA2[+] B cells; however IL-6 induces much higher IgA2 responses in appendix B cell cultures.[7] This would suggest that GALT is a major source of IgA precursor B cells which populate the effector regions of the small and large intestine with some preference for the IgA2 subclass.

It was in former years assumed that BALT are the normal counterpart for GALT, e.g., Peyer's patches (PP) in the GI tract. BALT was first described by Bienenstock and colleagues[8-10] in rabbits and several other species where organized lymphoid tissue was present at the branch points of the airways, most frequently between the bronchus and the artery. The BALT in rabbits and rats are covered by a stratified columnar epithelium which contains specialized epithelial cells which resemble M cells, and these cells most likely function in antigen uptake in a manner analogous to the dome region of the PP (see below).[11] In fact, it can be stated that the most extensive studies of BALT have been done in rabbits and rats and this has been reviewed in detail.[12] Recent work by Pabst and colleagues[13] have failed to demonstrated organized BALT structures in normal human lungs. It should be emphasized that these studies were done on lung tissue of patients without evident respiratory diseases, and on subjects of various ages who would normally encounter the microbial flora of the upper respiratory tract. A recent viewpoint[14] was taken that the human lungs contain lymphoid cells in other compartments, e.g., the intravascular pool, the interstitium and bronchoalveolar spaces, and may represent the cells which respond to antigenic insults in the lung.

We would offer an alternate but attractive possibility for the existence and function of human BALT. In humans, the most organized lymphoid tissue in the upper respiratory tract are the follicles present in the nasal cavity. These large, anatomically distinct masses are termed Waldeyer's ring. The most prominent follicles are the palatine and pharyngeal tonsils. These tonsils are covered by a stratified squamous epithelium, but certain regions have columnar epithelial cells and some of these bear characteristics of M cells. Of more importance, a number of studies have shown that high frequencies of B cells expressing surface IgA occur in human tonsils. Further, the majority of these sIgA[+] B cells express sIgA1 and mitogen stimulated tonsillar cultures produce mainly IgA1 antibodies. Further, it was recently shown that IL-6 also induces activated tonsillar B cells to become IgA1 secreting cells (Fujihashi et.al., manuscript in preparation). It is tempting to speculate that these are the B cell precursors for nasal mucosa and lamina propria regions of the upper respiratory tract. If indeed the tonsils represent the BALT- equivalent in humans, then the central question becomes- what is the relative importance of tonsils in mucosal immunity? In other words, does the BALT compare, for example, with the Peyer's patches (GALT) in terms of major mucosal inductive sites? If they are of importance, then vaccine strategies should begin to consider antigen presentation and uptake in human tonsillar tissues.

A recent review of NALT in the rat has suggested that in fact NALT may represent the major mucosal tissue in that species.[15] Certainly NALT develops earlier than BALT and resembles the PP in this regard. The NALT are covered by an M-cell lymphoepithelium, and this would suggest that it be classified as a mucosal inductive site. Nevertheless, it is too early to suggest that the rat NALT are normal counterparts for human tonsils. For example, NALT is more enriched for T cells, especially CD4[+] T cells, and does not appear to contain significant numbers of surface IgA[+] B cells,[15] a situation which is markedly different from human tonsils. At this point it may suffice to suggest that rats contain both NALT and BALT, and the relative contribution of each to mucosal immune responses to inhaled antigens/nasally administered vaccines remains to be determined. Finally, NALT may be a better description for the human Waldeyer's ring (tonsils) than BALT. However, until more functional studies are done with NALT in rats, mice, rabbits, and other species, we should avoid direct comparisons with human tonsillar tissues, or NALT.

Recent studies in SCID mice given human periperhal blood lymphocytes suggest that intestinal LP regions are selectively repopulated with human IgA plasma blasts.[16] This should now provide a model to test the hypothesis that tonsillar tissues (NALT) preferentially contain more IgA1 precursors destined for mucosal tissues, especially in the URT. Likewise, the appendix may contain greater numbers of mucosa-seeking IgA2 precursors. Experiments along these lines are currently underway in our group to determine if SCID mice reconstituted with representative tissues of human NALT and GALT exhibit preferences for IgA subclass responses and possibly result in repopulation of SCID mouse mucosal effector tissues in patterns which mimic the natural human host.

ANTIGEN UPTAKE AND INDUCTION OF IgA RESPONSES

The mucosal immune system can be divided into sites where antigen is first encountered and initial responses are induced (inductive sites) and into the larger surface areas where IgA plasma cells are found and where the production of S-IgA antibodies result in local immune protection (effector sites). A number of studies in mice and in humans have now provided compelling evidence that the stimulation of IgA precursor B cells and helper T (Th) cells in PP with orally administered antigens/vaccines leads to the dissemination of antigen-specific B and T cells to mucosal effector tissues such as LP regions of the intestinal, respiratory and genitourinary tracts and various secretory glands for subsequent antigen-specific S-IgA antibody responses.[2-4]

It is important to distinguish antigen uptake in mucosal inductive sites, e.g., PP from IgA effector regions such as LP of the gut, lungs and genitourinary tract and exocrine glandular tissues. IgA inductive sites are equipped with specialized uptake cells, termed microfold or M cells.[17,18] The current view is that microorganisms and complex antigens are sampled and passed intact through endocytotic vesicles and delivered to underlying lymphoid cells for further antigen processing and presentation (Fig. 2). However, recent evidence in humans suggest that M cells can express MHC class II, and thus could serve as an epithelial type of antigen presenting cell (APC).[11] These functions are not mutually exclusive; for example certain viruses (reovirus types 1 and 3 and rotavirus) or bacteria (*E coli*-RDEC) bind to receptors on M cells and may pass through in nondegradative endosomal/lysosomal pathways.[20,21] Other macromolecules may enter endosomes which fuse with class II+ secondary lysosomes with subsequent processing of peptides. APCs provide CD4+ Th cells with the triggering signal via peptide-MHC class II and results in the initial activation steps and formation of memory cells. B cells in germinal centers also receive triggering signals by antigen and develop into IgA precursor stages and into memory cells. Immediate homing to mucosal effector sites precedes the second signals required for actual immune responses.

In mucosal effector tissues, antigen uptake and presentation also occurs; however, important differences are noted. For example, vaccine antigen may be endocytosed by epithelial cells, and in certain situations the epithelial cells themselves express class II MHC and can process the antigen with subsequent association of immunogenic peptides with MHC class II (Fig. 2). In other situations, intact proteins can traverse tight junctions, and in this instance intact vaccine antigen could trigger B and T cell responses. For example, sIgA+ B cells may bind antigen and through endocytic pathways process and present peptides, together with MHC class II, to Th cells. Macrophages in lamina propria regions also could serve this function for more complex antigens. The simplest scenario would be presentation by class II+ sIgA+ B cells to Th2 cells, and this would complete signal two allowing full activation of Th2 cells with IL-4, IL-5, IL-6 and IL-10 cytokine release (see below). sIgA+ B cells would receive signal two from activated Th2 cells and derived cytokines (IL-4, -5 and -6), for subsequent B cell proliferation and differentiation into IgA producing plasma cells with specificity for the vaccine antigen (Fig. 2).

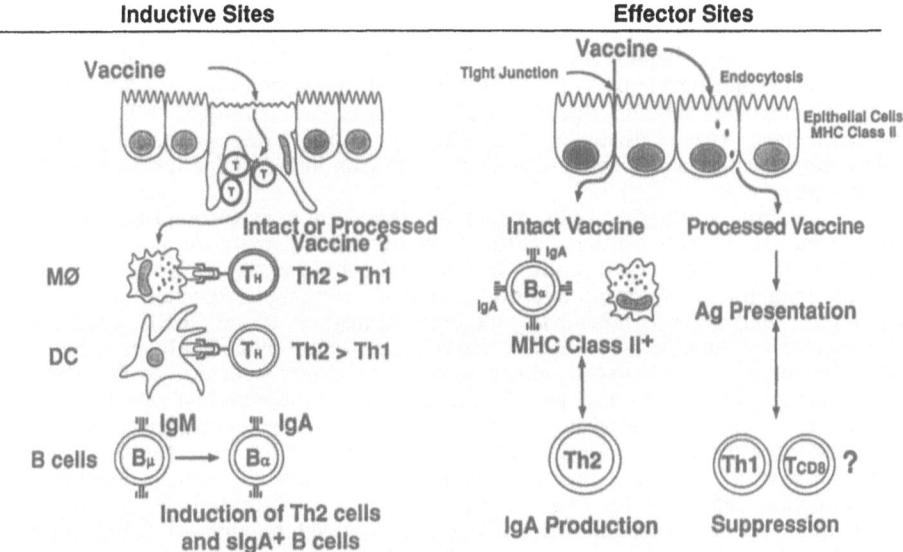

Figure 2. Major pathways for antigen/vaccine uptake in the mucosal immune system. In IgA inductive sites, M- cells in the epithelium endocytose/phagocytose intact antigens/vaccines and deliver them to underlying T and B cell areas. Antigens/vaccines are also taken up in mucosal effector sites and may be processed and presented by class II+ epithelial cells. Intact antigen may pass tight junctions and following processing by APCs may trigger Th2 cell regulated IgA responses.

MUCOSAL MEMORY: COMPARISON OF PROTEIN- AND CARBOHYDRATE-BASED VACCINES

The question of memory in the mucosal immune system has been the subject of debate for a long time, since many studies have shown that IgA responses are sometimes short-lived, and boosting may not always induce IgA responses faster, to higher levels or of increased avidity, all of which are properties of memory T and B cells involved in anamnestic responses. This has tended to confuse those who work with vaccines, since it is sometimes assumed that continual restimulation of mucosal surfaces will be required for protective S-IgA responses. This clearly may not be the case, since there is good evidence now for mucosal memory to certain proteins. However, the nature of the vaccine antigen, e.g., protein versus polysaccharide is a major determinant of memory B cell induction for anamnestic mucosal IgA responses.

The first encounter with an antigen leads to clonal expansion of antigen-specific B cells which develop into a germinal center. It is thought that memory B cells also arise in these germinal centers, including those present in the PP B cell zones. These germinal center B cells include memory cells and acquire unique properties including loss of sIgD and enhanced expression of Peanut Agglutinin (PNA) receptors (sIgM+, SIgD-, PNA^{Hi}). It is usually assumed that a single B cell can give rise to the progeny for a germinal center and hybridoma analyses have shown that an entire secondary antibody response is generated by a small number of clones (~20).[22] It is probably safe to conclude that this would also apply to anamnestic IgA responses in mucosal sites; however this assumption has not been experimentally addressed. During germinal center responses and memory B cell generation, switches to other isotypes occur and for IgA high frequencies of sIgA+ B cells occur in PP germinal centers.[23] Presumably, switches to IgA, e.g., μ→α or μ→γ→α occur in PP germinal centers; however direct proof of this is also lacking. It is known that PP germinal center B cells are sIgM+, PNA^{Hi} and Kappa^{lo}[24]; however significant numbers of sIgA[+], PNA[Hi] B cells are also found in PP germinal centers.[23,24] The preference for switching to IgA in GALT is thought to be intrinsic to the gut microenvironment.[25] The immediate fate of IgA memory cells and sIgA[+] B cells in the PP has been incompletely studied. It is known that Th cells and cytokines are required for germinal center formation[25] and for induction of memory B cells.[26-28] Further, CD4[+] Th cells are required for germinal center formation in PP.[29] It is plausible to suggest that memory B cells exit the germinal centers and pass directly into T cell areas enriched in CD4[+] Th cells.[28] It is known that memory cells do not begin to recirculate until 2-3 weeks after antigen priming, and memory B cells (and T cells) may therefore remain in IgA inductive sites such as the PP for long periods.

The above discussion only applies to protein antigens (protein-based vaccines) which result in Th cell-mediated antibody responses and memory B and T cells for secondary or anamnestic responses. In the classical sense, the first encounter with a systemic foreign protein (vaccine) results in antibody responses characterized by IgM and later IgG. This is also accompanied by induction of memory B and T cells, which respond to a second vaccination with higher affinity IgG antibodies in higher titers. This secondary response is due to the generation of B cell clones with surface Ig receptors of higher affinity, which arise through somatic mutations in the germinal centers. A similar scenario likely occurs following oral immunization with protein vaccines; however in these instances the primary and boosted response are both of the IgA isotype. The prediction from this would be that the first oral vaccine encounter induces low affinity S-IgA antibodies and following clonal expansion and somatic mutations, memory IgA B cells could respond to a second oral vaccine encounter with heightened, S-IgA antibodies of much higher affinity.

Studies in a number of laboratories have shown that IgA antibodies can be induced to polysaccharides. Early work indicated that IgA anti-polysaccharide responses had a requirement for T cells;[30] an effect which could be replaced by late acting T cell derived factor(s).[31] It is likely that T cell derived cytokines, e.g., IL-5 and/or IL-6, may serve as co-factors to facilitate B cell IgA responses. Two examples may be provided to show that IgA is readily induced to polysaccharides. Oral administration of pneumococcal polysaccharide vaccines result in IgA responses in external secretions and in serum.[32] Although few studies have been done in athymic, nude mice, evidence has been provided that these animals make IgA following immunization by systemic[33] or oral[34] routes to T-independent antigens. Perhaps the most definitive studies on the role of S-IgA anti-polysaccharide responses and protection have been carried out by Neutra and her colleagues.[35,36] In the initial work, mice were orally immunized with live cholera *Vibrios* and the PP B cells were used to generate IgA hybridomas with specificity for *V. cholerae* LPS (0 polysaccharide side chains).[35] When an IgA myeloma of this specificity was grown in the upper backs (piggy back model), it was shown that pIgA anti-LPS was transported into GI tract secretions, and more importantly the S-IgA antibodies were protective.[35] This work has more recently been applied to IgA anti-*Salmonella* LPS mAbs, and protection against an oral *Salmonella* challenge was shown.[36] It should be indicated that many studies have shown that IgA anti-LPS responses are induced to gram negative bacteria, and high levels of S-IgA (usually S-IgA2) occur in human external secretions.[37] One must conclude that IgA anti-bacterial polysaccharide responses can be (and often are) induced in the mucosal immune system.

This raises an important issue, e.g., antibody responses to polysaccharide antigens are short-lived and do not result in memory B cell responses. Thus, on the one hand, it is clear that S-IgA responses to capsular polysaccharide and to lipopolysaccharide are induced and are protective, at least at the level of monoclonal S-IgA responses.[29,30] However, it is equally clear that these S-IgA responses will quickly wane due to absence of memory B cells, and thus will require continued restimulation. This would explain why S-IgA anti-bacterial responses decline with time and in some instances do not result in protection when rechallenged. Thus, one explanation for the failure to show that mucosal IgA responses exhibit memory has been the choice of antigens used. In many of these studies, S-IgA responses to bacterial surface antigens, e.g., LPS and capsular polysaccharides, have been assessed and as indicated above, memory is not induced in these situations.

ORAL ADMINISTRATION OF VACCINES LEAD TO THE PREFERENTIAL INDUCTION OF TYPE 2 CD4+ T CELLS WHICH SUPPORT IgA RESPONSES

Functionally mature helper T (Th) cells express α/β T cell receptors (TCR) which recognize foreign peptides in association with MHC class II on APCs. These α/β TCR+, CD4+ Th cells can often be further subdivided into at least two subsets, Th1 and Th2, based upon unique profiles of cytokines produced and major functions in host immune responses.[38,39] For example, Th1 cells secrete IL-2, interferon gamma (IFN-γ) and tumor necrosis factor beta (TNF-β) and Th1 cells function in cell mediated immunity for protection against intracellular bacteria such as *Mycobacterium tuberculosis* and *Salmonella typhi*. Th1 cells may provide help for B cell responses and the IFN-γ produced supports IgG2a responses in mice.[38,39] The Th2 cells preferentially secrete IL-4, IL-5, IL-6 and IL-10 and provide effective help for B cell responses, most notably for IgG1 and IgG2b subclasses, and for IgA.[38,39] Other studies have also shown that T cells and certain cytokines, e.g., IL-5 and IL-6 are of particular importance for induction of committed sIgA+ B cells to differentiate into IgA producing plasma cells.[7,40] One would predict from this that higher frequencies of Th2 cells may occur in mucosal effector sites, and indeed this has been shown by our recent studies (see Chapter by Mega *et al*.) .[41,42]

In order to focus on antigen-specific Th cell responses, the sensitive cytokine-specific ELISPOT assay was employed for enumeration of increased numbers of Th1- and Th2-type cells present following antigen-specific priming of mucosal inductive sites by oral immunization.[43-45] Our initial studies revealed the surprising finding that oral immunization with a T cell-dependent antigen, sheep erythrocytes (SRBC) preferentially induced SRBC-specific Th2 type cells in PP, while systemic immunization (intraperitoneal; i.p.) resulted principally in Th1 type cell responses (See chapter by Xu-Amano *et al*.).[43] The results of our studies on oral immunization with SRBC have significant implications for design and delivery of oral vaccines. For example, oral vaccines which induce effective levels of mucosal S-IgA antibodies do so by induction of Th2 cells in GALT which then support mucosal antibody production. Likewise, vaccines which induce significant Th2 cell responses in PP will result in significant S-IgA responses in mucosal effector tissues, while vaccines which preferentially induce Th1 cell responses may not be as effective for provision of help for B cells undergoing IgA responses.

Therefore to test these assumptions, we have now used more relevant vaccine antigens, e.g., cholera toxin (CT), first as an oral immunogen and second as an mucosal adjuvant with tetanus toxoid (TT) to assess whether oral immunization induces Th2 cells that directly correlate with S-IgA responses in the GI tract of mice.[44,45] When CD4+ T cells from PP of mice orally immunized with CT were examined for antigen-specific Th1 and Th2 cell responses by *in vitro* CT-B stimulation, substantially higher numbers of IL-4 and IL-5 producing Th2 type cells occurred when compared with Th1 type cells secreting IFN-γ and IL-2.[44] Further, SP CD4+ Th cell cultures from mice orally immunized with CT also showed higher frequencies of Th2-type cells. These results suggest that CT, like other antigens such as SRBC, induce Th2 cell responses when given by the mucosal route, and may explain why significant IgA anti-CT responses occur following oral immunization.

It was of importance to determine if CT could enhance Th2 cell responses to other protein vaccines when given by the oral route. We next assessed PP CD4+ Th cells from mice orally immunized with TT and the mucosal adjuvant CT for frequencies of TT-specific Th1 and Th2 cell responses.[45] Clearly, higher numbers of TT-specific IL-5 producing Th2 type cells were induced in PP cultures when compared with IFN-γ -secreting Th1 type cells. Our results have shown that both TT and CT, when given orally to mice, preferentially induce Th2 cell responses. Furthermore, this immunization regimen also resulted in significant IgA anti-TT and anti-CT responses in lamina propria of the GI tract, a major IgA effector site.[44,45] The responses in LPLs were entirely of the IgA isotype, while splenic responses to CT and TT were largely IgG and IgA.

In summary, these studies provide the first direct evidence that oral immunization with T-cell dependent vaccine antigens preferentially induce Th2 cell responses in GALT.

9

A direct correction exists between antigen-specific Th2 cells in PP, in higher frequencies of polyclonal Th2-type cells in mucosal effector tissues, and antigen-specific IgA and total IgA SFC in these sites, respectively. It will be of obvious importance to construct rational mucosal vaccines for optimal induction of Th2 cell responses.

SUMMARY AND FUTURE DIRECTIONS

Oral immunization preferentially induces Th2 cell responses which directly correlate with antigen-specific IgA responses in mucosal effector sites. It is tempting to suggest that activated, antigen-specific Th2 cells leave the PP and home to IgA effector sites via the common mucosal immune system. It is likely that Th2 cells producing IL-5 and IL-6 direct antigen-specific $sIgA^+$ B cells to become IgA producing plasma cells. Nevertheless, additional studies will be required to establish that IgA responses to T cell dependent antigens depend upon Th2 cell-derived help.

What are the implications of these studies for current oral vaccines, including novel antigen delivery systems? The most obvious would be that vaccines should now be optimized for induction of Th2 cell responses in IgA inductive sites such as GALT. It would appear that Th2 cell responses are favored in this site for yet-to-be determined reasons; however certain types of oral vaccines may not induce the help required for mucosal IgA responses. For example, it remains to be determined if attenuated viruses (such as Sabin Poliovirus vaccine) induce $CD4^+$ Th2 cell responses. Further, one might predict that recombinant Gram negative bacteria expressing novel vaccine antigens, which normally induce T cell-mediated CMI (or Th1-type cells), may not provide optimal help for IgA responses.

ACKNOWLEDGMENTS

Work summarized in this review was supported by U.S. Public Health Service Contract AI 15128 and Grants DE 04217, DE 09837 and DE 08228. HK is the recipient of NIH Research Career Development Award DE 00237. We thank Ms. Debra H. Clisby for the preparation of this paper.

REFERENCES

1. M.E. Conley and D.L. Delacroix, Intravascular and mucosal immunoglobulin A: Two separate but related systems of immune defense? *Annu. Intern. Med.* 106:892 (1987).
2. J. Mestecky, and J.R. McGhee, Immunoglobulin A (IgA): Molecular and cellular interactions involved in IgA biosynthesis and immune response, *Adv. Immunol.* 40:153 (1987).
3. J.R. McGhee, J. Mestecky, M.T. Dertzbaugh, J.H. Eldridge, M. Hirasawa, and H. Kiyono, The mucosal immune system: From fundamental concepts to vaccine development, *Vaccine* 10:75 (1992).
4. J. Mestecky, The common mucosal immune system and current strategies for induction of immune responses in external secretions, *J. Clin. Immunol.* 7:265 (1987).
5. K. Kett, P. Brandtzaeg, J. Radl, and J.F. Haaijman, Different subclass distribution of IgA-producing cells in human lymphoid organs and various secretory tissues, *J. Immunol.* 136:3631 (1986).
6. P. Brandtzaeg, Immune functions of human nasal mucosa and tonsils in health and disease, *In:* Immunology of the Lung and Upper Respiratory Tract, Bienenstock, J. (ed.). McGraw-Hill Book Co., New York, NY, pp. 28-95 (1984).
7. K. Fujihashi, J.R. McGhee, C. Lue, K.W. Beagley, T. Taga, T. Hirano, T. Kishimoto, J. Mestecky, and H. Kiyono, Human appendix B cells naturally express receptors for an respond to interleukin 6 with selective IgA1 and IgA2 synthesis, *J. Clin. Invest.* 88:248 (1991).
8. J. Bienenstock, N. Johnson, and D.Y.E. Perey, Bronchial lymphoid tissue. I. Morphologic characteristics, *Lab. Invest.* 28:686 (1973).
9. J. Bienenstock, N. Johnson, and D.Y.E. Perey, Bronchial lymphoid tissue. II. Functional characteristics, *Lab. Invest.* 28:693 (1973).

10. O. Rudzik, R.L. Clancy, D.Y.E. Perey, R.P. Day, and J. Bienenstock, Repopulation with IgA-containing cells of bronchial and intestinal lamina propria after transfer of homologous Peyer's patch and bronchial lymphocytes, *J. Immunol.* 114:1599 (1975).

11. R. Scicchitano, A. Stanisz, P.B. Ernst, and J. Bienenstock, Chapter 10. A common mucosal immune system revisited. *in*: "Migration and Homing of Lymphoid Cells," Vol. II, Husband, A.J. (ed.), CRC Press, Inc., Boca Raton, Florida, pp. 1-34.

12. T. Sminia, G.J. van der Brugge-Gamelkoorn, and S.H.M. Jeurissen, Structure and function of Bronchus-Associated Lymphoid Tissue (BALT), *Critical Rev. Immunol.* 9 (issue 2): 119 (1989).

13. R. Pabst, and I. Gehrke, Is the bronchus-associated lymphoid tissue (BALT) an integral structure of the lung in normal mammals, including humans ? *Amer. J. Respir. Cell. Mol. Biol.* 3:131 (1990).

14. R. Pabst, Is BALT a major component of the human lung immune system ?, *Immunol. Today* 13:119 (1992).

15. C.F. Kuper, P.J. Koornstra, D.M.H. Hameleers, J. Biewenga, B.J. Spit, A.M. Duijvestijn, P.J.C. van Breda Vriesman, and T. Sminia, The role of nasopharyngeal lymphoid tissue, *Immunol. Today* 12:219 (1992).

16. C. Lue, H. Kiyono, K. Fujihashi, J.R. McGhee, and J. Mestecky, The use of the hu-PBL-SCID mouse model to study lymphocyte homing and responsiveness to recall antigens, *Reg. Immunol.* (in press).

17. D.E. Bockman, and M.D. Cooper, Pinocytosis by epithelium associated with lymphoid follicles in the bursa of Fabricius, appendix and Peyer's patches, *Amer. J. Anat.* 136:455 (1973).

18. R.L. Owens, and A.L. Jones, Epithelial cell specialization within human Peyer's patches: an ultrastructural study of intestinal lymphoid follicles, *Gastroenterology* 66:189 (1974).

19. L. Mayer, and R. Shlien, Evidence for function of Ia molecules on gut epithelial cells in man, *J. Exp. Med.* 166:1471 (1987).

20. J.L. Wolf, D.H. Rubin, R. Finberg, R.S. Kauffman, A.H. Sharpe, J.S. Trier, and B.N. Fields, Intestinal M cells: a pathway for entry of reovirus into the host, Science 212:471 (1981).

21 L. Inman and J. Cantey, Specific adherence of *Escherichia coli* (strain RDEC-1) to membraneous M cells of the Peyer's patch in *Escherichia coli* diarrhea in the rabbit, *J. Clin. Invest.* 71:1 (1983).

22. P. Blier, and A. Bothwell, A limited number of B cell lineages generates the heterogeneity of a secondary immune responses, *J. Immunol.* 136:3996 (1987).

23. E.C. Butcher, R.V. Rouse, R.L. Coffman, C.N. Nottenburg, R.R. Hardy, and I.L. Weissman, surface phenotype of Peyer's patch germinal center cells: implications for the role of germinal centers in B cell differentiation, *J. Immunol.* 129:2698 (1982).

24. D.A. Lebman, P.M. Griffin, and J.J. Cebra, Relationship between expression of IgA by Peyer's patch cells and functional IgA memory cells, *J. Exp. Med.* 166:1405 (1987).

25. P.D. Weinstein, and J.J. Cebra, The preference for switching to IgA expression by Peyer's patch germinal center B cells is likely due to the intrinsic influence of their microenvironment, *J. Immunol.* 147:4126 (1991).

26. Y.-J. Liu, G.D. Johnson, J. Gordon, and I.C.M. MacLennan, Germinal centers in T-cell-dependent antibody responses, *Immunol. Today* 12:17 (1992).

27. R. Rouse, J. Ledbetter, and I.L. Weissman, Mouse lymph node germinal centers contain a selected subset of T cells: the helper phenotype, *J. Immunol.* 128:2243 (1982).

28. W.J. Krall, and J. Braun, *In vivo* retroviral marking of antigen-specific B lymphocytes, *Sem. Immunol.* 4:19 (1992).

29 J. Mega, M.G. Bruce, K.W. Beagley, J.R. McGhee, T. Taguchi, A.M. Pitts, M.L. McGhee, R.P. Bucy, J.H. Eldridge, J. Mestecky, and H. Kiyono, Regulation of mucosal responses by CD4+ T lymphocytes: Effects of anti-L3T4 treatment on the gastrointestinal immune system, *Intern. Immunol.* 3:793 (1991).

30. M.F. Kagnoff, IgA anti-dextran B 1355 responses, *J. Immunol.* 122:866 (1979).

31. M.F. Kagnoff, L.S. Arner, and S.L. Swain, Lymphokine-mediated activation of a T cell dependent IgA anti-polysaccharide response, *J. Immunol.* 131:2210 (1983).

32. E. Abraham, and A. Robinson, Oral immunization with bacterial polysaccharide and adjuvant enhances antigen-specific pulmonary secretory antibody response and resistance to pneumonia, *Vaccine* 9:757 (1991).

33. P.K.A. Mongini, W.E. Paul, and E.S. Metcalf, IgG subclass, IgE and IgA Anti-Trinitrophenyl antibody production within Trinitrophenyl-ficoll-responsive B cell clones. Evidence in support of three distinct switching pathways, *J. Exp. Med.* 157:69 (1983).

34. J.L. Babb, H. Kiyono, S.M. Michalek, and J.R. McGhee, LPS regulation of the immune response: suppression of immune response to orally-administered T-independent antigen, *J. Immunol.* 127:1052 (1981).

35. L. Winner, J. Mack, R. Weltzin, J.J. Mekalanos, J.-P Kraehenbuhl, and M.R. Neutra, New model for analysis of mucosal immunity: Intestinal secretion of specific monoclonal immunoglobulin A from hybridoma tumors protects against *Vibrio cholerae* infection. *Infect. Immun.* 59:977 (1991).

36. P. Michetti, M.J. Mahan, J.M. Slauch, J.J. Mekalanos, and M.R. Neutra, Monoclonal secretory IgA protects mice against oral challenge with the invasive pathogen *Salmonella typhimurium*, *Infect. Immun.* 60:1786 (1992).

37. T.A. Brown, and J. Mestecky, Immunoglobulin A subclass distribution of naturally occurring salivary antibodies to microbial antigens, *Infect. Immun.* 49:459 (1985).

38. T.R. Mosmann, and R.L. Coffman, Th1 and Th2 cells: different patterns of lymphokine secretion lead to different functional properties, *Annu. Rev. Immunol.* 7:145 (1989).

39. N.E. Street, and T.R. Mosmann, Functional diversity of T lymphocytes due to secretion of different cytokines patterns, *FASEB J.* 5:171 (1991).

40. J.R. McGhee, J. Mestecky, C.O. Elson, and H. Kiyono, Regulation of IgA synthesis and immune response by T cells and interleukins, *J. Clin. Immunol.* 9:175 (1989).

41. T. Taguchi, J.R. McGhee, R.L. Coffman, K.W. Beagley, J.H. Eldridge, K. Takatsu, and H. Kiyono, Analysis of Th1 and Th2 cells in murine gut-associated tissues. Frequencies of CD4[+] and CD8[+] T cells that secrete IFN-γ and IL-5, *J. Immunol.* 145:68 (1990).

42. J. Mega, J.R. McGhee, and H. Kiyono, Cytokine-and Ig-producing cells in mucosal effector tissues: analysis of IL-5-and IFN-γ producing T cells, T cell receptor expression, and IgA plasma cells from mouse salivary gland-associated tissues, *J. Immunol.* 148:2030 (1992).

43. J. Xu-Amano, W.K. Aicher, T. Taguchi, H. Kiyono, and J.R. McGhee, Selective induction of Th2 cells in murine Peyer's patches by oral immunization, *Intern. Immunol.* 4:433 (1992).

44. J. Xu-Amano, K. Fujihashi, H. Kiyono and J.R. McGhee, Effect of cholera toxin on induction of Th1 and Th2 responses in mucosa-associated tissues, (Submitted for publication).

45. J. Xu-Amano, K. Fujihashi, R. Jackson, H. Kiyono and J.R. McGhee, Mucosal vaccine and Th1/Th2 responses: oral immunization of mice with tetanus toxoid induces antigen-specific Th2 type responses in mucosa-associated tissues, (submitted for publication).

PROSPECTS FOR HUMAN MUCOSAL VACCINES

Jiri Mestecky and Jerry R. McGhee

Departments of Microbiology and Medicine
Immunobiology Vaccine Center
University of Alabama at Birmingham
Birmingham, AL 35294-10005

INTRODUCTION

Introduction of vaccines together with antibiotics into human and veterinary practice represent the most successful achievements in preventive medicine. Current resurgence in academic and industrial efforts to develop new vaccines and improve existing ones has been facilitated by impressive advances in immunology and molecular biology technologies that make this goal feasible in the foreseeable future. For example, the ability to produce highly specific vaccines free of unnecessary components, express desired antigens in suitable microbial vectors, and extend the duration of the immune response by using novel antigen delivery systems, will undoubtedly foster a widespread use of such improved vaccines.

Development of vaccines for the parenteral route of immunization has so far received the major emphasis. This may be justifiable by the fact that an *exact* dose of desired antigen is introduced by one of the commonly acceptable routes and the magnitude of the ensuing immune response can be *precisely evaluated* in sera obtained from immunized subjects. Yet, extensive studies of the immune system indicate that, to a significant extent, this systemic immunity may not always provide an *optimal protection* against several types of infectious diseases. Despite the induction of corresponding antibodies and cell-mediated immune responses, as measured in the blood of systemically immunized individuals, protection against diseases such as cholera, influenza and probably AIDS, to name just a few, may be rather limited or nonexistent. Furthermore, alum is the only FDA-approved adjuvant for use in human vaccines, and this adjuvant is unlikely to be highly effective for boosting both cell-mediated and antibody responses.

Numerous lines of evidence which have emerged over the last two decades have clearly shown that the immune system can be divided into two functionally independent compartments: *systemic* - represented by the bone marrow, spleen and lymph nodes; and *mucosal* - represented by lymphoid tissues in mucosae and external secretory glands.[1,2] For the development of vaccines, this compartmentalization is essential: induction of an immune response in one of these two systems may not necessarily be reflected in the other. Certainly peripheral immunization induces poor mucosal immunity; however, mucosal immunization offers the advantage in that some delivery systems (see below) induce both mucosal and systemic immunity. Furthermore, these two systems do not display a parallel maturation pattern and the products of immunocytes (i.e., antibodies and cytokines) differ remarkably in their quality and quantity.[3,4]

Genetically Engineered Vaccines, Edited by
J.E. Ciardi *et al.*, Plenum Press, New York, 1992

Why Do We Need Mucosal Vaccines?

Diverse environmental antigens derived from ingested food, inhaled and ingested microorganisms and endogenous bacterial flora also provide a constant stimulation of the entire immune system. Due to their enormous contiguity (~400 m^2), mucosal surfaces represent the primary site of impingement with such antigens including agents causing the absolute majority of infectious diseases irrespective of their predominantly local (e.g., influenza) or systemic (e.g., meningitis) involvement.[3] This point is frequently underappreciated. And yet, recent statistical data indicate that approximately 80% of HIV infections worldwide are acquired through heterosexual contact, most apparently through unimpaired mucosal surfaces of genital mucosae![3-7] Furthermore, the presence of antibodies in the circulation may protect an individual against a systemic but not mucosal challenge with the live virus. This fact has been experimentally demonstrated in monkeys infected with SIV through the genital tract.[7]

It should not be surprising, as it seems to be to most, that the mucosal tissues and secretory glands comprise the largest accumulation of T cells, B cells and plasma cells in the body, far exceeding the numbers of such cells found in the bone marrow, spleen and lymph nodes.[1,8] This is true not only for antibody-forming cells, but also for cells secreting interferon-gamma.[9] The mucosal environment also includes a full complement of antigen-processing and -presenting cells functionally endowed with the ability to participate in the initiation of specific B and T cell-mediated immune responses.[8]

Experiments that addressed the origin of mucosal antibodies have led to the clear conclusion that an overwhelming proportion of such antibodies is produced locally in mucosal tissues and that in most species, including humans, antibodies derived from the circulation represent only a minor fraction.[10] Consequently, mucosal surfaces can be colonized by bacteria and viruses can infect epithelial cells in spite of the presence of corresponding antibodies in the circulation. For example, systemic immunization with inactivated poliovirus may prevent the development of poliomyelitis, but does not prevent initial infection in the gastrointestinal tract. Thus, these findings should be considered with increased awareness in the development of all vaccines destined to combat infections encountered through mucosae.[3]

COMMON MUCOSAL IMMUNE SYSTEM

Extensive studies concerning the origin of precursors of mucosal IgA plasma cells revealed that the organized lymphoepithelial structures found along the gastrointestinal and respiratory tracts are the main source of such cells (for reviews see 11,12). These precursors committed to IgA synthesis mature in mesenteric lymph nodes and through the thoracic duct enter the circulation. Subsequently, they lodge in the lamina propria of the intestinal, respiratory, and genital tracts and in the mammary, salivary, and lacrymal glands where terminal differentiation into IgA plasma cells occurs under the influence of locally produced cytokines such as IL-5 and IL-6, derived from T cells and mucosal epithelial cells.[13] In animals, the evidence for this IgA cycle is primarily based on the adoptive transfer of cells from the gut- and bronchus-associated lymphoid tissues (GALT and BALT) into recipients whose mucosal tissues and glands were populated by IgA plasma cells of the donor origin. Furthermore, specific IgA antibodies from several external secretions displayed identical spectrotypes, suggesting that IgA cells that populate various exocrine tissues were derived from a common clone.[14] Most importantly for vaccine development, Montgomery et al.[15] demonstrated that the oral administration of antigen to lactating rabbits led to the appearance of specific secretory IgA (S-IgA) in milk but not in serum of immunized animals. These results were further validated and extended in many subsequent studies performed in a large number of animal species using microbial antigens. Furthermore, such specific S-IgA antibodies also appeared in parallel in secretions of the intestinal, respiratory and genital tracts as well as in tears, saliva, and milk. It should be noted, however, that experiments performed in several laboratories strongly suggest that there are alternative sources(s) of mucosal IgA plasma cell precursors: surgical removal of PP, appendix, and sacculus rotundus, or chronic drainage of thoracic duct lymph resulted in the depletion of the intestinal IgA plasma cell population by no more than 50%.[16] Recent results concerning the properties and repopulation potential of Ly-1$^+$ (CD5) B cells from the peritoneal cavity of mice have provided new insights to explain this apparent paradox: Up to half of the IgA plasma cells found in the murine intestinal wall were derived from a distinct lineage of self-replenishing and long-living peritoneal lymphoid cells.[17-19] Interestingly, surface membrane IgA-positive (sIgA+) cells constitute no more than 1% of total peritoneal cells; therefore, a considerable proportion of IgA plasma cells in the murine gut originating from peritoneal Ly-1+ B cells is derived from the sIgA- population. Another important point concerns the antibody repertoire of these cells: they have been shown to produce autoantibodies and polyreactive antibodies to common surface antigens of microorganisms that colonize mucosal surfaces and to which the immune system is likely to be exposed shortly after birth.[20] Furthermore, intraperitoneal immunization of animals has been reported as an effective route for the induction of specific S-IgA antibodies in external secretions.[21-24]

Evidence for the existence of the common mucosal system in humans has been strengthened in recent years by several studies. In addition to the detection of specific S-IgA antibodies in remote secretions induced by natural exposure to antigens or oral immunization, analyses of IgA-secreting cells from peripheral blood and mucosal tissues provided strong evidence for this concept.[25] For example, cells secreting IgA antibodies specific for orally administered antigens have been detected and predominate in the peripheral blood, before the appearance of S-IgA in tears and saliva and in intestinal and minor salivary gland biopsies.[9,26] These results provide a sound physiological basis for the rational immunization protocols that exploit the potential of the common mucosal immune system in the design of vaccines that induce protective immunity at the portal of entry of most pathogens.

Alternative IgA Inductive Sites

Although GALT, represented by ileo-jejunal Peyer's patches (PP), and BALT have been considered as primary sources of precursors of mucosal IgA plasma cells, additional lymphoid structures elsewhere in the body may serve a similar purpose. As described above, Ly-1+ (CD 5+) B cells from the murine peritoneal cavity that are probably derived from peritoneal lymphoid "milky spots" populate the intestinal tract.[17-20] Whether cells from the human peritoneal cavity also supply mucosal tissues with IgA precursors is currently studied in our laboratory. Initial experiments suggest that approximately 60% of cells obtained from the ascitic fluid from patients with alcoholic cirrhosis or in dialysate fluids from patients on continuous ambulatory peritoneal dialyses are lymphocytes, mostly T cells. CD5+ B cells constitute ~ 2% of total cells. Intraperitoneal immunization with a soluble preparation of tetanus toxoid added to the peritoneal dialysis fluid resulted in the induction of specific IgG- and IgA-secreting cells in both peripheral blood and peritoneal fluid.

Other accumulations of lymphoid tissues such as palatine, lingual, and nasopharyngeal tonsils (Waldeyer's ring) are strategically positioned at the beginning of the digestive and respiratory tracts. They are continuously exposed to ingested and inhaled antigens and possess structural features similar to both lymph nodes and GALT.[27] A lymphoepithelium is present which contains M cells for selective antigen uptake. Thus, B and T cells, plasma cells as well as M cells and antigen-processing and -presenting cells have been shown to occur in abundance. Several observations, summarized by Brandtzaeg,[27] have suggested that these lymphoid tissues may serve as a source of precursors of IgA plasma cells found in the upper aerodigestive tract: the distribution of IgA1- and IgA2-producing cells in the nasal and gastric mucosae and in lacrimal and salivary glands resembles such distribution in tonsils.[28] Furthermore, the immune response to the oral poliovirus vaccine in tonsillectomized children is inferior when compared to that induced in children with intact tonsils, and such individuals display reduced levels of S-IgA.[29] Further studies using innovative approaches should be initiated to critically evaluate the role of the nasopharyngeal lymphoid tissues in human mucosal immunity; experiments in animals are of limited value due to the absence of analogous lymphoid structures in almost all species.

Although most of the investigations of the IgA inductive sites have primarily centered on PP and appendix, analogous follicular structures are also found in the large intestine with especially pronounced accumulations in the rectum.[30] The potential importance of the rectal lymphoid tissues as an IgA inductive site and as a source of IgA plasma cell precursors is suggested by several studies. The distribution of IgA1 and IgA2 cells in the lamina propria of the large intestine differs from other mucosal tissue by the pronounced predominance of the IgA2 plasma cells.[28,31] The fact that this is also the case in the female genital mucosal tissues (uterus, cervix, fallopian tubes, and vagina)[32] suggest the rectal lymphoid tissues may be the most important source of IgA precursors destined for the genital tract. Furthermore, recent results concerning the effectiveness of various immunization routes for the induction of SIV-specific antibodies in secretions of the female genital tract has emphasized the preference of intrarectal immunization.[33] Further, intrarectal challenge of macaques parenterally immunized with SIV were protected from infection[34]; however, in this study, no antibody analyses of genital tract secretions were performed.

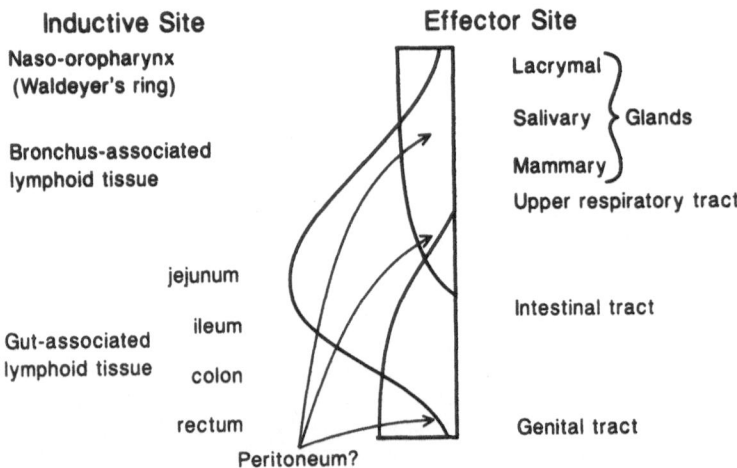

Figure 1. The common mucosal immune system: possible subcompartments for mucosal immunity in the gastrointestinal, upper respiratory, and genitourinary tracts. Limited evidence suggests that precursor cells from a given inductive site preferentially lodge in certain effector tissues. Thus, it appears that the upper aerodigestive tract is primarily supplied by IgA, precursor cells from Waldeyer's ring and BALT, while the genital tract receives precursors from the lower digestive tract. The possible contribution of peritoneal CD5+ B cells in humans remains to be evaluated.

Considered in the context of the common mucosal immune system, these studies suggest that further subcompartmentalization may exist and be controlled by the pronounced preference of housing of IgA plasma cell precursors. Thus, certain IgA inductive sites may be restricted to IgA and provide precursor lymphocytes for a particular effector site (Fig. 1). By extension, the oral route of immunization may be less effective in the induction of S-IgA in the genital tract than the introduction of antigens in the rectum. Thus, further comparative studies of the distribution of specific S-IgA antibodies in various external secretions induced by diverse mucosal immunization routes should be performed to address this point with implications important in the design of effective vaccines.

ANTIGEN DELIVERY SYSTEMS FOR THE INDUCTION OF MUCOSAL IMMUNE RESPONSES

Empirical experience with mucosal immunization has resulted in a generally accepted conclusion that considerably higher doses of antigens are required. This is due to the elimination of antigens, existence of effective mechanical (epithelial cells) and chemical (e.g., mucins) barriers, and degradation and denaturation of antigens by enzymes and acids. Thus, only minute quantities of fully potent antigens reach the mucosal lymphoid tissues. In addition, preexisting secretory antibodies complex with antigens at mucosal surfaces and further reduce antigen uptake.[35]

Consequently, several strategies recently reviewed elsewhere[3,4,36,37] and extensively discussed in this volume have been proposed to circumvent this problem. Here we summarize and critically evaluate the advantages and disadvantages of some of these approaches and provide their comparisons and applicabilities in selected situations.

Incorporation of Antigens in Protective Vehicles

To avoid their degradation and denaturation by pepsin and hydrochloric acid in the stomach, vaccine antigens have been incorporated into gelatin capsules, subsequently coated with substances that became soluble in the alkaline pH of the small intestine.[38,39] Similar approaches have been used by industry for controlled delivery and targeting of drugs for many years and this knowledge can be undoubtedly applied for the optimal delivery of oral vaccines.

Liposomes containing bacterial antigens have been used in animals and in humans for oral immunization against dental caries[40] (see Michalek et al. - this volume); improved immunogenicity of liposome-incorporated as compared to free antigen has been observed. Further studies of the ability of liposomes to deliver different types of antigens by the oral route are necessary, and the susceptibility of liposomes to solubilization by the bile acids should be investigated.

Biodegradable microspheres have been used for systemic and oral immunizations in several recent studies.[41-45] The microspheres used were composed of biodegradable and biocompatible materials such as poly DL-lactide-co-glycolide (DL-PLG) copolymers with antigens incorporated within such particles during their preparation. Biodegradation, which may range from several days to months depending on the lactide-glycolide proportion, proceeds by hydrolysis of ester bonds to yield catabolizable lactic and glycolic acids. The size of such microspheres can vary from a fraction of μm to several mm. Those with the size range 5-10 μm are absorbed from the gastrointestinal tract through PP where they are retained and subsequently release antigen. However, only a minute proportion is taken up. The incorporation of antigens into biodegradable microspheres has several advantages including the protection of antigen from proteolysis and possible co-incorporation of immunological adjuvants and cytokines that may further enhance the immune response. Furthermore, a single injection or ingestion of microspheres with a programmed short and long biodegradative times may induce overlapping primary and long-lasting secondary immune responses, thus eliminating the need for booster immunization. Small microspheres (1-5 μm) were not retained in PP and were found in the spleen and lymph nodes; thus, a concurrent systemic and secretory immune response may be induced by ingestion of microspheres of appropriate sizes. As a dry powder, microspheres containing antigens are stable and with the smaller number of antigens tested thus far, indicate that their immunogenicity can be preserved for many months. Nevertheless, several problems will have to be addressed to further increase the general usability of microsphere-based vaccines in both human and veterinary medicine. The organic solvents used during the preparation are likely to denature many of the candidate antigens. Most of the studies performed so far use a relatively stable and highly immunogenic antigen-staphylococcal enterotoxin B.[43] Although promising results have been obtained with more complex antigens, such as the influenza virus,[4,5] additional experiments will be necessary to select the most suitable organic solvents and procedures for microsphere preparation to preserve the immunogenicity of incorporated antigens. Furthermore, it is likely that the limited absorption of microspheres from the gastrointestinal tract noted thus far can be increased by modification of the surface properties of microspheres.

Cholera Toxin (CT) for Oral Vaccine Delivery and CT-B Subunit Conjugates

Cholera toxin, produced by *Vibrio cholerae* and an essentially identical toxin from *E. coli*, termed labile toxin (LT), bind to all nucleated cells and especially to intestinal epithelial cells through a specific GM 1 ganglioside receptor.[46] The CT (and LT) molecules consist of A and B subunits. The 28 kDa A subunit contains the toxin moiety, induces ADP-ribosylation of adenyl cyclase regulatory G proteins, and results in elevated cAMP and net secretion of anions from epithelial cells with ensuing diarrhea. The A subunit is post translationally cleaved into A1 (toxin) and A2 peptides, and the latter is associated with the B subunits. Five identical B subunits (11.6 kDa) are non-covalently associated and bind to GM 1 ganglioside receptors for insertion of the toxic A1 subunit into the cell.

CT is a potent oral immunogen and in addition promotes significant mucosal IgA and serum antibodies to proteins when co-administered by the oral route.[47,48] When proteins such as keyhole limpet hemocyanin (KLH) are given orally without CT to mice, systemic unresponsiveness (oral tolerance) is induced.[48] Furthermore, vaccine antigens such as influenza are somewhat immunogenic when given by the oral route; however, co-administration with CT significantly enhances S-IgA, anti-influenza responses in external secretions, as well as serum anti-influenza antibodies.[49,50]

Due to its inherent toxicity, CT is not a suitable adjuvant for use in humans. Recent studies performed in mice have shown that CT-B-protein conjugates induce mucosal antibody responses[51]; however, the results of some studies have been variable. Usually small amounts of CT must be added to CT-B-protein conjugates, in order to induce mucosal and serum antibody responses. The mechanisms involved in CT-induced adjuvant responses to orally administered proteins/vaccines are at present unknown. However, recent studies[52] have begun to elucidate potential mechanisms. For example, administration of CT (but not CT-B), together with fluorescein-conjugated dextran beads, resulted in a significant accumulation of the latter in mucosal epithelial cells and underlying tissue.[52] This would support the idea that CT increases epithelial cell permeability, and results in an influx of lumenal material including co-administered proteins/vaccines. Of interest was the finding that a primary dose of CT plus FITC-beads resulted in a marked influx into the gut tissues, while lower levels were seen after a second and even lower amounts after a third oral exposure. This would clearly support the concept that initial oral administration of CT results in significant local S-IgA responses, which subsequently neutralize the CT administered at later times and thus blocks bead (and antigen) penetration across the epithelium.

In summary, CT and CT-B have inherent advantages and significant disadvantages (Table 1). CT is the most potent nonliving mucosal antigen and can induce immunity to itself which protects against cholera. In addition, CT promotes mucosal and serum antibody responses to co-administered proteins, which would induce oral tolerance. Further, CT acts as an adjuvant and significantly boosts antibody responses to viral vaccines. The disadvantages include the unsuitability of CT for human use–microgram amounts given orally induce a massive diarrhea. Mucosal immune responses to CT and to CT-B would also negate the advantage of their use as oral adjuvants. Further, CT-B either co-administered or conjugated to vaccine is vastly inferior to CT for induction of mucosal and serum antibody responses. Finally, if CT increases mucosal permeability, as recent studies suggest,[52] then its use as an oral adjuvant could induce immune responses to bystander antigens present in the GI tract. Thus, CT would elicit antibodies to natural flora and autoimmune responses to self antigens present in the GI tract. This may not, however, be a problem with bystander antigens.[53] Many other non-CT related substances have been explored as mucosal adjuvants with highly variable effectiveness.[37]

Live Microbial Vaccines and Vectors

The ability of some microorganisms to colonize and infect mucosa of the intestinal tract, and the potential for inclusion of genes from unrelated microorganisms which code for protective proteins, represent an attractive possibility for design of novel vaccines effective in the protection of mucosal surfaces.[54] Although several bacterial species, including genetically modified *Vibrio cholerae*, various strains of *Salmonellae, Escherichia coli, Mycobacteria, Yersinia enterocolitica,* and *Lactobacilli* have been considered (for review see 55) most of the experimental work has been performed with extensively attenuated by genetic modifications strains of *Salmonellae*[56-60] and recently BCG.[61,62] In addition, vaccinia, polio and adeno viruses have also been considered as vectors suitable for mucosal immunization.[55]

Table 1. Oral vaccine delivery systems for mucosal and potentially systemic immunity.

System	Advantages	Disadvantages
Muramyl dipeptides and related molecules	Clearly enhance systemic and mucosal immunity to several types of antigens	May be toxic in humans; may also enhance immune response to luminal bystander antigens
CT and CT-B, CT-B conjugates; CT-B gene fusion genes	CT boosts significant mucosal/serum ab responses to co-administered proteins; CT-B is less effective but induces responses; the CT-B-fusion protein could potentially be optimal for oral immunization	Chemical coupling may be required; high levels of antibodies to CT or CT-B could reduce absorption and preclude the use of CTB for multiple immunizations with other antigens linked to it; CT is quite toxic in humans
Colonization of Peyer's patches with genetically engineered strains of *Salmonella, E. coli,* BCG or with adeno or polio viruses	Potential of expressing several antigens from unrelated microorganisms in a single vaccine strain; vigorous mucosal and systemic immune responses induced by *live* bacteria	Immune responses induced to carrier bacterium may diminish the effectiveness of colonization and enhancement of the immune response
Liposomes	Immune response may be enhanced by inclusion of targeting substances (increased uptake by Peyer's patches) and incorporation of oral adjuvants; protection of antigen from digestion	Low uptake, low stability in the intestinal environment (pH, bile salts, lipolytic enzymes)
Microencapsulation	Immune response may be enhanced by increased uptake by Peyer's patches (cationization of microcapsule surfaces); protects incorporated antigens from digestion; both mucosal and systemic immune responses may be induced depending on the size of microcapsules; programmed release (combination of fast and slow) of antigens may induce both primary as well as booster responses by single immunization; biocompatible (nonantigenic) and biodegradable materials are used in microcapsules; freeflowing powder, eliminates problems with vaccine storage	Degradation of sensitive antigens by organic solvents during the preparation of microcapsules; expense in preparation; limited uptake

This approach has several obvious advantages. Colonization and infection with live microorganisms are known to induce long-lasting, vigorous immune responses in both mucosal as well as systemic compartments. Furthermore, the possibility of introduction of genetic material coding for many different and unrelated antigens into a single microbial vector would reduce, otherwise, multiple immunization to a single dose and yet induce protection against several diseases. For example, oral immunization with BCG which expresses genes encoding for HIV glycoproteins and nontoxic tetanus toxin or many other candidate antigens of medical importance resulted in the induction of corresponding antibodies.[61] Furthermore, such vaccines would be easily administrable and inexpensive to produce – factors that are of paramount importance for large-scale immunizations in the Third World countries.

However, a number of related questions and problems associated with this approach must be considered. The most important aspect concerns the reusability of individual vectors. The immune responses are induced not only against the desired antigen expressed in a given microbial vector, but also to the vector itself.[63] As a matter of fact, the response to the vector is dominant. Such immune responses could limit the effectiveness of subsequent secondary or tertiary immunization with the same microorganism as recently demonstrated by Kantele et al.[63] However, others have shown that oral immunization with the vector may actually prime for antibody responses to the recombinant protein expressed in this vector.[64] In case of immunizations with BCG, potentially severe immunological reactions are likely to be encountered upon reexposure to this microorganism. Some of the antigens considered are poorly expressed in a given vector or are not secreted from the periplasmic space after the death of a bacterium. Furthermore, the antigens expressed in a bacterial vector are likely to differ radically from the structure of such antigens derived from the original source. For example, the secondary and tertiary structures of glycoproteins will be dissimilar to those expressed in *Salmonellae* or *E. coli* due to the inability of these bacteria to glycosylate the core protein and thus alter the epitopes on a given glycoprotein.

Finally, the genetic stability of such vectors and the possibility of the environmental contamination may pose additional problems. Nevertheless, genetically engineered live vectors offer unique opportunities for effective mucosal immunization and further research in this area should be vigorously pursued.

SUMMARY

The selective induction of antibodies in external secretions and mucosal T cell-mediated immunity are desirable for the prevention of various systemic as well as predominantly mucosa-restricted infections. An enormous surface area of mucosal membranes is protected primarily by antibodies that belong, in many species, to the IgA isotype. Such antibodies are produced locally by large numbers of IgA-containing plasma cells distributed in subepithelial spaces of mucosal membranes and in the stroma of secretory glands. In humans and in some animal species, plasma-derived IgA antibodies do not enter external secretions in significant quantities and systemically administered preformed IgA antibodies would be of little use for passive immunization. Systemic administration of microbial antigens may boost an effective S-IgA immune response only in a situation whereby an immunized individual had previously encountered the same antigen by the mucosal route. Immunization routes that involve ingestion or possibly inhalation of antigens lead to the induction of not only local but also generalized immune responses, manifested by the parallel appearance of S-IgA antibodies to ingested or inhaled antigens in secretions of glands distant from the site of immunization. Convincing evidence is available that antigen-sensitized and IgA-committed precursors of plasma cells and T cells from IgA inductive sites (e.g., BALT, GALT, and tonsils) are disseminated to the gut, other mucosa-associated tissues, and exocrine glands. However, due to the limited absorption of desired antigens from the gut lumen of orally immunized individuals, repeated large doses of antigens are required for an effective S-IgA response. Novel antigen delivery systems for the stimulation of such responses has been briefly reviewed here. These, of course, include genetically engineered bacteria and viruses, CT/CFB, liposomes and microspheres. Live attenuated or genetically manipulated bacteria expressing other microbial antigens have been used for selective colonization of GALT. Unique antigen packaging and the use of adjuvants suitable for oral administration hold promise for an efficient antigen delivery to critical tissues in the intestine and deserve extensive exploration. The oral immunization route appears to have many advantages over systemic immunization; however, one must consider alternate IgA inductive sites and compartmentalization within the Common Mucosal Immune System. In addition to providing immunity on mucosal surfaces, which are the most common sites of entry of infectious agents, the mucosal routes of administration are more acceptable and do not require stringent criteria applicable for injectable vaccines, storage problems may be simplified, and large populations of individuals can be immunized simultaneously without the assistance of highly trained health personnel.

ACKNOWLEDGMENTS

We thank Ms. Maria Bethune for preparation of this manuscript. Experimental work included in this article has been supported by grant AI-18745, DE08182, and Contract AI-15128 from NIH.

REFERENCES

1. J. Mestecky and J.R. McGhee. Immunoglobulin A (IgA): molecular and cellular interactions involved in IgA biosynthesis and immune response, *Adv. Immunol.* 40:153 (1987).
2. C.D. Alley and J. Mestecky, The Mucosal Immune System, *in*: "B Lymphocytes in Human Disease," G. Bird, and J.E. Calvert (eds.), Chapter 9, Oxford University Press, Oxford pp. 222-254 (1988).
3. J.R. McGhee and J. Mestecky, In defense of mucosal surfaces. Development of novel vaccines for IgA responses protective at the portals of entry of microbial pathogens, *Infect. Dis. Clin. North Amer.* 4:315 (1990).
4. J.R. McGhee, J. Mestecky, M.T. Dertzbaugh, J.H. Eldridge, M. Hirasawa, and H. Kiyono, The mucosal immune system. From fundamental concepts to vaccine development, *Vaccine* 10:75 (1992).
5. C. Miller and M.B. Gardner, Editorial, AIDS and mucosal immunity: Usefulness of the SIV macaque model of genital mucosal transmission, *J. AIDS* 4:1169 (1991).
6. B.D. Forrest, Women, HIV, and mucosal immunity, *Lancet* 337:835 (1991).
7. C.J. Miller, N.J. Alexander, S. Sutjipto, A.A. Lackner, A. Gettie, L.J. Lowenstine, M. Jennings, and P.A. Marx, Genital mucosal transmission of simian immunodeficiency virus: animal model for heterosexual transmission of human immunodeficiency virus, *J. Virol.* 63:4277 (1989).
8. P. Brandtzaeg, Overview of the mucosal immune system, *Curr. Top. Microbiol. Immunol.* 146:13 (1989).
9. M. Quiding, I. Nordström, A. Kilander, G. Andersson, L.A. Hanson, J. Holmgren, and C. Czerkinsky, Intestinal immune responses in humans, *J. Clin. Invest.* 88:143 (1991).
10. J. Mestecky, C. Lue, and M.W. Russell, Selective transport of IgA. Cellular and molecular aspects, *Gastroenterol. Clin. North Amer.* 20:441 (1991).
11. R. Scicchitano, A. Stanisz, P.B. Ernst, and J. Bienenstock, A common mucosal immune system revisited, *in*: "Migration and Homing of Lymphoid Cells," Vol. II., A.J. Husband (ed.), CRC Press, Boca Raton, FL., pp. 1-34 (1988).
12. J.M. Phillips-Quagliata, and M.E. Lamm, Migration of lymphocytes in the mucosal immune system, *in*: "Migration and Homing of Lymphoid Cells", Vol. II., A.J. Husband (ed.), CRC Press, Boca Raton, FL., pp. 53-75 (1988).
13. J.R. McGhee, J. Mestecky, C.O. Elson, and H. Kiyono, Regulation of IgA systhesis and immune response by T cells and interleukins, *J. Clin. Immunol.* 9:175 (1989).
14. P.C. Montgomery, A. Ayyildiz, I.M. Lemaitre-Cuelho, J.-P. Vaerman, and J.H. Rockey, Induction and expression of antibodies in secretions: the ocular immune system, *Ann. New York Acad. Sci.* 409:428 (1983).
15. P.C. Montgomery, J. Cohn, and E.T. Lally, The induction and characterization of secretory antibodies, *Adv. Exp. Med. Biol.* 45:453 (1974).
16. R.V. Heatley, J.M. Stark, P. Horsewood, E. Bandouvas, F. Cole, and J. Bienenstock, The effects of surgical removal of Peyer's patches in the rat on systemic antibody responses to intestinal antigen, *Immunology* 44:543 (1981).
17. F.G.M. Kroese, E.C. Butcher, A.M. Stall, and L.A. Herzenberg, A major peritoneal reservoir of precursors for intestinal IgA plasma cells, *Immunological Invest.* 18:47 (1989).
18. F.G.M. Kroese, E.C. Butcher, A.M. Stall, P.A. Lalor, S. Adams, and L. Herzenberg, Many of the IgA plasma cells in the murine gut are derived from self replenishing precursors in the peritoneal cavity, *Intern. Immunol.* 1:75 (1988).

19. N. Solvason, A. Lehuen, and J.F. Kearney, An embryonic source of Ly1 but not conventional B cells, *Intern. Immunol.* 3:543 (1991).

20. M.T. Kasaian, H. Tkematsu, and P. Casali, CD5[+] B lymphocytes, *Proc. Soc. Exp. Med. Biol.* 199:226 (1991).

21. N.F. Pierce, and J.L. Gowans, Cellular kinetics of the intestinal immune response to cholera toxoid in rats, *J. Exp. Med.* 142:1550 (1975).

22. K.J. Beh, A.J. Husband, and A.K. Lascelles, Intestinal response of sheep to intraperitoneal immunization, *Immunology* 37:385 (1979).

23. M.A. Thapar, E.L. Parr, and M.B. Parr, Secretory immune responses in mouse vaginal fluid after pelvic, parenteral or vaginal immunization, *Immunology* 70:121 (1990).

24. M.L. Dunkley, and A.J. Husband, Routes of priming and challenge for IgA antibody-containing cell responses in the intestine, *Immunol. Lett.* 26:165 (1990).

25. J. Mestecky, The common mucosal immune system and current strategies for induction of immune responses in external secretions, *J. Clin. Immunol.* 7:265 (1987).

26. C. Czerkinsky, A.-M. Svennerholm, M. Quiding, R. Jonsson, and J. Holmgren, Antibody-producing cells in peripheral blood and salivary glands after oral cholera vaccination of humans, *Infect. Immun.* 59:996 (1991).

27. P. Brandtzaeg, Immune functions of human nasal mucosa and tonsils in health, *in:* "Immunology of the Lung and Upper Respiratory Tract", J. Bienenstock (ed.), McGraw-Hill, New York, pp. 28-95 (1984).

28. K. Kett, P. Brandtzaeg, J. Radl, and J.J. Haaijman, Different subclass distribution of IgA-producing cells in human lymphoid organs and various secretory tissues, *J. Immunol.* 136:3631 (1986).

29. P.L. Ogra, Effect of tonsillectomy and adenoidectomy on nasopharyngeal antibody response to poliovirus, *New Eng. J. Med.* 284:59 (1979).

30. H. Strindel, Bartel's 'Tonsille des Mastdarmes' und ihre Stellung in der Pathologie des lymphatischen Apparates des Mastdarme, *Zbl. Chirurgie* 62:2594 (1935).

31. J. Mestecky and M.W. Russell, IgA subclasses, *Monogr. Allergy* 19:277 (1986).

32. W.H. Kutteh, K.D. Hatch, R.E. Blackwell, and J. Mestecky, Secretory immune system of the female reproductive tract: I. Immunoglobulin and secretory component-containing cells, *Obstet. Gynec.* 71:56 (1988).

33. T. Lehner, L. Bergmeier, C. Panagiotides, R. Brookes, and S. Adams, Oral, rectal, and vaginal routes of immunization by means of hybrid SIV p26-Ty virus-like particles in rhesus monkeys, VII International Conference on AIDS, June, 1991, Vol. 2, TH.A.65, p. 72.

34. M.P. Cranage, N. Cook, M. Dennis, J. Rose, P. Kitchin, A. Baskerville, and P.J. Greenaway, Parenterally administered formalin-inactivated SIVmac protects rhesus macaques from intrarectal challenge, 9th Annual Symposium on Nonhuman Primate Models for AIDS, November, 1991, Seattle, Washington, Abstract 29.

35. W.A. Walker, Role of the mucosal barrier in antigen handling by the gut, *in:* "Food Allergy and Intolerance", J. Brostoff, and S.J. Challacombe (eds.), Balliére Tindall, London, pp. 209-222 (1987).

36. C.A. Gilligan, and P.A. Liwan, Oral vaccines: design and delivery, *Internat. J. Pharmaceut.* 75:1 (1991).

37. J. Mestecky, and J.H. Eldridge, Targeting and controlled release of antigens for the effective induciton of secretory antibody responses, *Curr. Opin. Immunol.* 3:492 (1991).

38. R.H. Müller (ed.), Colloidal carriers for controlled drug delivery and targeting, CRC Press, Boca Raton, FL, pp. 1-379 (1991).

39. J.N.C. Healey, Enteric coatings and delayed release, *in:* "Drug Delivery to the Gastrointestinal Tract", J.G. Hardy, S.S. Davis, and C.G. Wilson (eds.), Ellis Horwood Limited, Halsted Press, New York, pp. 83-96 (1989).

40. D. Wachsmann, J.P. Klein, M. Schöller, and R.M. Frank, Local and systemic immune response to orally administered liposome-associated soluble *S. mutans* cell wall antigens, *Immunology* 54:189 (1985).

41. F.A. Klipstein, R.F. Engert, and W.T. Sherman, Peroral immunization of rats with *Escherichia coli* heat-labile enterotoxin delivered by microspheres. *Infect. Immun.* 39:1000 (1983).

42. D.T. O'Hagan, H. Jeffery, M.J.J. Roberts, J.P. McGee, and S.S. Davis, Controlled release microparticles for vaccine development, *Vaccine* 9:768 (1991).

43. J.H. Eldridge, C.J. Hammond, J.A. Meulbroek, J.K. Staas, R.M. Gilley, and T.R. Tice, Controlled vaccine release in the gut-associated lymphoid tissues. I. Orally administered biodegradable microspheres target the Peyer's patches, *J. Controlled Release* II:205 (1990).

44. D.T. O'Hagan, D. Rahman, J.P. McGee, H. Jeffery, M.C. Davies, P. Williams, S.S. Davis, and S.J. Challacombe, Biodegradable microparticles as controlled release antigen delivery systems, *Immunology* 73:239 (1991).

45.   Z. Moldoveanu, J.K. Staas, R.M. Gilley, R. Ray, R.W. Compans, J.H. Eldridge, T.R. Tice, and J. Mestecky, Immune responses to influenza virus in orally and systemically immunized mice, *Curr. Top. Microbiol. Immunol.* 146:91 (1989).
46.   P. Cuatrecasas, Gangliosides and membrane receptors for cholera toxin, *Biochemistry* 12:3558 (1973).
47.   C.O. Elson, Cholera toxin and its subunits as potential oral adjuvants, *Curr. Top. Microbiol. Immunol.* 146:29 (1989).
48.   C.O. Elson and W. Ealding, Cholera toxin feeding did not induce oral tolerance in mice and abrogated oral tolerance to an unrelated protein antigen, *J. Immunol.* 133:2892 (1984).
49.   Y. Hirabayashi, H. Kurata, H. Funato, T. Nagamine, C. Aizawa, S. Tamura, K. Shimada, and T. Kurata, Comparison of intranasal inoculation of influenza HA vaccine combined with cholera toxin B subunit with oral or parenteral vaccination, *Vaccine* 8:243 (1990).
50.   K. Kikuta, Y. Hirabayashi, T. Nagamine, C. Aizawa, Y. Ueno, A. Oya, T. Kurata, and S. Tamura, Cross-protection against influenza B type virus infection by intranasal inoculation of the HA vaccines combined with cholera toxin B subunit, *Vaccine* 8:595 (1990).
51.   M.W. Russell and H.-Y. Wu, Distribution, persistence, and recall of serum and salivary antibody responses to peroral immunization with protein antigen I/II of *Streptococcus mutans* coupled to the cholera toxin B subunit, *Infect. Immun.* 59:4061 (1991).
52.   N. Lycke, U. Karlsson, A. Sjölander, and K.-E. Magnusson, The adjuvant action of cholera toxin is associated with an increased intestinal permeability for luminal antigens, *Scand. J. Immunol.* 33:691 (1991).
53.   J.G. Nedrud and N. Sigmund, Cholera toxin as a mucosal adjuvant: III. Antibody responses to nontarget dietary antigens are not increased, *Reg. Immunol.* 3:217 (1990/1991).
54.   R. Curtiss, S.M. Kelly, P.A. Gulig, and K. Nakayama, Selective delivery of antigens by recombinant bacteria, *Curr. Top. Microbiol. Immunol* 146:35 (1989).
55.   F. Schodel and M. Hofnung, Oral immunization using recombinant bacteria, *Res. Microbiol.* 141:745-1019 (1990).
56.   J. X. Bao and J.D. Clements, Prior immunologic experience potentiates the subsequent antibody response when *Salmonella* strains are used as vaccine carriers, *Infect. Immun.* 59:3841 (1991).
57.   N.F. Fairweather, S.N. Chatfield, A.J. Makoff, R.A. Strugnell, J. Bester, D.J. Maskell, and G. Dougan, Oral vaccination of mice against tetanus by use of a live attenuated *Salmonella* carrier, *Infect. Immun.* 58:1323 (1990).
58.   F. Schödel, D.R. Milich, and H. Will, Hepatitis B virus nucleocapsid/pre-S2 fusion proteins expressed in attenuated *Salmonella* for oral vaccination, *J. Immunol.* 145:4317 (1990).
59.   R. Curtiss, III, J.E. Galan, K. Nakayama, and S.M. Kelly, Stabilization of recombinant avirulent vaccine strains *in vivo*, *Res. Microbiol* 141:797 (1990).
60.   D.A. Herrington, L.Van De Verg, S.B. Formal, T.L. Hale, B.D. Tall, S.J. Cryz, E.C. Tramont, and M.M. Levine, Studies in volunteers to evaluate candidate *Shigella* vaccines: further experience with a bivalent *Salmonella typhi-Shigella sonnei* vaccine and protection conferred by previous *Shigella sonnei* disease, *Vaccine* 8:353 (1990).
61.   C.K. Stover, V.F. de la Cruz, T.R. Fuerst, J.E. Burlein, L.A. Benson, L.T. Bennett, G.P. Bansal, J.F. Young, M.H. Lee, G.F. Hatfull, S.B. Snapper, R.G. Barletta, W.R. Jacobs, Jr., and B.R. Bloom, New use of BCG for recombinant vaccines, *Nature* 351:456 (1991).
62.   A. Aldovini and R.A. Young, Humoral and cell-mediated immune responses to live recombinant BCG-HIV vaccine, *Nature* 351:479 (1991).
63.   A. Kantele, J.M. Kantele, H. Avrilommi, and P.H. Mäkelä, Active immunity is seen as a reduction in the cell response to oral live vaccine, *Vaccine* 9:428 (1991).
64.   J.X. Bao and J.D. Clements, Prior immunologic experience potentiates the subsequent antibody response when *Salmonella* strains are used as vaccine carriers, *Infect. Immun.* 59:3814 (1991).

BACTERIAL DISEASES OF THE ORAL TISSUES

Francis L. Macrina

Department of Microbiology and Immunology
Virginia Commonwealth University
Richmond, VA 23298-0678

OVERVIEW OF DENTAL CARIES

Human oral diseases of bacterial origin can be divided into two categories: those of the hard tissues (dental decay) and those of the soft tissues (periodontal disease). Coronal dental caries is associated with a group of *Streptococcus* species collectively referred to as the mutans streptococci. Owing to a variety of factors, the mutans streptococci have assumed a prominent role as the agents of smooth surface dental caries. First, they have been intensely investigated by molecular taxonomic techniques resulting in an understanding of their phylogenetic relationships to one another and to streptococci in general. Second, their ability to fulfill Koch's Postulates in rodent animal models has been clearly established. Third, the mutans streptococci have yielded to molecular analyses providing the means to genetically and biochemically dissect those factors involved in oral colonization and caries virulence. Such genetic and biochemical analyses have been followed by appropriate evaluation in *in vitro* and *in vivo* model systems. These approaches are illuminating the process of caries pathogenesis while, at the same time, opening avenues for the prevention and control of caries including the development of novel vaccines. Dental decay of the tooth root surface is recognized as being of bacterial etiology but a precise definition of the agent(s) involved is presently unavailable. Although the application of the approaches used to study coronal caries has not yet been implemented in exploring root surface caries, there are no *a priori* reasons to believe that similar advances cannot be made in our understanding of this oral tissue disease.

OVERVIEW OF PERIODONTAL DISEASES

Our knowledge of bacterial diseases of the soft tissues of the human oral cavity is limited compared to all aspects of caries microbiology. Diseases of the periodontium represent a collection of pathologic processes, likely to have differing etiologies. Gingival inflammation may present itself in chronic or acute forms. Our knowledge of the bacterial etiology of the acute form indicates this disease is associated with gram-negative anaerobes and spirochetal bacteria. The chronic form of gingivitis has a nonspecific bacterial

Genetically Engineered Vaccines, Edited by
J.E. Ciardi *et al.*, Plenum Press, New York, 1992

etiology. The microbiology of adult and early onset periodontitis is beginning to come into focus with the indictment of causative organisms such as *Porphyromonas gingivalis*, *Actinobacillus actinomycetemcomitans*, *Prevotella intermedia* and *Fusobacterium nucleatum*. Molecular taxonomic methods are being applied to potentially periodontipathic organisms, but much work is needed to better appreciate the phylogeny of the organisms which populate the subgingival plaque. Until recently, there has been a lack of facile animal models and *in vitro* systems in which to evaluate pathogenesis and to test Koch's Postulates. This has been a problem in establishing the unequivocal definition of etiologic agents. Many of the organisms of the subgingival plaque are aerosensitive anaerobes and/or are nutritionally fastidious making their growth and laboratory manipulation difficult. Finally, the development of genetic systems in these organisms has been slow-moving for a variety of reasons. This has prevented the rapid identification and analysis of colonization and virulence factors. Nonetheless, work in these latter areas is beginning to establish a momentum that is likely to lead to breakthroughs in our understanding of specific agents of disease and their virulence determinants.

This chapter briefly reviews the pathology and bacterial etiology of dental caries and periodontal diseases. I also will review our understanding of some of the more important bacterial factors that are established or suspected virulence factors. In this context the focus will be on *S. mutans*-mediated smooth surface decay and on the periodontal pathology associated with *P. gingivalis* and *A. actinomycetemcomitans*.

SMOOTH SURFACE DENTAL CARIES

Dental caries has long been recognized as a multifactorial disease involving the inciting microbial agent, dietary sucrose, and susceptible teeth.[1,2] The interaction of these three factors to produce caries is also dependent on time. Based on epidemiological data and on the fulfillment of Koch's Postulates in animals,[1] *Streptococcus mutans* and some related species of streptococci (the mutans streptococci) are recognized as the etiologic agents of smooth surface dental decay. *S. mutans* is the species most commonly isolated in humans followed by *S. sobrinus*. Other related species include *S. cricetus*, *S. rattus*, *S. ferus*, and *S. macacae*.[3] The processes leading to the formation of dental caries by *S. mutans* may be summarized as follows. Attachment and accumulation of the organisms to the tooth surface is crucial. Initial attachment is believed to be independent of dietary sucrose and recent evidence supports this position. Specifically, a surface protein or proteins of *S. mutans* appear to be involved in the initial cellular adherence to the saliva-coated hydroxyapatite surface.[4,5] *S. mutans* establishment in the oral cavity is also likely to involve adhesive interactions between *S. mutans* and other oral microorganisms.[6] Sucrose-independent colonization of the tooth surface eventually gives way to the tenacious attachment and significant accumulation of cells mediated by polymer formation which is dependent on the presence of dietary sucrose. Such polymers include water soluble and water insoluble glucans as well as fructans. Using mutant strains of *S. mutans*, it has been possible to implicate the synthesis of certain of the polymers made from sucrose in caries pathogenesis. Indeed the role for insoluble glucans in cariogenicity was first suggested from the studies of mutants isolated over 20 years ago.[7] The ultimate pathology in *S. mutans*-mediated cariogenicity is due to bacterial acid production from the fermentation of dietary substrates, including sucrose.[1] The acidogenic and aciduric nature of *S. mutans* established in the dental plaque matrix translates to the long-term exposure of the enamel surface to metabolic acids. This effect is believed to be enhanced by the plaque itself interfering with the buffering effects of saliva.

MOLECULAR CLONING OF POTENTIAL VIRULENCE FACTORS FROM *S. MUTANS*

The cloning of genes postulated to be involved in caries virulence has been driven by a number of issues. Because many streptococcal genes are readily expressed in *Escherichia*

Table 1. Properties of Isogenic Mutants Constructed by Allelic Exchange.

Mutated Gene	Product/Activity	Evaluation	Virulence	Reference
gtfB/C	insoluble and soluble glucan	rodent caries model	reduced	Munro et al.[9]
gtfD	soluble glucan	rodent caries model	(wild type)[A]	Munro et al.[9]
ftf	fructan	rodent caries model	reduced	Schroeder et al.[14]
gtfA	sucrose phosphorylase	rodent caries model	wild type	Barletta et al.[10]
scrA/B	sucrose transport	rodent caries model	wild type	Macrina et al.[11]
spaP1	surface protein antigen	salivary induced agglutination	reduced	Bleiweis et al.[12]
		coated hydroxy -apatite adherence	reduced	Bleiweis at al.[12]
		rodent caries model	wild type	Bowen et al.[5]

[A] Little or no contribution to virulence inferred from *gtfD* mutation failure to further reduce virulence in a *gtfB/C* mutant *coli*, cloned genes immediately provide the means to overproduce and study the gene product. Of equal, if not greater, importance is the ability to create a defective copy of the gene in question using standard recombinant DNA methods in *E. coli*.[8] This defective gene then can be introduced into the *S. mutans* cell, usually by transformation resulting in the construction of a so-called allelic exchange mutant where the wild-type gene is replaced by the defective copy. This recombinational event is easily detected because the insertional inactivation of the cloned gene uses a directly selectable marker such as an antibiotic resistance gene.[8] Design of such constructs is straightforward, but it requires some information about the structure of the gene targeted for inactivation. Recombination into the chromosome is via additive integration, and the genotype of the mutant can be definitively evaluated using Southern blot analysis to confirm the predicted recombinational products. To ensure the validity of the phenotype of mutants constructed in such a fashion, one must be aware of potential pitfalls. Specifically, untoward effects of the insertion, such as polarity on downstream genes, must be explored by evaluating mutants with insertions in downstream regions and/or by scoring the original mutant for a wide variety of phenotypes to confirm the absence of other measurable effects of the insertion. Where possible, the addition *in trans* of a wild-type copy of the gene (e.g., on a plasmid) into the allelic exchange mutant with an observed restoration of the original phenotype provides definitive proof of the specific effect of the mutation. Properly constructed and verified allelic exchange mutants provide powerful tools for the analysis of virulence factors in bacteria. Such mutants can be evaluated *in vitro* to assess biochemical and physiological implications of the loss of the gene in question. Animal testing of allelic exchange mutants provides the means to assess the relative importance of specific genes and their gene products as colonization and virulence factors. Thus, such mutants can provide information that can lead to the design of rational intervention strategies, including the identification of potential immunogens for antibacterial vaccines. Indeed, the method of allelic exchange itself can be used to create strains of reduced virulence which, in turn, can be considered for use as whole cell vaccine candidates.

The application of recombinant DNA to the mutans streptococci has resulted in the isolation of a variety of genes thought to be involved in colonization or virulence. Such cloned genes in many instances have provided the raw material for the construction of allelic exchange mutants which have been comparatively studied *in vitro* and *in vivo*. Table 1 is a compilation of some of the results obtained from such experimental approaches. The table is not comprehensive in terms of listing all cloned sequences from the mutans streptococci; rather, it focuses especially, but not exclusively, on examples where *in vivo* virulence testing has been used to evaluate the mutant strain.

The study of mutants of *S. mutans* created by allelic exchange is beginning to provide a framework for pathogenesis. As expected from early mutagenesis studies,[7,13] insoluble glucans are important for virulence probably by promoting tenacious adherence of the cariogenic bacteria to the teeth.[9] Soluble glucans appear less important in contributing to virulence. Fructan synthesis has been shown to have a role in virulence but whether this involves fructans as extracellular storage compounds, adherence factors, or both, remains to be clearly established.[14] The SpaP1 surface protein antigen (also called antigen B, antigen I/II or SpaA) is an abundant surface protein that appears to be involved in the adherence of cells to saliva-coated hydroxyapatite.[5] However, when tested in a rodent caries model, which uses animals fed a diet containing 56 % sucrose, SpaP1-deficient mutants constructed by allelic exchange were still cariogenic when compared to wild type.[5] The increasing number of available cloned genes from the mutans streptococci bodes well for continued efforts in constructing strains with single as well as multiple mutations which then can be used to evaluate virulence. Strategies should build on available genetic information as well. For example, chemically induced mutants isolated by a variety of means indicate a role for lactate dehydrogenase,[15] intracellular polysaccharide,[16] dextranase[17] and the SpaA protein[17] in virulence. The role of SpaA (see above synonyms) needs clarification in light of a seemingly conflicting report of its importance in cariogenicity.[5] This issue is likely to be reconciled as the above- and below-mentioned caveats and pitfalls are sorted out in further studies.

In addition to the genetic caveats of allelic exchange mentioned above, investigators should be cautious in the interpretation of results owing to other pitfalls. Of crucial importance is the extrapolation or comparison of results when bacteriologic or animal model system variables exist. Different strains of *S. mutans* are known to possess differing degrees of virulence in the same animal models (S. Michalek, pers. communication). Results can differ significantly depending on whether gnotobiotic or specific pathogen-free animals are used.[13] Physiologic parameters such as control of saliva flow in the animals can also strongly influence caries scores.[5] Finally, the necessity of including sucrose in the animal's diet in order to induce caries formation needs to be considered. For example, it is reasonable to expect that systems which use 5% dietary sucrose[14] will yield differing results from those that employ 56% sucrose.[5]

ROOT SURFACE CARIES

Lesions affecting the cementum and dentin of the tooth surface are called root surface caries.[18] These lesions are generally found in adults originating at the cemental-enamel junction and are clearly associated with dental plaque. Epidemiologically, filamentous bacteria (*Actinomyces*) are associated with human root surface lesions. This association has been strengthened by the ability of certain *Actinomyces* species to cause root surface caries in rodent animal models. Based on animal model studies, *A. viscosus* as well as other species such as *naeslundii* have been implicated. A role for streptococcal species including *salivarius, mutans,* and *sanguis* has also been suggested in the initiation and progression of root surface lesions. Bacterial factors involved in root surface caries pathology are not defined. Elements of initial attachment and subsequent accumulation need to be identified. Intuitively, the ultimate destruction of the tooth surface in this process is most likely to be the dissolution of the tooth surface mediated by prolonged contact of bacterial metabolic by-products with the tissue surface. The use of animal models along with the application of molecular genetic methods currently being used to explore coronal surface caries should further our understanding of this disease. Such information can then be applied to the design of strategies for intervention or prevention of root surface caries.

BACTERIOLOGY OF PERIODONTAL DISEASES

Periodontal disease is a phrase used to describe a number of different pathologic processes all of which have a bacterial etiology.[19] The complexity of the human subgingival microflora population, along with the aerosensitivity and nutritional fastidiousness of many of the potentially periodontipathic organisms has made the definition of etiologic agents and their virulence factors difficult. Also contributing to this issue has been the lack of good animal models in which specific infectious processes can be evaluated. Modern bacteriologic techniques and the application of molecular taxonomic, genetic and recombinant DNA techniques have helped move the field ahead in the past decade.

Gingivitis is defined as an inflammation of the gingival tissues.[20] It is marked by gingival swelling and bleeding and can involve exudate formation in the gingival crevice. Gingivitis is further characterized as not involving detachment of any of the periodontal structures or loss of alveolar bone. Thus, gingivitis is considered to be a reversible process; health can be restored by proper oral hygiene. Although generally recognized as being microbially non-specific in origin, a few species have been observed to be elevated consistently and significantly in human epidemiological studies (Table 2). Bacterial virulence factors have not been identified in this process, and the pathology seen in large part is probably due to host immune responses which have destructive effects on the gingival tissues.[20]

While non-specific gingivitis is recognized as a chronic process, an acute form of gingival inflammation has been long recognized.[20] Necrotizing ulcerative gingivitis (NUG) is an acute disease. Predisposition to NUG appears related to factors such as physiologic and/or psychologic stress and the pathology of the disease is associated with a specific microbial flora (see Table 2). NUG is characterized by necrotic ulcerative lesions that often result in destruction of tips of the gingival papillae. The affected areas are painful and subject to bleeding. Successful treatment of the disease involves professional debridement of plaque and necrotic tissue followed by carefully implemented personal oral hygiene and plaque control. Antibiotic therapy is recommended when systemic symptoms (e.g., fever) are apparent.

Table 2. Summary of bacterial agents implicated in human periodontal diseases.

Disease	Primary Agent(s)	Evidence[A]
gingivitis - non-specific	*Actinomyces naeslundii* II *Campylobacter concisus* *Streptococcus anginosus* *Streptococcus sanguis*	epidemiological studies animal studies
Necrotizing ulcerative gingivitis	*Prevotella intermedia* spirochetes *Fusobacterium* sp.	epidemiological studies clinical response
Adult periodontitis[B]	*Porphyromonas gingivalis* *Prevotella intermedia*	epidemiological studies animal studies immune response clinical response
Localized early-onset periodontitis	*Actinobacillus actinomycetemcomitans*	epidemiological studies immune response clinical response

[A] Epidemiological studies refer to data collected using healthy or diseased human subjects. Animal studies refer to use of one or more of the indicted organisms in virulence assessment using an *in vivo* system. Clinical response refers to an arrest or reversal of the disease process being associated with removal or inhibition of the organism(s) in question. Immune response refers to a specific antibody response to the organism(s) in question.

[B] Other organisms which may play a role in initiation or progression of adult periodontitis include *Wolinella sp.*, *Eubacterium sp.*, *Selenomonas sp.*, *Eikenella sp.*, *Prevotella sp.*, *Bacteroides forsythus*, and spirochetes.

Adult peiodontitis[20] refers to an inflammatory process which is an extension of gingivitis into the deeper structures of the periodontium. This chronic process is characterized by the formation of a periodontal pocket. This pathologically deepened gingival sulcus eventually is accompanied by alveolar bone loss, detachment of the periodontal ligament, and subsequent tooth mobility. Although causative microorganisms for this process have been sought over the years, only recently have a limited number of species been indicted in the etiology of periodontitis (see Table 2). Factors involved in bacterial virulence are also starting to emerge but a clear picture of pathogenesis will require a systematic dissection of the inciting organisms along with appropriate *in vivo* and *in vitro* virulence testing of wild-type and mutant strains. Moreover, it is recognized that host factors play a role in contributing to the disease process by mounting inappropriate immune responses which have destructive effects on the periodontium.

Localized early onset periodontitis (also called juvenile periodontitis) is characterized by periodontal pocket formation principally around first molars and incisors.[20] It is a disease process that occurs in teenagers and young adults and it results in the loss of periodontal attachment structures as well as alveolar bone loss. In localized early onset periodontitis, these symptoms usually occur in the absence of plaque or calculus formation and without pre-existing gingivitis. There seems to be a genetic predisposition to localized early onset periodontitis based on familial occurrence of the disease and mounting evidence provides strong support that the gram-negative, capnophilic, coccobacillus, *A. actinomycetemcomitans* is the principal etiologic agent of this disease. Among other factors, the genetic predisposition of this localized disease may be due to chemotactic disorders in neutrophils. Finally, it should be noted that a generalized form of early onset periodontitis is recognized which, unlike the localized form, is non-self limiting in nature. The generalized form can result in extensive destruction of the periodontium in young adults. The bacterial etiology of the generalized form is not yet clear but, at present, *P. gingivalis* and other flora seem associated with the disease.

P. GINGIVALIS AND *A. ACTINOMYCETEMCOMITANS* VIRULENCE

Because both *A. actinomycetemcomitans* and *P. gingivalis* are becoming increasingly recognized in the etiology of specific periodontal diseases (see above) the exploration of the basis of their virulence is moving forward using the molecular techniques that have yielded success with other pathogens. Potential virulence factors of *P. gingivalis* have been reviewed[21] and include fimbriae, hemagglutinin, capsule, lipopolysaccharide, proteases (including a native collagenase), and a variety of non-specific cytotoxins (e.g., ammonia, acids). Likewise, potential virulence factors of *A. actinomycetemcomitans* include collagenase, fibroblast inhibition factor, epitheliotoxin, bone resorbing factor(s), leukocyte migration inhibition factor, lipopolysaccharide, and a leukotoxin.[22,23] Moreover, recent evidence indicates that *A. actinomycetemcomitans* is capable of entering and surviving within a human oral epithelial cell line, suggesting the presence of an invasion-like virulence factor as well.[24]

Such information on potential virulence factors is leading the way toward the assessment of pathogenicity. Being able to clone, identify, and characterize genes of interest should give rise to the construction of gene sequences that can be used in preparing allelic exchange mutants. Consideration should be given to problems associated with gene expression, especially in the case of *P. gingivalis*, since this group of organisms (formerly genus *Bacteroides*) is distantly[25] related to *E. coli*, the organism commonly used in recombinant DNA work. The use of expression vectors to achieve transcription and translation of some *P. gingivalis* genes may be necessary. For example, genetic determinants encoding hemagglutinin(s)[26,27] and for a protease[28] have been cloned and expressed in *E. coli* with little difficulty. On the other hand, the *P. gingivalis* gene for superoxide dismutase has also been cloned in *E. coli* but its expression appears to be governed by a plasmid vector promoter.[29] The structural gene for the *P. gingivalis* fimbrial subunit also does not seem to be readily expressed in *E. coli* .[30] Problems of this type can be solved by using genetic expression vectors. The determination of the nucleotide sequence of cloned fragments also can be helpful in defining specific genes. Introduction of specific mutated genes into *P. gingivalis* provides a challenge at present. Whether or not DNA can be introduced into this species by electroporation remains to be established. If this methodology proves unsuccessful, then the use of conjugational mobilization systems to introduce mutated genes would be a promising avenue. Such systems have been developed for other gram-negative anaerobes, such as *Bacteroides fragilis,* and have been used successfully to construct allelic exchange mutants.[31] The *B. fragilis* tools could be tried directly in *P. gingivalis* or similar systems adapted for use in *Porphyromonas*.

Genes cloned from *A. actinomycetemcomitans* include a leukotoxin with killing specificity for human neutrophils and monocytes,[32,33] surface proteins,[34] and proteins reactive to antisera from patients with periodontitis.[35] Thus far, *A. actinomycetemcomitans* genes cloned in *E. coli* have been transcribed and translated with no apparent difficulties. Manipulation of cloned genes in the *Actinobacillus* host cell should be readily accomplished in light of the recent development of an electroporation system and useful *E. coli-A. actinomycetemcomitans* shuttle plasmids.[36]

CONCLUDING THOUGHTS

The study of the biochemical genetic basis of colonization and virulence factors of many oral microflora has obvious implications for the control and prevention of the diseases reviewed here. One should consider, however, that there are other compelling reasons for our investigation of the human oral microflora. Oral streptococci of the viridans group account for more than 50% of all cases of human native valve endocarditis[37]; moreover, cases of endocarditis caused by *A. actinomycetemcomitans* are not uncommon. Dental plaque is also a repository for obligately anaerobic bacteria like *Porphyromonas, Bacteriodes,* and *Fusobacterium* which are common causes of pulmonary[38] and brain abscesses.[39] *Candida* yeast infections of oral origin present a significant threat to the immunocompromised patient.[40] As we probe the mechanisms of microbial virulence involved in oral infections, it is likely that we will achieve a better understanding of how to approach and control extra-oral infections caused by the indigenous oral microorganisms. The oral microbial population also holds important questions of microbial ecology and evolution which await exploration. The extent to which horizontal gene transfer occurs between oral microflora and other indigenous microflora is unclear. The dissemination of drug resistance genes and toxin genes among procaryotes has important human health implications. Antibiotic resistance genes which have relatives existing in a wide variety of procaryotes (e.g., macrolide-lincosamide resistance determinants) have been demonstrated in oral microflora[41] and such organisms may be a significant reservoir for resistance genes. A toxin gene found in *A. actinomycetemcomitans* has been determined to be a member of a family of pore-forming cytotoxins which appear to be disseminated in a diverse group of gram-negative pathogens.[42] The study of such systems promises to contribute to our understanding of virulence mechanisms as well as to the evolutionary pathways operative in procaryotic pathogens.

ACKNOWLEDGMENTS

Work from my laboratory was supported by NIH grants DE04224 and DE09035. I thank R.J. Genco for his helpful comments regarding Table 2.

REFERENCES

1. W.J. Loesche, Role of *Streptococcus mutans* in human dental decay, *Microbiol. Rev.* 50:353 (1986).
2. S. Hamada and H.D. Slade, Biology, immunology and cariogenicity of *Streptococcus mutans, Microbiol. Rev.* 44:331 (1980).
3. A. Coykendall, Classification and identification of the viridans streptococci, *Clin. Microbiol. Revs.* 2:315 (1989).
4. E. Kishimoto, D.I. Hay, and R.J. Gibbons, A human salivary protein which promotes adhesion of *Streptococcus mutans* serotype c strains to hydroxyapatite, *Infect. Immun.* 57:3702 (1989).
5. W.H. Bowen, K. Schilling, E. Giertsen, S. Pearson, S.F. Lee, A. Bleiweis, and D. Beeman, Role of a cell surface associated protein in adherence and dental caries, *Infect. Immun.* 59:4606 (1991).
6. R.J. Lamont, D.R. Demuth, C.A. Davis, D. Malamud, and B. Rosan, Salivary-agglutinin-mediated adherence of *Streptococcus mutans* to early plaque bacteria, *Infect. Immun.* 59:3446 (1991).
7. J.D. de Stoppellaar, K. Konig, A. Plasschaert, and J. van der Hoeven, Decreased cariogenicity of a mutant of *Streptococcus mutans, Arch. Oral. Biol.* 16:971 (1971).

8. F.L. Macrina, M.T. Dertzbaugh, M.C. Halula, E.R. Krah, and K.R. Jones, Genetic approaches to the study of oral microflora: a review, *CRC Crit. Rev. in Oral Biol. and Med.* 1:207 (1990).

9. C. Munro, S.M. Michalek, and F.L. Macrina, Cariogenicity of *Streptococcus mutans* V403 glucosyltransferase and fructosyltransferase mutants constructed by allelic exchange, *Infect. Immun.* 59:2316 (1991).

10. R.G. Barletta, S.M. Michalek, and R. Curtiss, III, Analysis of the virulence of *Streptococcus mutans* serotype c mutants in the rat model system, *Infect. Immun.* 56:322 (1988).

11. F.L. Macrina, K.R. Jones, C.-A. Alpert, B.M. Chassy, and S.M. Michalek, Repeated DNA sequence involved in mutations affecting transport of sucrose into *Streptococcus mutans* V403 via the phosphoenolpyruvate phosphotransferase system, *Infect. Immun.* 59:1535 (1991).

12. A.S. Bleiweis, S.F. Lee, L.J. Brady, A. Progulske-Fox, and P.J. Crowley, Cloning and inactivation of the gene responsible for a major surface antigen on *Streptococcus mutans*, *Arch. Oral Biol.* 35 Suppl.:15S (1990).

13. J.M. Tanzer, M.L. Freedman, R.J. Fitzgerald, and R.H. Larson, Diminished virulence of glucan synthesis defective mutants of *Streptococcus mutans*, *Infect. Immun.* 10:197 (1974).

14. V.A. Schroeder, S.M. Michalek, and F.L. Macrina, Biochemical characterization and evaluation of virulence of a fructosyltransferase-deficient mutant of *Streptococcus mutans* V403, *Infect. Immun.* 57:3560 (1989).

15. K.P. Johnson, S.M. Gross, and J.D. Hillman, Cariogenic potential in vitro in man and in vivo in the rat of lactate dehydorgenase mutants of *Streptococcus mutans*, *Arch. Oral Biol.* 25:707 (1980).

16. J. Tanzer and M. Freedman, Genetic alterations of *Streptococcus mutans* virulence, *Adv. Exp. Med. Biol.* 107:661 (1978).

17. R. Curtiss, R. Goldschmidt, R. Pastan, M. Lyons, S. Michalek, and J. Mestecky, Cloning virulence determinants from *Streptococcus mutans* and the use of recombinant clones to construct bivalent oral vaccine strains to confer protective immunity against *S. mutans*-induced dental caries, *in* "Molecular Microbiology and Immunology of *Streptococcus mutans*", S. Hamada, S. Michalek, H. Kiyono, J. McGhee, and L. Menaker, eds., pp. 173-180, Elsevier Science Publishers, Amsterdam (1986).

18. L.E. Wolinsky, Caries and Cariology, *in* "Oral Microbiology and Immunology", M.G. Newman and R. Ninsegard, eds., pp. 389-309, W.B. Saunders Co., Philadelphia (1988).

19. J. Zambon, Microbiology of Periodontal Disease, *in* "Contemporary Periodontics", R.J. Genco, H.M. Goldman, and D.W. Cohen, eds., pp. 147-160, C.V.Mosby Co., St. Louis (1990).

20. R.J. Genco, Classification and clinical and radiographic features of periodontal disease, *in* "Contemporary Periodontics", R. Genco, H.M. Goleman, and D.W. Cohen, eds., pp. 63-81, C.V. Mosby Co., St. Louis.

21. T.J.M. van Steenbergen, A.J. van Winkelhoff, and J. de Graff, Black pigmented oral anaerobic rods: classification and role in periodontal disease. *in* "Periodontal Disease - Pathogens and Host Immune Response", S. Hamada, S.C. Holt, and J.R. McGhee, eds., pp. 41-52, Quintessence Publishing Co. Ltd, Chicago (1991).

22. J. Slots, and S.E. Schonfeld, *Actinobacillus actinomycetemcomitans* in localized juvenile periodontitis, *in* "Periodontal Disease - Pathogens and Host Immune Response", S. Hamada, S. Holt, and J.R. McGhee, eds., pp. 53-64, Quintessence Publishing Co., Ltd., Tokyo (1991).

23. J. Zambon, *Actinobacillus actinomycetemcomitans* in human periodontal disease, *J. Clin. Periodontol.* 12:1 (1985).

24. D.H. Meyer, P.K. Sreenivasan, and P.M. Fives-Taylor, Evidence for invasion of a human oral cell line by *Actinobacillus actinomycetemcomitans*, *Infect. Immun.* 59:2719 (1991).

25. W.G. Weisburg, Y. Oyaizu, H. Oyaizu, and C.R. Woese, Natural relationship between *Bacteroides* and flavobacteria, *J. Bacteriol.* 164:230 (1985).

26. B.C. McBride, A. Joe, and U. Singh, Cloning of *Bacteroides gingivalis* surface antigens involved in adherence, *Arch. Oral Biol.* 35 Suppl.:59S (1990).

27. A. Progulske-Fox, S. Tumwasorn, and S.C. Holt, The expression and function of a *Bacteroides gingivalis* hemagglutinin gene in *Escherichia coli, Oral. Microbiol. Immunol.* 4:121 (1989).

28. M.A. Arnott, G. Rigg, H. Shah, D. Williams, A. Wallace, and I.S. Roberts, Cloning and expression of a *Porphyromonas (Bacteroides) gingivalis* protease gene in *Escherichia coli, Arch. Oral Biol.* 35 Suppl.:97S (1990).

29. J. Choi, N. Takahashi, T. Kato, and H. Kuramitsu, Isolation, expression, and nucleotide sequence of the sod gene from *Porphyromonas gingivalis, Infect. Immun.* 59:1564 (1991).

30. D.P. Dickinson, M.A. Kubiniec, F. Yoshimura, and R.J. Genco, Molecular cloning and sequencing of the gene encoding the fimbrial subunit protein of *Bacteroides gingivalis, J. Bacteriol.* 170:1658 (1988).

31. E.P. Guthrie, and A.A. Salyers, Use of targeted insertional mutagenesis to determine whether chondroitin lyase II is essential for chondroitin sulfate utilization by *Bacteroides thetaiotaomicron, J. Bacteriol.* 166:966 (1986).

32. E.T. Lally, I.R. Kieba, D.R. Demuth, J. Rosenbloom, E.E. Golub, N.S. Taichman, and C.W. Gibson, Identification and expression of the *Actinobacillus actinomycetemcomitans* leukotoxin gene, *Biochem. Biophys. Res. Commun.* 159:256 (1989).

33. D. Kolodrubetz, T. Dailey, J. Ebersole, and E. Kraig, Cloning and expression of the leukotoxin gene from *Actinobacillus actinomycetemcomitans, Infect. Immun.* 57:1465 (1989).

34. G.J. Sunday, J.J. Zambon, and R.J. Genco, Molecular cloning and expression of antigens from *Actinobacillus actinomycetemcomitans* in *Escherichia coli, Arch. Oral Biol.* 35 Suppl.:85S (1990).

35. S. Arakawa, S. Hata, I. Ishikawa, and N. Tsuchida, Gene cloning of an *Actinobacillus actinomycetemcomitans* Y4 antigen which reacts with peripheral blood sera in patients with advanced destructive periodontitis, *Arch. Oral Biol.* 35 Suppl.:93S (1990) .

36. P.K. Reenivasan, D.J. LeBlanc, L.N. Lee, and P. Fives-Taylor, Transformation of *Actinobacillus actinomycetemcomitans* by electroporation, utilizing constructed shuttle plasmids, *Infect. Immun.* 59:4621 (1991).

37. R. Bayliss, C. Clark, C. Oakley, W. Somerville, A. Whitfield, and S. Young, The microbiology and pathogenesis of infective endocarditis, *Br. Heart. Journal* 50:513 (1983).

38. S.M. Finegold, Lung Abscess. *in* "Principals and Practices of Infectious Disease", G. Bartlett, S. Gorbach, G. Mandell, R. Douglas, and J. Bennett, eds., pp. 560-564, Churchill-Livingston, New York (1990).

39. C. Chun, J. Johnson, M. Hofstetter, and M. Raff, Brain abscess: a study of 45 consecutive cases, *Medicine* 65:415 (1986).

40. J.E. Edwards, Candidiasis. *in* "Principals and Practices of Infectious Disease", G. Bartlett, S. Gorbach, G. Mandell, R. Douglas, and J. Bennett, eds., pp. 1943-1958, Churchill-Livingston, New York (1990).

41. D.A. Odelson, J.L. Rasmussen, C.J. Smith, and F.L. Macrina, Extrachromosomal systems and gene transmission in anaerobic bacteria, *Plasmid* 17:87 (1987).

42. R.A. Welch. Pore-forming cytolysins of gram-negative bacteria, *Molec. Microbiol.* 5:521 (1991).

ORAL VIRUS INFECTIONS: THE POTENTIAL FOR GENE TRANSFER IN

TREATMENT AND PREVENTION

Cherrilee Steele and Edward J. Shillitoe

Department of Microbiology
Dental Branch
University of Texas Health Science Center at Houston
Houston, TX 77225

INTRODUCTION

The oral cavity is the site of a small number of virus infections, which have been studied closely in recent years. Herpes simplex virus type-1 (HSV-1) can cause primary or recurrent infections of the oral mucosa and lips. Recurrent infections are due to a reactivation of latent virus that is residing in a local sensory nerve ganglion, and are usually more troublesome than primary infections. Viruses that are in a latent state are protected from many antiviral drugs since their metabolism is so limited (Garcia-Blanco and Cullen, 1991) and they avoid the immune system in several ways (Oldstone, 1991). Thus no effective vaccine for HSV-1 has become available, despite a large amount of promising research (Cohen, this volume; Rooney et al., this volume). Human papillomaviruses (HPVs) represent the other major group of oral viruses and are found in many oral lesions (Chang et al., 1991). It is possible that oral cancer is due, in many cases, to the effects of papillomaviruses (Steele and Shillitoe, 1991). Relatively little is known about the immune responses to HPV since they cannot be studied in cell culture, but good progress may be expected in the near future since the genes for the major antigens have now been cloned (Galloway and McDougall, 1989).

Although progress toward a vaccine for HSV-1 or HPVs has been slow, more and more information has accumulated about the molecular biology of these and many other viruses. This has led to the prospect of controlling infections by the deliberate manipulation of the target cells at the molecular level. Such approaches have become known as "intracellular immunization" when they are applied to non-infected cells for the sake of protection, and as "gene therapy" when applied in cases where an infection already exists. The term "gene transfer" covers both situations. This manipulation of cells or organisms has enormous potential for the correction of a number of deficiency diseases or metabolic diseases as well as infectious disease, but this review will be limited to those approaches that have promise for the control of oral virus infections.

INTRACELLULAR EXPRESSION OF ANTIVIRAL MOLECULES

In theory it is possible to protect cells from viruses by giving them genes to express antiviral antibodies within their own cytoplasm. To date, research using this approach has been sparse but encouraging. Biocca et al. (1990) demonstrated that if the genes for IgM molecules are modified by removal of hydrophobic leader sequences and are transferred to epithelial cells, the antibody heavy and light chains can be synthesized and assembled in the cytoplasm. Furthermore, if the leader sequence of the light chain was replaced with a fragment that encodes a nuclear localization signal the antibodies could be targeted to the nucleus of the cell. Although no biological effect of the antibody was demonstrated it clearly could be important in protection against viruses. Another example of intracellular expression of antibody molecules is a study by Werge et al., (1990) using intracellularly expressed monoclonal antibodies in transfected NIH 3T3 cells to inhibit oncogenic transformation by the *ras* oncogene. Although they were able to induce expression of the antibody in the cell in reasonable concentration they did not observe any reduction in $p21^{ras}$ activity, possibly due to separation of the antibody and its target in different cell compartments. However the antigen-binding ability of the antibody was confirmed, and immunological effects of intracellular antibody still seem to be feasible.

An elaborate scheme for the use of this type of system to protect cells from the human immunodeficiency virus (HIV) was proposed by Faraji-Shadan et al. (1990). They suggested that genes encoding antibody to the viral reverse transcriptase enzyme could be used. Immunoglobulin mRNA could be isolated from B lymphocytes that were immunized *in vitro* and be amplified by the reverse polymerase chain reaction (PCR). The cDNA obtained from the PCR reaction could then be transferred to stem cells of HIV-infected individuals and the cells would make intracellular antibody and be protected from the virus. The patients could then be irradiated to remove their existing, infected lymphocytes, and receive a transplant of their own, virus-free immune stem cells. Although no stage of this complicated approach appears to have been actually performed, it is based on phenomena that are confirmed to exist and uses technology that is in routine use in some laboratories.

Cells often show some natural resistance to viruses by expression of the interferon (IFN) family of proteins, and overexpression of these proteins in the correct circumstances should increase the level of protection. Human IFN-induced MxA and MxB genes are related to the murine Mx1 protein which confers resistance to influenza infection. It has been shown that the transfer of the gene for MxA to cultured cells results in an increased resistance to infection by influenza A virus and to vesicular stomatitis virus in cells that are normally susceptible to these viruses (Pavlovic et al., 1990). Another protein that is increased in cells by IFN is the (2'-5') oligo A synthetase. When the gene that encodes this protein was transferred to CHO cells and expressed constitutively, the cells became resistant to picornaviruses (Chebath et al., 1987). The IFN system can evidently be exploited in several ways so as to increase resistance of cells to a number of RNA viruses. However the major oral viruses, HSV-1 and the HPVs, are DNA viruses that are relatively resistant to IFN, and so it is not clear that this approach will lead to increased treatment or protection against oral virus infections.

Although most work using gene transfer to induce protection against viruses has been performed in cell culture, it appears that protection can also be effective in entire animals. Arnheiter et al. (1990) produced transgenic mice that express intracellular Mx1 protein under the control of an interferon-responsive promotor. These constructs were injected into the pronuclei of mouse embryos prior to transplantation into pseudopregnant females. Offspring were divided into three groups; negative or controls, low-responders and high-responders. After challenge with live virus, it was found that the animals in the high-responder group were resistant to infectious doses up to 50,000 LD_{50} whereas the majority of the low-responder animals were only resistant in the 5-50 LD_{50} range. All mice negative for the expression of the Mx1 gene were sensitive and easily infected at dosages similar to that of the control populations. Interestingly, but not surprisingly, it was observed that the simultaneous stimulation of IFN at the time of viral challenge increased resistance in the low-responders up to 1,000-fold.

Another approach to the protection of cells from viruses is based on the observation that in cancer, expression of a mutant gene often confers a dominant phenotype on the cells in which it is expressed. The effect is produced by interference between the mutant molecules and the function of the wild-type gene product, or by allowing the cells expressing the mutated genes to have some selective growth advantage. Friedman et al. (1988) used the first of these approaches in an effort to protect cells against HSV-1 by the use of the viral VP16 protein. This protein is responsible for activation of the immediate early genes and so is essential for replication and pathogenicity of the virus. Sequences that encode 78 amino acids were deleted from the acidic transactivation domain of the gene, and this truncated VP16-coding sequence was then cloned into an expression vector for stable transfection into mouse L-cells. Through an interference mechanism with the wild- type virus, these mutant molecules dominantly competed for cis-acting sequences of the immediate-early region. This inhibited the transcription of immediate-early genes and reduced the amount of virus that was produced by each infected cell. Furthermore, this interference was specific for HSV-1 and not other types of herpes viruses. It was this study that led Baltimore to coin the useful phrase "intracellular immunization" (Baltimore, 1988). Some immunologists have objected that the word "immunization" should not be used in this context since the protection is not mediated by the immune system, but nevertheless the cells are immune in a functional sense, and "intracellular immunization" is a term that has started to appear quite frequently in the scientific literature.

With these experiments in mind, immunization and therapy for infection by HIV and AIDS by the use of dominant negative mutant molecules seems possible. It is known that the first step in the infection of a cell by HIV is binding of the gp120 molecule of the virus to the CD4 receptor of the target cell. Attempting to interfere with this process, Buonocore and Rose (1990) constructed cells in which mutant CD4 molecules were expressed that were soluble and retained in the endoplasmic reticulum or the cells. When the cells were infected by HIV, the viral gp120 protein was blocked from its transport to the cell surface, and this prevented the fusion of infected cells that typically occurs in HIV infection. It was suggested that if this could be reproduced in T lymphocytes, those cells could be protected from HIV.

In other investigations, viral regulatory and replicative proteins have been targeted. For example Trono et al. (1990) were able to interfere with replication of HIV in COS and H9 cells using mutated *gag* genes. The rationale for these experiments stems from the observation that the GAG proteins are known to form large multimers which are required for proper virion structure. Therefore, the transfer and expression of mutant proteins may interfere with the production of virus particles even if these interfering molecules are only expressed at low levels. The results confirmed that the expression of these mutant molecules interfered with the assembly of wild type virus when cells were either simultaneously infected and transfected, or when established transfectants were subsequently infected. Furthermore, they showed that the mechanism of inhibition was dependent upon the site of the deletions in the *gag* constructs. Some mutations interfered with virus assembly of release while others limited infectivity, viral adsorption or the uncoating process.

ANTISENSE OLIGONUCLEOTIDES

In the 1980s it became clear to many investigators that cellular gene expression could be inhibited by the use of RNA molecules that are complementary to target mRNA sequences. When "sense" RNA hybridizes with "antisense" RNA, the sense RNA becomes unable to function (reviewed by Chrisey, 1991). Initially this approach was used to inhibit cellular genes, such as some of the developmental genes of Drosophila, but more recently it has been developed to treat infections by viruses.

Zamecnik and Stephenson (1978) were the first investigators to demonstrate that the replication of a virus could be inhibited by complementary antisense oligodeoxynucleotides to target sequences. In their experiments, synthetic antisense oligodeoxynucleotides were used to inhibit the replication of the Rous sarcoma virus in chicken embryos. Since that time these molecules have been shown to be effective in inhibiting the expression of a wide variety of cellular, viral and parasitic genes *in vitro*. HSV-1 can be inhibited to some extent by the presence of antisense molecules that are directed to its immediate early genes, ICP-4 and ICP-5, although to date very high levels of oligonucleotides have been necessary (Smith et al., 1986). Another drawback of synthetic oligonucleotides is their instability *in vitro* and *in vivo*. After successful internalization of these molecules into the appropriate cell, the half-life is extremely short since the presence of numerous cellular nucleases rapidly degrade these structures. In fact the average half-life of a 20 base-pair oligomer, in cells, is approximately 15 minutes. Therefore, the molecules must find their way through the cellular milieu, bind to their target and cause inactivation or degradation of the target sequences very rapidly. Even more effective degradation could be expected if such molecules were injected to the circulation or into some body cavity.

Much emphasis, therefore, has been directed toward modifying these molecules so as to make them resistant to nucleases and other forms of attack. One modification consists of phosphorothioate modifications of the phosphodiester backbone and there is evidence that this will increase the *in vivo* survival of these molecules for up to 48 hours (Agrawal et al., 1991). Treatment of animals or patients with synthetic antisense molecules therefore is probably feasible, although neither animal nor clinical human trials have yet been reported.

INTRACELLULAR EXPRESSION OF ANTISENSE RNA

An alternate strategy to repeated application of antisense synthetic oligonucleotides is the effort to develop methods which permit the stable, endogenous expression of antisense RNA molecules within a selected population of cells. Easily obtainable, expression vectors can be constructed using any identified DNA fragment or cDNA clone. Integration and stable expression of antisense DNA templates were shown in several studies during the early 1980s (Pestka et al., 1984, Izant and Weintraub, 1985) to inhibit the expression and translation of target mRNA molecules. Certainly, as with the antisense oligomers, several hurdles are present and must be overcome before this technology can be realized for clinical therapy of disease. The obstacles include cellular targeting, regulation of antisense expression within specific cell types and the processing, localization and stability of the products.

One of the major impediments in successful antisense-mediated inhibition is inconsistency in the stable and sufficient levels of expression from the vector templates. Sullenger et al. (1990) demonstrated the importance of variables in specific vectors and cloning sites used for expression. In their studies, various regions of the

Moloney murine leukemia virus (MoMLV) genome were cloned into expression vectors which then expressed chimeric tRNA-MoMLV antisense transcripts. Although all of these plasmid constructs consistently expressed a similar level of antisense RNA within transfected cells, there was significant variation in the ability of the expressed messages to interfere with MoMLV replication. Furthermore, when different vectors were used to express the same sequences, the level of antisense transcription was extremely variable. However, with the best combination of all of these factors these workers did achieve 97% reduction in viral replication. Another important feature in the success or failure of antisense-mediated inhibition is the selection of the appropriate target site on the genome that is to be inhibited. Rittner and Sczakiel (1991) performed a careful comparison of target sites on the HIV genome and showed a very marked difference in effectiveness. To some extent the likely targets could have been predicted in advance, but some sites that might have seemed suitable were only moderately effective, and it appears that experimental screening of a target is the only method that can be relied on to reveal sites of susceptibility.

Since many oral cancers contain HPV, it seems reasonable to assume that inhibition of these viruses would inhibit the growth of oral cancer cells. This might become the basis of gene therapy for oral cancer, or of the intracellular immunization against malignant change in patients with pre-malignant lesions. To test this, we are investigating antisense-mediated inhibition of human papillomaviruses in various oral and cervical carcinoma cell lines through two of the possible approaches--antisense oligodeoxynucleotides and intracellular expression of antisense RNA. Current data indicate that treatment of oral cancer cells with oligonucleotides directed to the 5' end of the E6 and E7 genes of HPV-18 causes an inhibition in the growth rate of those cells that contain DNA of HPV-18, but has no effect on cells without the viral sequences. Furthermore, these effects are dose dependent and synergistic. In the presence of endogenously expressed antisense E6/E7 or E7 transcripts, not only is there a substantial decrease in proliferative capacity of HPV positive cells, but the induced transfectants show a significant decrease in other characteristics indicative of malignantly transformed cells, such as lower plating efficiency and ability to form foci in soft agar (Steele et al., 1992).

Although most studies on the effects of antisense RNA have been performed in cells, one study has shown effectiveness in animals. Han et al. (1991) generated a line of transgenic mice whose chromosomes contained the packaging sequences of MoMLV in an antisense orientation relative to appropriate regulatory elements. When the mice were infected with MoMLV 31% of nontransgenic, control animals developed virus-induced leukemia, while none of the animals that expressed the antisense RNA became leukemic. Thus antisense therapy may be feasible *in vivo* if the appropriate sequences can be expressed in the appropriate tissues.

RIBOZYMES

Recent studies of viruses that infect plants have led to the finding of a particular class of RNA molecules that are able to fold into specific conformations, and then cleave themselves into multiple smaller molecules. These molecules have become known as "hammerhead ribozymes," because of the apparent shape in which the essential sequences are often shown (Symons, 1989). Analysis of the region around the cleavage site has shown that only a very small set of conserved sequences is essential for cleavage, and the target sequences can be located on a different molecule from the ribozyme sequences. Thus if a potential target sequence can be identified in a cellular RNA molecule a synthetic ribozyme can be designed to cut the RNA at that site. Ribozymes are expected to be more efficient than other antisense molecules, since they can dissociate from the target after cleavage is complete, and attack another target.

The majority of research on ribozymes has been *in vitro*, but some experiments have now shown their effectiveness in cells. Sarver et al., (1990) designed a ribozyme that targeted the RNA transcripts that encode the p24 antigen of HIV. Expression of the ribozyme resulted in almost complete destruction of the targeted RNA and decreased the levels of HIV proviral DNA by up to 100 fold. Based on these results it was suggested that the actual RNA genome of HIV might be a target for destruction by ribozymes, and that these molecules could become effective forms of therapy for AIDS. To date, however, no animal or clinical trials of ribozyme-based treatments have been reported.

VIRAL VECTORS

A major problem in the adaptation of the antiviral methods described here will be to find a method of delivery of new genetic constructs into a patient's tissues. The techniques for gene transfer that are used in cell culture, such as precipitation, electroporation and development of transgenic animals, cannot be modified for human clinical use. Most investigators feel that the solution to the problem of routine use of gene transfer will come from the development of new virus vectors. Several viruses are available, although none of them meets all the requirements for a clinically useful product. Retrovirus vectors have been used for transfer to cells *in vitro* and might be effective *in vivo*. However, retroviruses become integrated into the cell genome at random sites, which would lead to side effects if an essential cellular gene was interrupted, or an undesirable gene was activated. Adenoviruses have been used for transfer of DNA to the respiratory mucous membrane of animals (Rosenfeld et al., 1991). This group of viruses has some potential for clinical use, but some adenoviruses are pathogenic to humans and others are oncogenic, and this will inhibit their development. HSV-1 has been used to transfer DNA to neural cells since it has the advantage of being able to enter and survive in nonreplicating cells (Geller et al., 1990). However, it is also a human pathogen and has many genes of unknown function. The adeno-associated viruses are an attractive possibility as vectors for gene therapy since they are nonpathogenic to humans, and appear to integrate at a specific, nonessential chromosomal site (Kotin et al., 1990). Although there appear to be many choices for possible viral vectors, most viruses have not been examined in detail for this particular purpose, and the development and comparison of new vectors have only just begun.

CONCLUSIONS

Although in the past the major efforts to protect against virus infections have focused on vaccines and antiviral drugs, it is clear that there are a large number of new approaches, based on gene transfer technology, that must be considered. There are substantial technical difficulties which must be overcome before any of these approaches can become practical. Nevertheless, this is an important area of intense research and numerous possibilities are close to becoming realized. Only a decade ago, the genetic manipulation of human cells was almost unimaginable, yet now it is possible to introduce new genes and inhibit others. Virus-resistant animals have been created through genetic engineering and human gene therapy trials for correction of enzyme deficiencies have begun. The requirements for genetic protection against viruses are clear, and the technology with which to perform the procedures are available. If progress continues at the current rate, then within another decade there might be genetic treatments available for the management of several virus infections, including those that affect the oral cavity.

REFERENCES

Agrawal, S., Temsamani, J., and Tang,J.Y., 1991, Pharmacokinetics, biodistribution, and stability of oligodeoxynucleotide phosphorothioates in mice, *Proc Natl Acad Sci*. 88:7595.

Arnheiter, H., Skuntz, S., Noteborn, M., Chang, S., and Meier, E., 1990, Transgenic mice with intracellular immunity to influenza virus, *Cell*. 62:51.

Baltimore, D., 1988, Intracellular immunization, *Nature*. 335:395.

Biocca, S., Neuberger,M.S., and Cattaneo,A. 1990, Expression and targeting of intracellular antibodies in mammalian cells, *EMBO J*. 9:101.

Buonocore, J. and Rose,J.K., 1990, Prevention of HIV-1 glycoprotein transport by soluble CD4 retained in the endoplasmic reticulum, *Nature*. 345:625.

Chang, F., Syrjanen, S., Kellokoski, J., and Syrjanen, K., 1991, Human papillomavirus (HPV) infections and their associations with oral disease, *J Oral Pathol Med*. 20:305.

Chebath, J., Benech, P., Revel, M., and Vigneron, M., 1987, Constitutive expession of (2'-5') oligo A synthetase confers resistance to picornavirus infection, *Nature*. 330:587.

Chrisey, L.A., 1991, An indexed bibliography of antisense literature, 1978-1990, *Antisense Res and Dev*. 1:65.

Faraji-Shadan, F., Stubbs, J.D., and Bowman, P.D., 1990, A putative approach for gene therapy against human immunodeficiency virus (HIV), *Med Hypoth*. 32:81.

Friedman, A.D., Triezenberg, S.J., and McKnight, S.L., 1988, Expression of a truncated viral trans-activator selectively impedes lytic infection by its cognate virus, *Nature*. 335:452.

Galloway, D.A. and McDougall, J.K., 1989, Human papillomaviruses and carcinomas, *Adv Virus Res*. 37:125.

Garcia-Blanco, M.A. and Cullen, B.R., 1991, Molecular basis of latency in pathogenic human viruses, *Science*. 254:815.

Geller, A.E., Keyomarsi, K., Bryan, J., and Pardee, A.B., 1990, An efficient deletion mutant packaging system for defective herpes simplex virus vectors: Potential applications to human gene therapy and neuronal physiology, *Proc Natl Acad Sci USA*. 87:8950.

Han, L., Yun, J.S., and Wagner, T.E., 1991, Inhibition of Moloney murine leukemia virus-induced leukemia in transgenic mice expressing antisense RNA complementary to the retroviral packaging sequences, *Proc Natl Acad Sci USA*. 88:4313.

Izant, J.G. and Weintraub, H., 1985, Constitutive and conditional suppression of exogenous and endogenous genes by anti-sense RNA, *Science*. 229:345.

Kotin, R.M., Siniscalco, M., Samulski, R.J., Zhu, X., Hunter, L., Laughlin, C.A., McLaughlin, S., Muzyczka, N., Rocchi, M., and Berns, K.I., 1990, Site-specific integration by adeno-associated virus, *Proc Natl Acad Sci USA*. 87:2211.

Oldstone, M.B.A., 1991, Molecular anatomy of viral persistence, *J Virol*. 65:6381.

Pavlovic, J., Zurcher, T., Haller, O., and Stacheli, P., 1990, Resistance to influenza virus and vesicular stomatitis virus conferred by expression of human MxA protein, *J Virol*. 64:3370.

Pestka, S., Daugherty, B.L., Jung, V., Hotta, K., and Pestka, R.K., 1984, Anti-mRNA: specific inhibition of translation of single mRNA molecules, *Proc Natl Acad Sci USA*. 81:7525.

Rittner, K. and Sczakiel G., 1991, Identification and analysis of antisense RNA target regions of the human immunodeficiency virus type 1, *Nucl Acids Res*. 19:1421.

Rosenfeld, M.A., Siegfried, W., Yoshimura, K., Yoneyama, K., Fukayama, M., Stier, L.E., Paakko, P.K., Gilardi, P., Stratford-Perricaudet, L.D., Perricaudet, M., Jallat, S., Pavirani, A., Lecocq, J.P., and Crystal, R.G., 1991, Adenovirus-mediated transfer of a recombinant alpha-1-antitrypsin gene to the lung epithelium in vivo, *Science*. 252:431.

Sarver, N., Cantin, E.M., Chang, P.S., Zaia, J.A., Ladne, P.A., Stephens, D.A., and Rossi, J.J., 1990, Ribozymes as potential anti-HIV-1 therapeutic agents, *Science*. 247:1222.

Smith, C.C., Aurelian, L., Reddy, M.P., Miller, P.S., and Ts'o, P.O.P., 1986, Antiviral effect of an oligo(nucleoside methylphosphonate) complementary to the splice junction of herpes simplex virus type 1 immediate early pre-mRNAs 4 and 5, *Proc Natl Acad Sci USA*. 83:2787.

Steele, C., Sacks, P., Adler-Storthz, K., and Shillitoe, E.J., 1992, Effect on cancer cells of plasmids that express antisense RNA of human papillomavirus type-18. (Submitted for publication).

Steele, C. and Shillitoe,E.J., 1991, Viruses and oral cancer, *Critical Revs Oral Biol Med.* 2:153.

Sullenger, B.A., Lee, T.C., Smith, C.A., Ungers, G.E., and Gilboa, E., 1990, Expression of chimeric tRNA-driven antisense transcripts renders NIH 3T3 cells highly resistant to Moloney murine leukemia virus replication, *Mol Cell Biol.* 10:6512.

Symons, R.H., 1989, Self-cleavage of RNA in the replication of small pathogens of plants and animals, *ITBS.* 14:445.

Trono, D., Feinberg, M.B., and Baltimore, D., 1990, HIV-1 Gag mutants can dominantly interfere with the replication of the wild-type virus, *Cell.* 59:113.

Werge, T.M., Biocca, S., and Cattaneo, A., 1990, Intracellular immunization: Cloning and intracellular expression of a monoclonal antibody to the p21-ras protein, *FEBS.* 274:193.

Zamecnik, P.C., and Stephenson, M.L., 1978, Inhibition of Rous sarcoma virus replication and cell transformation by a specific oligodeoxynucleotide, *Proc Natl Acad Sci USA.* 75:280.

BACTERIAL MUCOSAL VACCINES

John J. Mekalanos

Department of Microbiology and Molecular Genetics
Harvard Medical School
200 Longwood Avenue
Boston, Massachusetts 02115

INTRODUCTION

Vaccines are known to be among the most cost effective ways to control infectious diseases. Indeed, the widespread use of vaccines offers the potential of complete eradication of some infectious agents that have a single obligate host and environmental reservoir. Discontinued use of the vaccine that allowed eradication then brings more savings in costs for generations to come.

Sadly, such spectacular success has been achieved with only one disease (small pox). While other diseases offer the potential for eradication with effective vaccination programs (e.g., polio), the lack of sufficient health infrastructure in the world community makes such a repeat performance highly unlikely in the near future. Thus, one driving force in future vaccine production strategies is to construct inexpensive vaccines that are more easily distributed and administered. These are exactly the promises offered by mucosal vaccines.[1] Because they can be simply administered by the oral or nasal routes, they would not require the professional health care infrastructure demanded by injectable vaccines. More importantly, for some diseases a mucosal immune response may be the only effective way of producing protection.

In this short chapter, I will discuss general aspects of bacterial vaccine development with emphasis on old as well as new technologies applicable to mucosal vaccine development.

VACCINES PAST AND PRESENT

From a general point of view, we can divide bacterial vaccines into three different types: subunit vaccines, inactivated microorganisms, and live attenuated microorganisms. Progress is being made in facilitating the production of all three types of vaccines and considerable effort has focused on the conversion of all three types to mucosal vaccines.

Subunit Vaccines

Subunit vaccines are composed of a single protein or component of the microorganism. These include proteins such as toxins, carbohydrates, viral capsids, or even synthetic peptides. Where the protein was initially toxic, treatment with chemical agents such as formalin not only detoxified the protein but in many cases improved the immunogenicity and stability of the protein. Such "toxoids", although widely used, are potentially unsafe because chemical toxoiding procedures vary from lot to lot leading to poor inactivation or

Genetically Engineered Vaccines, Edited by
J.E. Ciardi *et al.*, Plenum Press, New York, 1992

43

even re-activation to toxicity. Thus, additional costs are incurred by elaborate systems to assure the safety and shelf-life of toxoid preparations.

Molecular biology offers to greatly impact this area of subunit vaccines. For example, it is now practical to genetically engineer detoxified versions of toxins by identifying active site residues, modifying these proteins by deletion or insertion of new amino acids, and then checking to see if these modified proteins retain the stability and immunogenicity of the old toxoids.[2] Experience, however, with diphtheria toxin showed that the formalin treatment of the toxin was very important for its immunogenicity and mutated forms of the toxin that lacked toxicity were poor immunogens until treated with formalin. Nonetheless, mutagenesis of subunit vaccine proteins is clearly an elegant approach and will certainly be very important for integration into live vaccines where immunostimulation due to chemical treatment may be replaced by immunomodulators associated with the carrier vaccine organism.

One of the other advantages to subunit vaccines concerns the ability to manipulate these chemically in ways that alter the nature of the natural immune response. For example, type B capsular polysaccharide of *Hemophilus influenzae* (Hib) is a T-independent antigen that is poorly immunogenic in the newborn. By chemically coupling capsule to various toxoids or membrane protein preparations, the capsular antigen can obtain "T-help" allowing neonatal immunization against Hib disease for the first time.[3] Similarly, modification of the peptides by crosslinking to lipid moieties has resulted in changes in both the processing of antigens as well as their immunogenicity.[4] However, whether a modified subunit vaccine such as a capsular conjugate can be delivered mucosally is only beginning to be investigated.[7]

The current limitation in subunit vaccine technology as applied to mucosal immunization is defining "delivery systems" that will allow small amounts (micrograms of protein) to be administered mucosally and still be as immunogenic as adjuvant formulated injectable vaccines. Technological advances have been made in this area with the application of microencapsulation and the use of mucosal adjuvants such as cholera toxin B subunit, muramyl dipeptide, ISCOMS, and lipid A derivatives.

More is known about immunogenicity of antigens delivered orally than any other mucosal route. In general terms, the ability of an antigen preparation to induce a mucosal immune response correlates best with the tendency for that antigen to be transported by specialized M-cells associated with the follicle-associated epithelium of Peyer's patches.[5] Many mucosal adjuvants act simply by increasing this uptake.

For example, microencapsulation of antigens in biodegradable polymers has been found to simulate antigen uptake into M-cells by mechanisms that depend primarily on the size of the particle.[6] The use of liposomes and ISCOMS (immuno-stimulating complexes) may offer similar advantages in particle formulations and applications.[7] In contrast, soluble proteins like cholera toxin B subunit and the related LT-B subunit have the ability to bind to surface receptors on M-cells thus increasing the efficiency of their uptake. Accordingly, much effort has been made to fuse the LT-B and cholera B subunits to various antigens in order to produce chimeric proteins suitable for use as oral or mucosal immunogens.[8] In general this approach has worked on an experimental scale but no licensed products have come from this technology as of yet.

Inactivated Microorganisms

The second type of vaccine to consider consists of inactivated virulent microorganisms. The clear advantage of this type of vaccine is that it is inexpensive and relatively easy to prepare. The complexity of these vaccines is potentially advantageous in that multiple protective antigens are probably present although ill defined. The heat stability of this type of vaccine is in theory another advantage but little information has been gathered that confirms this notion.

The best bacterial example of this type of vaccine is the very effective killed whole cell pertussis vaccine which although parenterally administered protects against the mucosal pathogen *Bordetella pertussis*.[9] The effacacy of this vaccine may find its origin in the *in vitro* expression of a wide array of protein antigens (i.e., pertussis toxin, filamentous hemagglutinin, pertactin, pili, adenylate cyclase toxin, outer membrane proteins, etc.) which are also virulence factors of the organism. The potency of this vaccine drops as the content of these antigens decrease. The presence of immunomodulating proteins such as pertussis toxin may also contribute to this vaccine's potency. In contrast, the limited and short term protection afforded by parenterally administered killed whole cell typhoid and cholera vaccines probably depends exclusively on cellular components such as lipopolysacchrides. Like, *B. pertussis*, *Vibrio cholerae* and *Salmonella* species have coordinate regulatory systems that control the expression of virulence factors.[10] Current parenteral vaccines for cholera and typhoid lack these protein antigens because culture conditions have not been completely defined for the GMP manufacture of bacterial cells expressing these virulence factors. By analogy to the whole cell pertussis vaccine, the the potency of these parenteral vaccines might be dramatically increased if these proteins were present and represented protective antigens.

Conversion of inactivated organisms into good mucosal vaccines may be as simple as administering them by a mucosal route in sufficient quantity to stimulate a mucosal immune response. Such an approach has been successfully tested in field trials with the two killed whole cell cholera vaccines developed by Jan Holmgren, Ann-Mari Svennerholm and colleagues.[11] These vaccines composed of 10^{11} heat and formalin killed *V. cholerae* organisms alone or mixed with cholera toxin B subunit, when given orally in three doses produced high level protection against cholera in the first year and 50% effacacy even three years after vaccination.[11]

The whole cell component of this promising vaccine is clearly "low tech" in the sense that no genetic engineering or specialized fermentation has been applied to improve expression of protective antigens such as the TCP pilus and other ToxR-regulated virulence factors.[12] Molecular genetic advances in this area have begun to make an impact on this vaccine beginning with the genetic engineering of cholera B subunit production to levels 10-100 times those seen under standard fermentation conditions.[13] Similar genetic engineering of TCP expression or the use of constitutive mutants expressing ToxR-regulated antigens may eventually facilitate the production of a whole cell component that displays a protein profile similar to that seen during infection by *V. cholerae*. However, it remains to be determined what if any technology will improve the immunogenicity of whole inactivated organisms and proteins expressed when administered by the oral route.

In general inactivated organisms are much less immunogenic when administered by the mucosal route than live, attenuated organisms. This difference in immunogenicity probably reflects the fact that killed organisms are transported by M-cells far less efficiently than living organisms of the same type.[14] It is not at all clear why this is the case. It may have to do with expression of surface properties by live bacteria *in vivo* that are not expressed during laboratory cultivation (or are inactivated by the methods used to kill the harvested bacteria). As in the case of subunit vaccines, microencapsulation and adjuvant formulations may have an impact on the efficiency of mucosal vaccines based on inactivated microorganisms.

Live, Attenuated Microorganisms

The third category to consider includes vaccines composed of living, attenuated organisms. One of the major advantages of live vaccines is that a single relatively small innoculum can in theory expand by replication to a larger and therefore more immunogenic dose of vaccine. In fact this promise has not yet been fulfilled for any live attenuated bacterial vaccine licensed or in development without paying the price of reactogenicity. A more realistic advantage of live vaccines is that they presumably express most of the natural target immunogens of the natural infection and do so in a way that promotes their processing and presentation in a way that is most similar to the natural infection.

While viral vaccines dominate this category, two field-tested vaccines, BCG and Ty21a do provide us with information that has demonstrated the feasibility of this approach in bacterial vaccine development. BCG, an attenuated strain of *Mycobacterium bovis*, is administered parenterally and provides some protection against tuberculosis presumably through cellular rather than humoral immune mechanisms.[15] Ty21a is an attenuated derivative of *Salmonella typhi* Ty2 which when administered orally in high multiple doses, provides significant protection against typhoid fever by stimulating cellular as well as humoral immunity.[16]

Converting live attenuated vaccines to mucosal vaccines is dependent on not only mucosal administration but also on the prerequisite that the vaccine organism does multiply or at least persist in mucosal tissues. In fact, some organisms naturally "target" to lymphoid tissue of the mucosal epithelium, adhering to and multiplying within these sites that are so important to antigen sampling and initiation of the local mucosal immune response. Live, attenuated vaccines made from such organisms offer the potential of a living, replicating, antigen delivery system, specialized in stimulating mucosal immune responses.

At present, this type of vaccine includes examples that are the most advanced mucosal vaccines to date and will therefore be the focus of the remainder of this chapter.

LIVE, ATTENUATED BACTERIAL MUCOSAL VACCINES UNDER DEVELOPMENT

Salmonella typhi

Over the last decade, attenuated species of the genus *Salmonella* have enjoyed the attention of the research community interested in delivery of heterologous antigens by a live, attenuated carrier organism. This interest was based on the recognition that salmonellae are efficient enteric pathogens that gain entry into mucosal tissue by direct interaction with the follicular lymphoid tissues of the gut. This interaction with the gut-associated lymphoreticular tissue (or GALT) leads ultimately in surviving animals to a strong humoral immune response dominated by secretory IgA. Because salmonellae are also facultative intracellular parasites, they are, in theory, capable of inducing a vigorous cell-mediated immune responses as well.

Most of the work defining attenuating mutations has utilized the species *S. typhimurium* and the BALB/c mouse model. Among the mutations used to attenuate *Salmonella* species are SmD (producing streptomycin dependency), galE (producing galactose toxicity *in vivo*), pur and aro (producing auxotrophy for compounds not available in animal tissues), crp and cyc (producing global changes in genes expression under catabolite control), phoP and phoP^c (producing global changes in the expression of virulence genes), pagC (producing a macrophage survival defect).[16-22] Very few comparison studies have been done to a clear advantage of one of these attenuating mutations over another. Most importantly, it seems clear that more than one mutation will be needed to attenuate *S. typhi* for use in humans.

The only licensed, live, attenuated *Salmonella* based vaccine is Ty21a, a galE mutant of *S. typhi* which also carries other undefined auxotrophic mutations. Ironically, the galE mutation alone is clearly not responsible for the attenuation of Ty21a since genetically engineered galE mutants of *S. typhi* are still able to cause typhoid in human volunteers.[23] Many more human studies will be required to determine what is a safe but effective combination of attenuating mutations.

A variety of different heterologous antigens have been expressed in attenuated salmonellae including lipopolysaccharide,[24] toxin B subunits,[25] enzyme targets,[26] and viral antigens.[27] In general, an immune response to these "carrier antigens" can be demonstrated suggesting that this approach is of considerable merit.

Vibrio cholerae

There has been over a century's worth of effort to develop an effective cholera vaccine. This effort has played a major role in our understanding of the importance of the local immune response in combating mucosal infections. The recognition that exposure of the gut mocosal to *V. cholerae* infection leads to long term immunity, has prompted the development of several generations of live, attenuated oral cholera vaccines.

The earliest versions of this approach depended on chemical mutagenesis,[28] but the most current hope for success relies on the methods of recombinant DNA.[29,30] Deletion of the cholera toxin A subunit gene (ctxA) is the common theme in the construction of a variety of vaccine candidates that have undergone various degrees of clinical testing.

Initial human volunteer studies with strains JBK70, CVD101, and derivatives of these strains showed that a single dose of a genetically engineered strain could indeed induce immunity.[31] However, like earlier chemical mutants,[28] these recombinant vaccines caused reactions (e.g., moderate diarrhea) in the volunteers that were severe enough to preclude their use. These studies suggested that *V. cholerae* expressed other virulence properties besides cholera toxin that could induce diarrhea. Interestingly, testing of an additional ctxA mutant in volunteers shed some light on this proposal. Volunteers that ingested strain O395-N1 showed significantly less reactions, particularly diarrhea, than other isogenic derivatives such as CVD101.[32] This observation led the discovery that O395-N1 produced less of an activity named "ZOT" for zonula occuludens toxin than CVD101.[33] It is still not clear whether ZOT is responsible for residual reactogenicity in strains such as CVD101 and O395-N1, but the recent identification of the zot structural gene[34] should make it possible to test this idea by construction of ctxA zot double deletion mutants.

Currently, the most advanced recombinant vaccine, is *V. cholerae* CVD103-HgR. This strain when given in a single large doses of 5×10^9 bacteria produces high levels of seroconversion and protection that lasts at least one month with little if any severe side effects.[35] This vaccine is a double ctxA hlyA double deletion mutant that has been constructed in the 569B strain background. Because the deletion of the hlyA gene had no significant effect on the virulence of other prototypic cholera vaccines such as CVD104 and CVD105,[31] it can be concluded that the properties that allow significant attenuation of CVD103-HgR are probably associated more with its parental strain than with any of its recombinantly engineered properties. The parental 569B strain is an unusual *V. cholerae* strain in that it colonizes poorly, is deleted for the regulatory gene ToxS, and is virtually nonmotile. CVD103-HgR does produce ZOT suggesting that an unknown property such as its relatively poor colonization is probably responsible for its apparent attenuation. Accordingly, double mutants of ZOT deficient strains such as O395-N1 carrying secondary mutations in recA or irgA may be useful to test in volunteers since both of these mutations cause a measurable decrease in the colonization of *V. cholerae* in animal models.[36]

Finally, it should be noted that both cholera toxin and ZOT are encoded by a large genetic element that undergoes duplication and amplification in *V. cholerae*.[34,37] The CTX genetic element is a type of transposon that can recombine into nontoxigenic strains by a illegitimate recombinational event.[36] Thus, it is possible that a live attenuated cholera vaccine can regain the capacity to produce cholera toxin and ZOT after re-acquiring these genes in natural settings by coming in contact with toxigenic *V. cholerae* that carry DNA mobilization systems. Reversion of attenuated vaccines to virulence by recombination is not typically considered a likely event but should be carefully considered whenever the widespread use of a live vaccine is being considered.

Bacille Calmette-Guerin (BCG)

BCG is a strain of *Mycobacterium bovis* which is used as a live vaccine for tuberculosis.[15] Although it has varied in its effectiveness as a vaccine, it is still the most widely used vaccine in the world. Although originally used as an oral vaccine, BCG is now used as an injectable vaccine because of its propensity to cause lymphadenitis. Its use as an oral vaccine also requires much higher doses because of the acid sensitivity of the organism. The possibility to enteric coat lyophilized BCG so that oral delivery can be re-examined has prompted much discussion on whether BCG can be used as a mucosal vaccine. It is further hoped that the known adjuvant properties and ability to induce cell-mediated immunity, might make BCG exceptional as a carrier vaccine for heterologous antigens as well.

Impressive progress has been made in the effort to engineer expression of heterologous antigens in BCG. Genetic systems exist for introduction of gene constructs into this organism and expression via heat-shock and other promoters has been successful.[38] However, so far there has been no reports on the induction of a secretory IgA response to either BCG antigens or heterologous antigens expressed by BCG. It remains to be determined whether BCG can be used as an effective live delivery system for inducing local immune responses on mucosal surfaces.

Shigella Species

Considerable effort has been made to devise live vaccines effective in controlling shigellosis. Efforts to construct *E. coli-Shigella* hybrid strains have successfully defined virulence genes but have not yet been totally fruitful in the search for a non-reactogenic, protective live vaccine.[39] Recent work focusing on the construction of aroA mutations in *S. flexneri* 2a has been promising in studies involving monkeys and volunteers.[40] An exciting and elegant approach has been the construction of virulence factor deficient mutants. In this regard, derivatives of various *Shigella* species have been constructed that carry mutations in genes encoding cell to cell spread (icsA), iron uptake (iuc, iut), catalase (katF), virulence regulation (ompB) and shiga toxin production.[41] However, so far there have been no reports of expression of heterologous antigens in Shigella species although this should be technically easy to accomplish once a suitable vaccine has been developed.

Another promising candidate for live, oral vaccine development is *Yersinia enterocolitica*. Genetically engineered versions of this organism has been shown to express cholera B subunit and to induce antibody responses to CT-B after gut infections.[42] However, a means of attenuating this organism for gastoenteritis in humans has not yet been defined and the worry of *Y. enterocolitica*-induced reactive arthritis may further preclude the use of this organism as a human carrier vaccine.

CONCLUSIONS

There are still many hurdles we must leap before the promise of multivalent, live bacterial oral vaccines can be realized. Issues such as reactogenicity must be addressed with thought given to the fact that some individuals within humans populations are immunocompromised. The stability of vaccines must be considered in light of changes that may occur in the vaccine on acquisition of virulence genes encoded by transposable and other genetic elements (plasmids and bacteriophage). We must understand whether heterologous antigens expressed by these vaccines need to be surface localized to induce good IgA responses and if so we need to accelerate the definition of vehicles for accomplishing this.[43,44] The expression systems that allow the right level of heterologous antigen to be expressed at the appropriate stage in the infection cycle must also be defined for each candidate system. Uptake by M-cells and subsequent events must be understood in greater detail so that each organisms can be evaluated in objective terms for their ability to induce IgA responses to heterologous antigens. Finally, the delivery and immunogenicity of both living and dead oral vaccines may become more efficient if we can produce advances in the area of microencapsulation and mucosal adjuvant formulation.

REFERENCES

1. J. Mestecky and J.R. McGhee, Oral immunization: Past and Present, *Curr. Top. Microbiol. Immunol.*, 146:4 (1989).
2. R.K. Tweten, J.T. Barbieri, and R.J. Collier, Diphtheria toxin, effect of substituting aspartic acid for glutamic acid 148 on ADP-ribosyltransferase activity, *J. Biol. Chem.*, 260:10392 (1985).
3. R. Schneerson, O. Barrera, A. Sutton, and J.B. Robbins, Preparation, characterization, and immunogenicity of *Haemophilus influenzae* type b polysaccharide-protein conjugates, *J. Exp. Med.*, 152:361 (1980).
4. K. Deres, H. Schild, K.B. Wiesmuller, G. Jung, and H.G. Rammensee, *In vivo* priming of virus-specific cytotoxic T lymphocytes with synthetic lipopeptide vaccine, *Nature*, 342:561 (1989).
5. R.L. Owen, Sequential uptake of horseradish peroxidase by lymphoid follicle epitheium of Peyer's patches in the normal unobstructed mouse intestine: an ultrastructural study, *Gastroenterology*, 72:440 (1977).
6. J. Eldridge, C.J. Hammond, J.A. Meulbroek, J.K. Staas, R.M. Gilley, and T.R. Tice, Controlled vaccine release in the gut-associated lymphoid tissues. I. Orally administered biodegradable microspheres target the Peyer's patches, *J. Controlled Release*, 11:205 (1990).
7. T. Bruyere, D. Wachsmann, J.P. Klein, M. Scholler, and R.M. Frank, Local response in rat to liposome-associated *Streptococcus mutans* polysaccharide-protein conjugate, *Vaccine*, 5:39 (1987).
8. M.T. Dertzbaugh, and C.O. Elson, Cholera toxin as a mucosal adjuvant, *in* "Topics in Vaccine Adjuvant Research," D.R. Spriggs, W.C. Koff eds., CRC Press, Boca Raton, FL, pp, 119-131 (1991).
9. M.J. Brennan, D.L. Burns, B.D. Meade, R.D. Shahin, and C.R. Manclark, Recent advances in the development of pertussis vaccines, *in* "Vaccines: New Approaches to Immunological Problems," R.W. Ellis eds., Butterwork-Heinemann, Boston, MA, pp, 23-44 (1992).
10. J.J. Mekalanos, Environmental signals controlling expression of virulence determinants in bacteria, *J. Bacteriol.*, 174:1 (1992).

11. J.D. Clemens, D.A. Sack, J.R. Harris, F. Van Loon, J. Chakraborty, F. Ahmed, M.R. Rao, M.R. Khan, M. Yunus, N. Huda, B.F. Stanton, B.A. Kay, S. Walter, R. Eeckels, A. Svennerholm, and J. Holmgren, Field trial of oral cholera vaccines in Bangladesh: results from three-year follow-up, *The Lancet*, 335:270 (1990).

12. R. Taylor, C. Shaw, K. Peterson, P. Spears, and J.J. Mekalanos, Safe, live *Vibrio cholerae* vaccines?, *Vaccine*, 6:151 (1988).

13. J. Sanchez, and J. Holmgren, Recombinant system for overexpression of cholera toxin B subunit in *Vibrio cholerae*, *Proc. Natl. Acad. Sci.*, 86:481 (1989).

14. N.F. Pierce, J.B. Kaper, J.J. Mekalanos, W.C. Cray Jr., and K. Richardson, Determinants of the immunogenicity of live virulent and mutant *Vibrio cholerae* 01 in rabbit intestine, *Infect. Immun.*, 55:477 (1987).

15. R.G. Barletta, B. Snapper, J.D. Cirillo, N.D. Connell, D.D. Kim, W.R. Jacobs, and B.R. Bloom, Recombinant BCG as a candidate oral vaccine vector, *Res. Microbiol.*, 141:931 (1990).

16. M.M. Levine, R.E. Black, C. Ferruccio, and R. Germanier, Chilean typhoid committee. Large-scale field trial of Ty21a live oral typhoid vaccine in enteric-coated capsule formulation, *The Lancet*, i:1049 (1987).

17. M.M. Levine, D. Hone, C. Tacket, C. Ferreccio, and S . Cryz, Clinical and field trails with attenuated *Salmonella typhi* as live oral vaccines and as "carrier" vaccines, *Res. Microbiol.*, 141:807 (1990).

18. C.E. Hormaeche, H.S. Joysey, L. Desilva, M. Izhar, and B.A.D. Stocker, Immunity induced by live attenuated Salmonella vaccines, *Res. Microbiol.*, 141:757 (1990).

19. J.O. Hassan, and R. Curtiss III, Control of colonization by virulent *Salmonella typhimurium* by oral immunization of chickens with avirulent Δcya Δcrp S. typhimurium, *Res. Microbiol*, 141:839 (1990).

20. S.I. Miller, J.J. Mekalanos, and W.S. Pulkkinen, *Salmonella* vaccines with mutations in the *phoP* virulence regulon, *Res. Microbiol*, 141:817 (1990).

21. S.I. Miller, and J.J. Mekalanos, Constitutive expression of the PhoP regulon attenuates *Salmonella* virulence and survival within macrophages, *J. Bacteriol*, 172:2485 (1990).

22. S.I. Miller, A.M. Kukral, and J.J. Mekalanos, A two component regulatory system (*phoP* and *phoQ*) controls *Salmonella typhimurium* virulence, *Proc. Natl, Acad. Sci.*, 86:5054 (1989).

23. D.M. Hone, S.R. Attridge, B. Forrest, R. Morona, D. Daniels, J.T. LaBrooy, R.C.A. Bartholomeusz, D.J.C. Shearman, and J. Hackett, A *galE via* (Vi antigen-negative) mutant of *Salmonella typhi* Ty2 retains virulence in humans, *Infect. Immun.*, 56:1326 (1988).

24. S.B. Formal, L.S. Baron, D.J. Kopecko, O. Washington, C. Powell, and C.A. Life, Construction of a potential bivalent vaccine strain: introduction of *Shigella sonnei* form 1 antigen genes into the *galE Salmonella typhi* Ty21a typhoid vaccine strain, *Infect. Immun.*, 34:746 (1981).

25. J.D. Clements, F.L. Lyon, K.L. Lowe, A.L. Farrand, and S. El-Morshidy, Oral immunization of mice with attenuated *Salmonella enteritidis* containing a recombinant plasmid which codes for production of the B subunit of heat-labile *Escherichia coli* enterotoxin, *Infect. Immun.*, 53:685 (1986).

26. K. Nakayama, S.M. Kelly, and R. Curtiss III, Construction of an Asd⁺ expression-cloning vector: stable maintenance and high level expression of cloned genes in a *Salmonella* vaccine strain, *Bio/Tech*, 6:693 (1988).

27. F. Schödel, G. Enders, M. Jung, and H. Will, Recognition of a hepatitis B virus nucleocapsid T-cell epitope expressed as a fusion protein with the subunit B of *Escherichia coli* heat labile enterotoxin in attenuated salmonellae, *Vaccine*, 8:569 (1990).

28. M.M. Levine, R.E. Black, M.L. Clements, C. Lanata, S. Sears, T. Honda, C.R. Young, and R.A. Finkelstein, Evaluation in man of attenuated *Vibrio cholerae* El Tor Owaga strain Texas Star-SR as a live oral vaccine, *Infect. Immun.*, 43:515 (1984).

29. J.J. Mekalanos, D.J. Swartz, and G.D.N. Pearson, Cholera toxin genes: nucleotide sequence, deletion analysis and vaccine development, *Nature*, 306:551 (1983).

30. J.B. Kaper, H. Lockman, M. Baldini, and M.M. Levine, Recombinant nontoxinogenic *Vibrio cholerae* strains as attenuated cholera vaccine candidates, *Nature*, 308:605 (1984).

31. M.M. Levine, J.B. Kaper, D. Herrington, C. Tacket, G. Losonsky, J.G. Morris, B. Tall, and R. Hall, Volunteer studies of deletion mutants of *Vibrio cholerae* 01 prepared by recombinant techniques, *Infect. Immun.*, 56:161 (1988).

32. D.A. Herrington, R.H. Hall, G. Losonsky, J.J. Mekalanos, R.K. Taylor, and M.M. Levine, Toxin, Toxin-coregulated pili, and the *toxR* Regulon are essential for *Vibrio cholerae* pathogenesis in humans, *J. Exp. Med.*, 168:1487 (1988).

33. A. Fasano, B. Baudry, D.W. Pumplin, S.S. Wasserman, B.D. Tall, J.M. Ketley, and J.B. Kaper, *Vibrio cholerae* produces a second enterotoxin, which affects intestinal tight junctions, *Proc. Natl. Acad. Sci.*, 88:5242 (1991).

34. B. Baudry, A. Fasano, J. Ketley, and J.B. Kaper, Cloning of a gene (*zot*) encoding a new toxin produced by *Vibrio cholerae, Infect. Immun.*, 60:428 (1992).

35. M.M. Levine, J.B. Kaper, D. Herrington, J. Ketley, G. Losonsky, C.O. Tacket, B. Tall, and S. Cryz, Safety, immunogenicity, and efficacy of recombinant live oral cholera vaccines, CVD 103 and CVD 103-HgR, *The Lancet*, ii:467 (1988).

36. G.D.N. Pearson, V.J. DiRita, M.B. Goldberg, S.A. Boyko, S.B. Calderwood, and J.J. Mekalanos, New attenuated derivative of *Vibrio cholerae, Res. Microbiol.*, 141:893 (1990).

37. J.J. Mekalanos, Duplication and amplification of toxin gene in *Vibrio cholerae, Cell*, 35:253 (1983).

38. A. Aldovini, and R.A. Young, Humoral and cell-mediated immune responses to live recombinant BCG-HIV vaccines, *Nature*, 351:476 (1991).

39. K.L. Kotloff, D.A. Herrington, T.L. Hale, J.W. Newland, L. Van De Verg, J.P. Cogan, P.J. Snoy, J.C. Sadoff, S.B. Formal, and M.M. Levine, Safety, immunogenicity, and Efficacy in monkeys and humans to invasive *Escherichia coli* K-12 hybrid vaccine candidates expressing *Shigella flexneri* 2a somatic antigen, *Infect. Immun.*, 60:2218 (1992).

40. A.A. Lindberg, A. Kärnell, B.A.D. Stocker, S. Katakura, H. Sweiha, and F.P. Reinholt, Development of an auxotrophic oral live *Shigella flexneri* vaccine, *Vaccine*, 6:146 (1988).

41. A. Fontaine, J. Arondel, and P.J. Sansonetti, Construction and evaluation of live attenuated vaccine strains of *Shigella flexneri* and *Shigella dysenteriae* 1, *Res. Microbiol.*, 141:907 (1990).

42. M. Sory, P. Hermand, J. Vaerman, and G.R. Cornelis, Oral immunization of mice with a live recombinant *Yersinia enterocolitica* 0:9 strain that produces the cholera toxin B subunit, *Infect. Immun.*, 58:2420 (1990).

43. D. O'Callaghan, A. Charbit, P. Martineau, C. Leclerc, S. van der Werf, C. Nauciel, and M. Hofnung, Immunogenicity of foreign peptide epitopes expressed in bacterial envelope proteins, *Res. Microbiol.*, 141:963 (1990).

44. J.L. Harrison, I.M. Taylor, and C.D. O'Connor, Presentation of foreign antigenic determinants at the bacterial cell surface using the TraT lipoprotein, *Res. Microbiol.* 141:1009 (1990).

A GENERAL OVERVIEW OF VIRAL VACCINE DEVELOPMENT

E. Kanta Subbarao and Brian R. Murphy

Respiratory Viruses Section, Laboratory of Infectious Diseases
National Institute of Allergy and Infectious Diseases
National Institutes of Health, Bethesda, MD 20892

INTRODUCTION

Research toward the development of vaccines for the prevention of infectious diseases of the oral cavity can draw upon the extensive experience that has been accumulated in the development of vaccines for the prevention of viral respiratory tract diseases. The parallels between immunization of oral and respiratory tissues are obvious since they both involve prevention of disease of mucosal surfaces. We present an overview of approaches that have been employed in the development of vaccines to prevent respiratory tract diseases.

There are some general considerations that influence the development of a vaccine strategy against viral pathogens. First, it is essential to define the protective antigens of the pathogen and to determine whether antibodies or T cells play the predominant role in protection. Second, the goal of immunization must be defined. While it may not be feasible to prevent infection with certain pathogens, such as respiratory syncytial virus, it may be possible to prevent disease. Third, the age of the subject in whom the vaccine is to be used must be defined since pre-existing passively transferred maternal antibody can interfere with the response of very young infants to a vaccine. The choice of the type of vaccine (e.g., live versus non-living) and the dose or the route of administration of the vaccine may need to be modified to overcome the effects of the passively acquired antibody. Lastly, the site that is to be protected by the vaccine (i.e., mucosal protection vs. protection from systemic involvement) affects the choice of route of vaccine administration. In general, a topical route of immunization that induces an IgA antibody response is preferred for prevention of mucosal disease.

For respiratory viruses that infect mucosal surfaces, such as influenza and the paramyxoviruses, the surface glycoproteins are the antigens that induce protective antibodies.[1,2] The strategy employed with the respiratory viruses is directed at the prevention of disease, not infection. In the case of respiratory syncytial virus and parainfluenza virus type 3, the age group at risk for the development of serious illness includes very young infants, the majority of whom have circulating maternal antibodies.[2] Since passively acquired antibodies can suppress the antibody response to the protective antigens in live or inactivated virus vaccines as well as neutralize the infectivity of a live virus vaccine, a successful vaccine would have to overcome these

Genetically Engineered Vaccines, Edited by
J.E. Ciardi *et al.*, Plenum Press, New York, 1992

effects of pre-existing antibodies. The elderly and persons with cardiopulmonary disease are particularly at risk for the development of serious illness from influenza virus infections, so a vaccine strategy for prevention of influenza must be chosen with these subjects in mind. There are three major categories of vaccines for prevention of viral respiratory tract disease: non-living (whole virus or subunit) vaccines, vectored vaccines, and live attenuated vaccines and these will be discussed separately.

NON-LIVING VACCINES

Inactivated virus or subunit vaccines can be presented to the immune system safely since there is no risk of infection. However, immunity induced by non-living vaccines tends to be shorter-lived than that induced by infection, and mucosal antibody responses are very weak following parenteral immunization.[3] Non-living vaccines can be derived from whole or disrupted virions or from viral proteins expressed in eucaryotic or procaryotic cells.

Whole Virus Vaccines

Three viral pathogens of the respiratory tract for which inactivated vaccines have been used are influenza viruses, measles virus, and respiratory syncytial virus (RSV). Two distinctly different experiences resulted from the use of these vaccines. The influenza vaccine was safe, immunogenic, efficacious and generally well tolerated.[4] The inactivated whole virus vaccine retained mild reactogenicity in young seronegative children and has been replaced by disrupted virion vaccines that are in use today. The influenza vaccine has been fairly well accepted, but its shortcomings include a short-lived protective immunity, the need for annual re-vaccination (both due in part to the antigenic changes of the virus), and the induction of low levels of mucosal antibodies.

The experience with the inactivated measles and RSV vaccines was quite different. Both of these vaccines were formalin-inactivated preparations that are no longer in use because they induced a paradoxical, exaggerated clinical response to natural infection in vaccines.[5-9] The formalin-inactivated RSV vaccine induced moderately high levels of serum antibodies that were weak in neutralizing activity. Immunization with inactivated RSV vaccine did not prevent infection and those infected developed serious lower respiratory tract disease requiring hospitalization with a much higher frequency than controls.[5,6] Similarly, the killed measles vaccine was immunogenic and protective for several years, but immunity against infection waned. When these vaccinees were naturally infected with measles virus, they experienced an exaggerated and atypical clinical response.[7-9] This disease potentiation following soon after administration of inactivated RSV vaccine and years after killed measles vaccine is thought to be due to an abnormal and unbalanced immune response such that heightened pathologic reactions occur at sites of virus replication.[10,11]

Subunit Vaccines

Subunit vaccines can be derived from disrupted virions or from the expression of protective viral antigens in eucaryotic or procaryotic cells followed by purification of the expressed proteins. Two subunit vaccines have been developed for RSV: (1) a fusion (F) glycoprotein,[12] which is immunoaffinity-purified from RSV-infected cells, and (2) a chimeric protein containing the fused ectodomains of the F and G glycoproteins purified from baculovirus vector infected insect cell cultures from which the chimeric protein is purified.[13] The RSV F and G glycoproteins were selected for inclusion in a subunit vaccine because they are the major protective antigens of RSV.[2]

Although both of these vaccines elicited an antibody response in experimental animals, the antibodies were qualitatively different from those induced by natural infection[14] since they possessed only a low level of neutralizing activity. The subunit vaccines provided complete pulmonary resistance to RSV challenge in cotton rats,[12,13] but passive transfer of the qualitatively altered (high ELISA and low neutralizing titer) antibody induced by the FG vaccine did not protect animals from RSV challenge.[14] Protection in the passively immunized animals correlated only with levels of neutralizing antibody. Both the F and FG vaccines have caused potentiated disease in experimental animals.[14]

Clearly, licensed subunit vaccines such as the disrupted influenza A virus vaccines and the yeast-derived hepatitis B virus vaccines are effective, indicating that this vaccine strategy has been successful against certain pathogens. However, the undesirable potentiation of disease seen with RSV subunit vaccines indicates that each potential vaccine needs to be fully evaluated in preclinical studies.

VECTORED VACCINES

A vectored vaccine consists of a live replicating virus or bacteria that expresses a protective antigen of another pathogen and that replicates sufficiently in the vaccinee to induce an immune response to the expressed foreign protein. For such a strategy to be successful, the vector would have to be a viable but attenuated organism, preferably one that has a long history of use as a successful vaccine, that could not revert to virulence, and that could adequately express the foreign protein. The endogenously expressed protein antigens would be authentic and native and capable of inducing both humoral and cellular immune responses. If the site to be protected by the vaccine is a mucosal surface, the vector should be one that can be administered at or replicate at the local mucosal site or be a potent immunogen if administered at a distant site.[15]

The two vectored vaccines expressing respiratory virus antigens that we have tested in experimental animals are the vaccinia-RSV and the adenovirus-RSV recombinant vaccines. Vaccinia-RSV recombinant viruses expressing the F or G glycoprotein of RSV induced neutralizing serum antibody. Infection with vaccinia-RSV recombinants stimulated protective immunity in rodents and owl monkeys similar to infection with RSV and, therefore, seemed to be promising vaccine candidates.[16-18] However, these vaccines induced only moderate titers of ELISA antibody and low-to-moderate levels of neutralizing antibody in chimpanzees.[19] The chimpanzees were poorly protected from challenge, showing no effect on duration of clinical illness and only marginal restriction in duration and magnitude of virus shedding. Protection was not seen in the upper respiratory tract indicating that these intradermally administered vaccines, like the inactivated viruses given parenterally, are poor inducers of resistance on mucosal surfaces.[18] The chimpanzees did develop a very high titer of neutralizing antibodies following challenge, suggesting that the vaccine, though not protective, did efficiently prime the immune system.[19] The reasons for the failure of the vaccinia-RSV recombinants to protect the chimpanzees remain incompletely defined. An additional and significant observation that resulted from the studies of the vaccinia-RSV recombinants in cotton rats was that passively transferred RSV antiserum suppressed the antibody response to RSV glycoproteins but not to the vaccinia virus antigens. Those animals whose antibody responses were suppressed by passive antibody showed an increased susceptibility to RSV infection.[20] These immunosuppressive effects of pre-existing antibody on the immune response to intradermally administered vaccinia-RSV virus recombinants could be partially overcome by topical (intranasal) administration of the recombinants.[21] These observations have great relevance for RSV because RSV causes severe disease in two-month-old infants and immunization will have to be successful in the presence of maternal antibody.

RSV glycoproteins have also been engineered into the E3 region of adenovirus types 4, 5, and 7 under the control of endogenous adenovirus promoters.[22,23] The reasons for choosing adenoviruses as vectors were: (1) Adenovirus 4 and 7 vaccines have been used safely in military recruits for years; (2) the adenovirus vaccine replicates in the gastrointestinal tract of the vaccinee and therefore should be able to induce a generalized mucosal antibody response; (3) the adenovirus recombinants are high-level expression vectors in which relatively large pieces of DNA can be inserted in place of non-essential parts of the adenovirus genome, and the expressed protein is authentically processed in adenovirus infected cells; and (4) vectors based on different adenovirus serotypes can be developed for sequential immunization to boost immunity. As in with the vaccinia-RSV recombinants, evaluation of adenovirus-RSV recombinants in experimental animals produced conflicting results. In cotton rats and dogs, moderate to high levels of antibodies were induced and the animals were completely protected from challenge.[22] The immunogenicity of the RSV-recombinant vaccine in the one chimpanzee studied was disappointing since sequential immunization with three different adenovirus-RSV recombinants failed to induce a significant RSV-neutralizing antibody response.[22]

Alternatively, avipoxviruses have been used as vectors since these viruses replicate efficiently in avian hosts and are naturally restricted in replication in mammalian cells, in which they undergo an abortive cycle of replication but efficiently express the foreign proteins of interest.[24] Thus, protective viral immunogens are expressed in an authentic fashion in the avipox-recombinant-infected cells, but infectious virus is not produced. This reduces the risk of dissemination of the vector in the vaccinee or transmission to other non-vaccinated contacts, which is a major concern due to the high prevalence of immunosuppressive HIV infections in certain target populations.

Currently, the use of vectored vaccines in humans remains problematic with many obstacles to overcome. However, it is likely that vectored vaccines for veterinary medicine will precede their application in humans.

LIVE ATTENUATED VACCINES

The main advantage of a live virus vaccine is that the immune response induced more closely mimics that of a natural infection, including a mucosal antibody response when administered topically. The ideal live virus vaccine would demonstrate a satisfactory level of attenuation, replicate to a lower level than wild-type virus, induce resistance to wild-type virus challenge, and not lose the attenuation phenotype following replication. Because the chief concern about live virus vaccines is the genetic stability of their attenuation phenotype, it is preferable for live virus vaccines to have multiple mutations that contribute to the attenuation phenotype, thus decreasing the likelihood of restoration of virulence during replication in the vaccinee. Live virus vaccines fall into one of two major categories: "Jennerian" and classical vaccines.

The "Jennerian" Approach to Immunization

The "Jennerian" approach is modelled on the use of cowpox (vaccinia) virus by Edward Jenner to protect humans against the antigenically related smallpox virus and involves immunization with an antigenically related animal virus that is attenuated in humans to protect against a human pathogen.[3] There are two examples of using this approach in respiratory virus vaccine development: the use of the bovine parainfluenza virus type 3 (BPIV3) as a vaccine to protect against disease caused by human parainfluenza 3 (HPIV3), and the use of avian influenza A viruses as donors of attenuating genes in avian-human influenza A reassortant virus vaccines.

The BPIV3 virus was chosen as a vaccine candidate for several reasons. First, monoclonal antibody and post-infection serum studies identified a high degree of antigenic relatedness of BPIV3 and HPIV3 viruses.[25] Second, experimental animals immunized with BPIV3 were protected from HPIV3.[25] Third, nucleotide sequence analysis showed significant divergence between BPIV3 and HPIV3 genes, indicating independent evolution of BPIV3 and HPIV3 in their respective species. During the evolution of the BPIV3 in bovine hosts, it is likely that it acquired mutations that favor replication in this species but that restrict replication in humans.[25] The vaccine has been evaluated both in non-human primates and humans and showed properties of restricted replication and induction of a protective immune response to HPIV3, making it a promising candidate for a live parainfluenza 3 virus vaccine.[25,26] Further evaluation of this vaccine in humans is now in progress.

Avian influenza A viruses are restricted in replication in non-human primates and are attenuated in humans.[27] It is also known that the protective immune response in human influenza A virus infections is directed against the hemagglutinin (HA) and neuraminidase (NA) proteins.[1] Avian-human influenza A reassortant viruses containing the HA and NA genes from a human influenza A virus and the six internal attenuating genes from the avian influenza A/Mallard/NY/78 virus were found to be restricted in replication in humans and yet induced resistance against human influenza A virus infections. Thus, the six transferable avian influenza A virus genes conferred the attenuation phenotype. The reassortant viruses were immunogenic, protective, and well tolerated in adults,[27] but were unacceptably reactogenic in young children, causing fevers of $>39.4\,°C$ in 24% of the vaccine recipients.[28] Further development of avian-human influenza A reassortant vaccines derived from the A/Mallard/NY/78 avian influenza A parent was abandoned due to the reactogenicity in children. Currently, studies using gull influenza A viruses as donors are being actively pursued. Since the entire avian influenza A virus was not used as the vaccine but, rather, an avian-human influenza A reassortant virus; this immunization strategy has been referred to as a "modified" Jennerian approach.

The Classical Approach to Immunization

The classical approach to live virus vaccine development involves the derivation of a mutant virus from a wild-type (*wt*) virus which has acquired one or more mutations that specify the attenuation (*att*) phenotype. In addition, other phenotypes can develop during the process of attenuation, such as the cold-adaptation (*ca*) or the temperature-sensitive (*ts*) phenotype. Genes bearing *ca* and *ts* mutations have been identified as attenuating genes, suggesting that these "markers of attenuation" are indeed the genetic determinants of attenuation in some viruses.[29] Temperature-sensitive viruses are attractive candidates as vaccines for use in the respiratory tract because one can take advantage of the temperature differential in the upper and lower respiratory tracts. The virus could replicate in the cooler upper respiratory tract (and serve as an immunogen), but would be sufficiently restricted in replication at the higher temperature of the lower respiratory tract and, hence, would be incapable of causing disease.[30]

There are three major methods by which live attenuated respiratory virus vaccines have been developed: (1) mutagenesis of a virus with selection of variants bearing *ts* mutations; (2) repeated passage of virus at low temperatures, resulting in the acquisition of new phenotypes, such as the *ca*, *ts*, and *att* phenotypes; and (3) repeated passage at low temperature in cells from a heterologous host, resulting in a virus that is neither *ca* nor *ts* but has the *att* phenotype. Examples of each of these methods for selecting mutants will be presented.

The *ts* RSV vaccine was derived in the early 1970s by chemical mutagenesis of the wild-type RSV strain A2. It was found to be immunogenic and highly (although incompletely) attenuated, but was found to have some genetic instability.[31-34] The *ca* influenza A virus is an example of the second method and consisted of the serial passage of the wild-type influenza A/Ann Arbor/6/60 at low temperatures.[35] The passaged virus acquired three phenotypes that distinguish it from wild-type virus: the *ca* phenotype (i.e., the ability to replicate efficiently at 25°C); the *ts* phenotype (i.e., the restriction of replication of the virus at 39°C compared to the wild-type virus); and the *att* phenotype, restricted replication in the respiratory tract of mice, hamsters, ferrets and humans. Reassortant viruses deriving their internal genes from the A/AA/6/60 *ca* virus and the HA and NA from newly emerged wild-type influenza A viruses of different HA and NA subtypes are satisfactorily infectious, attenuated, immunogenic, and efficacious and are not reactogenic in young children. The complete sequence of the A/AA/6/60 *ca* virus has been determined and the mutations present in each of the internal genes have been identified.[36] The gene segment that specifies each of the phenotypes has also been determined,[29] and the mechanism of attenuation specified by each of the attenuating genes is an area of active study. Another example of this method is the *ca* human parainfluenza type 3 vaccine candidates that have also been derived by serial passage of a wild-type isolate at 20° to 22°C and that demonstrate the *ca*, *ts*, and *att* phenotypes.[37] These candidate vaccines are currently being evaluated in non-human primates and human volunteers. The third method to generate live respiratory viruses involves the passage of a wild-type virus in a heterologous host (bovine kidney tissue) at low temperature. The RSV vaccine candidate derived from this passage was neither *ca* nor *ts*, but was highly, though incompletely, attenuated.[38,39] Since the subunit vaccines have been recognized to potentiate disease and the vectored RSV vaccines showed a low efficacy in chimpanzees, further efforts to develop live attenuated *ts* and/or *ca* RSV vaccines seem warranted.

Recombinant DNA technologies employing site-directed mutagenesis are also being used to develop attenuated mutants for a variety of viral pathogens.[3] This new technology has been used to generate an influenza A virus vaccine candidate that contains a chimeric NA gene with influenza B virus non-coding 3'- and 5'-end sequences and the coding sequence of an influenza A virus N1 NA gene.[40] This chimeric virus was attenuated and efficacious in mice.

We have attempted to summarize a large body of information on the development, evaluation, and problems associated with different respiratory virus vaccines. A basic understanding of the natural history of the infection and of the biology of the pathogen is crucial to the development of a useful vaccine. The usefulness of an animal model for evaluation of candidate vaccines cannot be overstated. It is hoped that this information might help in the production of vaccines against oral pathogens.

REFERENCES

1. B.R. Murphy and R.G. Webster, Orthomyxoviruses, *in*: "Virology," B.N. Fields, D.M. Kline, *et al.*, eds., Raven Press, Ltd., New York (1990).
2. K. McIntosh and R.M. Chanock, Respiratory syncytial virus, *in*: "Virology," B.N. Fields, D.M. Kline, *et al.*, eds., Raven Press, Ltd., New York (1990).
3. B.R. Murphy and R.M. Chanock, Immunization against viruses, *in*: "Virology," B.N. Fields, D.M. Kline, *et al.*, eds., Raven Press, Ltd., New York (1990).
4. M.W. Shaw, N.H. Arden, and H.F. Maassab, New aspects of influenza viruses, *Clin. Microbiol. Rev.* 5:74 (1992).
5. H.W. Kim, J.G. Canchola, C.D. Brandt, G. Pyles, R.M. Chanock, K. Jensen, and R.H. Parrott, Respiratory syncytial virus disease in infants despite prior administration of antigenic inactivated vaccine, *Am. J. Epidemiol.* 89:422 (1969).

6. A.Z. Kapikian, R.H. Mitchell, R.M. Chanock, R.A. Shvedoff, and C.E. Stewart, An epidemiologic study of altered clinical reactivity to respiratory syncytial (RS) virus infection in children previously vaccinated with an inactivated RS virus vaccine, *Am. J. Epidemiol.* 89:405 (1969).

7. P.R. Nader, M.S. Horwitz, and J. Rousseau, Atypical exanthem following exposure to natural measles: Eleven cases in children previously inoculated with killed vaccine, *J. Pediatr.* 72:22 (1968).

8. T.F.M. Scott and D.E. Bonanno, Reactions to live-measles-virus vaccine in children previously inoculated with killed-virus vaccine, *N. Engl. J. Med.* 277:248 (1967).

9. F. Buser and B. Montagnon, Severe illness in children exposed to natural measles after prior vaccination against the disease, *Scand. J. Infect. Dis.* 2:157 (1970).

10. B.R. Murphy, G.A. Prince, E.E. Walsh, H.W. Kim, R.H. Parrott, V.G. Hemming, W.J. Rodriguez, and R.M. Chanock, Dissociation between serum neutralizing and glycoprotein antibody responses of infants and children who received inactivated respiratory syncytial virus vaccine, *J. Clin. Microbiol.* 24:197 (1986).

11. H.W. Kim, S.L. Leikin, J. Arrobio, C.D. Brandt, R.M. Chanock, and R.H. Parrott, Cell-mediated immunity to respiratory syncytial virus induced by inactivated vaccine or by infection, *Pediatr. Res.* 10:75 (1976).

12. E.E. Walsh, C.B. Hall, M. Briselli, M.W. Brandriss, and J.J. Schlesinger, Immunization with glycoprotein subunits of respiratory syncytial virus to protect cotton rats against viral infection, *J. Infect. Dis.* 155:1198 (1987).

13. M.W. Wathen, R.J. Brideau, D.R. Thomsen, and B.R. Murphy, Characterization of a novel human respiratory syncytial virus chimeric FG glycoprotein expressed using a baculovirus vector, *J. Gen. Virol.* 70:2625 (1989).

14. M. Connors, P.L. Collins, C.-Y. Firestone, A.V. Sotnikov, A. Waitze, A.R. Davis, P.P. Hung, R.M. Chanock, and B.R. Murphy, Cotton rats previously immunized with a chimeric RSV FG glycoprotein develop enhanced pulmonary pathology when infected with RSV, a phenomenon not encountered during immunization with vaccinia-RSV recombinants or RSV, *Vaccine* in press (1992).

15. P.L. Collins, M. Connors, R.M. Chanock, and B.R. Murphy, Expression of respiratory syncytial virus genes by recombinant expression vectors, *in*: "Animal Models of Respiratory Syncytial Virus Infections," B. Meignier, *et al.*, eds., Merieux Foundation Publication, France (1991).

16. E.J. Stott, L.A. Ball, K.K. Young, J. Furze, and G.W. Wertz, Human respiratory syncytial virus glycoprotein G expressed from a recombinant vaccinia virus vector protects mice against live-virus challenge, *J. Virol.* 60:607 (1986).

17. R.A. Olmsted, N. Elango, G.A. Prince, B.R. Murphy, P.R. Johnson, B. Moss, R.M. Chanock, and P.L. Collins, Expression of the F glycoprotein of respiratory syncytial virus by a recombinant vaccinia virus: Comparison of the individual contributions of the F and G glycoproteins to host immunity, *Proc. Natl. Acad. Sci. (USA)* 83:7462 (1986).

18. R.A. Olmsted, R.M.L. Buller, P.L. Collins, W.T. London, J.A. Beeler, G.A. Prince, R.M. Chanock, and B.R. Murphy, Evaluation in non-human primates of recombinant vaccinia viruses expressing the F or G glycoproteins of respiratory syncytial virus, *Vaccine* 6:519 (1988).

19. P.L. Collins, R.H. Purcell, W.T. London, L.A. Lawrence, R.M. Chanock, and B.R. Murphy, Evaluation in chimpanzees of vaccinia virus recombinants that express the surface glycoproteins of human respiratory syncytial virus, *Vaccine* 8:164 (1990).

20. B.R. Murphy, R.A. Olmsted, P.L. Collins, R.M. Chanock, and G.A. Prince, Passive transfer of respiratory syncytial virus (RSV) immune serum suppresses the immune response to the RSV fusion (F) and large (G) glycoproteins expressed by recombinant vaccinia viruses, *J. Virol.* 62:3907 (1988).

21. B.R. Murphy, P.L. Collins, L. Lawrence, J. Zubac, R.M. Chanock, and G.A. Prince, Immunosuppression of the antibody response to respiratory syncytial virus (RSV) by pre-existing serum antibodies: Partial abrogation by topical infection of the respiratory tract with vaccinia virus-RSV recombinants, *J. Gen. Virol.* 70:2185 (1989).

22. K.-H. Hsu, M.D. Lubeck, A.R. Davis, R.A. Bhat, B.H. Selling, B.M. Bhat, S. Mizutani, B.R. Murphy, P.L. Collins, R.M. Chanock, and P.P. Hung, Immunogenicity of recombinant adenovirus-respiratory syncytial viruses using Ad4, Ad5, and Ad7 vectors in dogs and a chimpanzee, *J. Infect. Dis.* submitted (1992).

23. P.L. Collins, A.R. Davis, M.D. Lubeck, S. Mizutani, P.P. Hung, G.A. Prince, R.H. Purcell, R.M. Chanock, and B.R. Murphy, Evaluation of protective efficacy of recombinant vaccinia viruses and adenoviruses that express respiratory syncytial virus glycoproteins, *in*: "Vaccines 90: Modern Approaches to New Vaccines Including Prevention of AIDS," F. Brown, R.M. Chanock, H. Ginsberg, and R.A. Lerner, eds., Cold Spring Harbor Laboratory, New York (1990).

24. D. Baxby and E. Paoletti, Potential use of non-replicating vectors as recombinant vaccines, *Vaccine* 10:8 (1992).

25. K.L.V. Coelingh, C.C. Winter, E.L. Tierney, W.T. London, and B.R. Murphy, Attenuation of bovine parainfluenza virus type 3 in non-human primates and its ability to confer immunity to human parainfluenza virus type 3 challenge, *J. Infect. Dis.* 157:655 (1988).

26. M.L. Clements, R.B. Belshe, J. King, F. Newman, T.U. Westblom, E.L. Tierney, W.T. London, and B.R. Murphy, Evaluation of bovine, cold-adapted human, and wild-type human parainfluenza type 3 viruses in adults and chimpanzees, *J. Clin. Microbiol.* 29:1175 (1991).

27. S.D. Sears, M.L. Clements, R.F. Betts, H.F. Maassab, B.R. Murphy, and M.H. Snyder, Comparison of live attenuated H1N1 and H3N2 cold-adapted and avian-human influenza A reassortant viruses and inactivated virus vaccine in adults, *J. Infect. Dis.* 158:1209 (1988).

28. M.C. Steinhoff, N.A. Halsey, L.F. Fries, M.H. Wilson, J. King, B.A. Burns, R.K. Samorodin, V. Perkis, B.R. Murphy, and M.L. Clements, The A/Mallard/6750/78 avian-human, but not the A/Ann Arbor/6/60 cold-adapted, influenza A/Kawasaki/86 (H1N1) reassortant virus vaccine retains partial virulence for infants and children, *J. Infect. Dis.* 165:1023 (1991).

29. M.H. Snyder, R.F. Betts, D. DeBorde, E.L. Tierney, M.L. Clements, D. Harrington, S.D. Sears, R. Dolin, H.F. Maassab, and B.R. Murphy, Four viral genes independently contribute to attenuation of live influenza A/Ann Arbor/6/60 (H2N2) cold-adapted reassortant virus vaccines, *J. Virol.* 62:488 (1988).

30. D.D. Richman and B.R. Murphy, The association of the temperature-sensitive phenotype with viral attenuation in animals and humans: Implications for the development and use of live virus vaccines, *Rev. Infect. Dis.* 1:413 (1979).

31. D.S. Hodes, H.W. Kim, R.H. Parrott, E. Camargo, and R.M. Chanock, Genetic alteration in a temperature-sensitive mutant of respiratory syncytial virus after replication *in vivo*, *Proc. Soc. Exp. Biol. Med.* 145:1158 (1974).

32. M.A. Gharpure, P.F. Wright, and R.M. Chanock, Temperature-sensitive mutants of respiratory syncytial virus, *J. Virol.* 3:414 (1969).

33. H.W. Kim, J.O. Arrobio, C.D. Brandt, P. Wright, D. Hodes, R.M. Chanock, and R.H. Parrott, Safety and antigenicity of temperature sensitive (ts) mutant respiratory syncytial virus (RSV) in infants and children, *Pediatrics* 52:56 (1973).

34. P.F. Wright, T. Shinozaki, W. Fleet, S.H. Sell, J. Thompson, and D.T. Karzon, Evaluation of live, attenuated respiratory syncytial virus vaccine in infants, *J. Pediatr.* 88:931 (1976).

35. H.F. Maassab, Adaptation and growth characteristics of influenza virus at 25°C, *Nature* 213:612 (1967).

36. N.J. Cox, F. Kitame, A.P. Kendal, H.F. Maassab, and C. Naeve, Identification of sequence changes in the cold-adapted, live attenuated influenza vaccine strain, A/Ann Arbor/6/60 (H2N2), *Virology* 167:554 (1988).

37. R.B. Belshe and F.K. Hisson, Cold adaptation of parainfluenza virus type 3: Induction of three phenotypic markers, *J. Med. Virol.* 10:235 (1982).

38. W.T. Friedewald, B.R. Forsyth, C.B. Smith, M.A. Gharpure, and R.M. Chanock, Low-temperature-grown RS virus in adult volunteers, *JAMA* 204:690 (1968).

39. H.W. Kim, J.O. Arrobio, G. Pyles, C.D. Brandt, E. Camargo, R.M. Chanock, and R.H. Parrott, Clinical and immunological response of infants and children to administration of low-temperature adapted respiratory syncytial virus, *Pediatrics* 48:745 (1971).

40. T. Muster, E.K. Subbarao, M. Enami, B.R. Murphy, and P. Palese, An influenza A virus containing influenza B virus 5' and 3' noncoding regions on the neuraminidase gene is attenuated in mice, *Proc. Natl. Acad. Sci. (USA)* 88:5177 (1991).

AN UPDATE ON THE "JENNERIAN" AND MODIFIED "JENNERIAN" APPROACH TO VACCINATION OF INFANTS AND YOUNG CHILDREN AGAINST ROTAVIRUS DIARRHEA

Albert Z. Kapikian,[1] Timo Vesikari,[2] Tarja Ruuska,[2] H. Paul Madore,[3] Cynthia Christy,[3] Raphael Dolin,[3] Jorge Flores,[1] Kim Y. Green,[1] Bruce L. Davidson,[4] Mario Gorziglia,[1] Yasutaka Hoshino,[1] Robert M. Chanock,[1] Karen Midthun,[5] and Irene Pérez-Schael[6]

[1]National Institutes of Health, Bethesda, MD 20892; [2]University of Tampere, Tampere, Finland 33520; [3]University of Rochester, Rochester, NY 14642; [4]Wyeth-Ayerst Research, Philadelphia, PA 19101; [5]Johns Hopkins University, Baltimore, MD 21205; and [6]Central University of Venezuela, Caracas, Venezuela 1010A

IMPORTANCE OF DIARRHEAL DISEASES

It is not generally appreciated that diarrheal diseases take an enormous toll in regard to morbidity and mortality in infants and young children in the developing countries of the world. Diarrheal illnesses are ranked first among infectious diseases in the number of episodes and deaths in developing countries in Asia, Africa, and Latin America: it is estimated that 3-5 billion cases of diarrhea and 5-10 million diarrhea-associated deaths occur annually in developing countries, predominantly in infants and young children.[1] Although deaths from diarrheal illnesses occur infrequently in developed countries, the morbidity from diarrheal illnesses in the pediatric age group is substantial.[2] The importance of diarrheal diseases was highlighted recently in a poster heralding "The World Summit for Children," a 2-day meeting of world leaders held at the United Nations in New York City, September 29-30, 1990, that aimed at a commitment to improving the well-being of children worldwide. The poster indicated that during the 2-day time period of the meeting, diarrheal illness would cause the death of 22,000 children worldwide, a toll greater than that from any other disease noted for this age group.

Up until the 1970s, the cause of a major portion of diarrheal illnesses was not known.[3-5] It was disappointing that despite the "golden age" of virology in the 1950s and 1960s, when scores of enteric viruses such as the echoviruses were discovered, none emerged as important etiologic agents of gastroenteritis. However, the discovery of the Norwalk virus in 1972 and of rotavirus in 1973, both without the benefit of *in vitro* tissue culture systems, ushered in a new era in the study of the etiology of gastroenteritis viruses.[6,7] The Norwalk virus was found to be an important cause of nonbacterial gastroenteritis in adults and older children, while rotaviruses were found to be major etiologic agents of severe diarrheal illness of infants and young children.[5,8]

IMPORTANCE OF ROTAVIRUSES

Rotaviruses are the single most important etiologic agents of severe diarrhea of infants and young children in the developed as well as developing countries being responsible for approximately 35-50% of such illnesses.[5] The impact of rotavirus

Genetically Engineered Vaccines, Edited by
J.E. Ciardi *et al.*, Plenum Press, New York, 1992

infections in the United States is substantial, as 90% of infants and young children have undergone a rotavirus infection by the end of the third year of life.[5] The disease burden annually in the United States for rotavirus diarrhea in the under-5 year age group is estimated to reach over 1 million cases of severe diarrhea and 150 deaths.[2] Estimates of the frequency of rotavirus diarrhea in a large Texas county, by extrapolation from the mean annual number of cases and probable cases of rotavirus gastroenteritis admitted to a large pediatric hospital over a 6-year period, indicated that the annual incidence of hospitalization for such rotavirus gastoenteritis in the first and second years of life was 11.1 and 5.8 per 1000 children, respectively, with a further decline with increasing age.[9] In addition, it was estimated in this study that 1 of every 46 children would be hospitalized for rotavirus gastroenteritis by 18 years of age (the upper age limit for admission to this pediatric hospital), with almost all of the risk occurring during the first 5 years of life. By extrapolation from these figures to the United States as a whole, it was estimated that 110,000 children are hospitalized annually with presumptive rotavirus gastroenteritis, resulting in 583,000 hospital days at a cost of $352 million.[9]

Rotaviruses are consistently shown to be the major etiologic agents of severe diarrhea in most developing countries.[5] Although rotavirus infections occur with similar frequency in both developing and developed countries, the morbidity associated with such infections is quite different.[5] It was estimated that annually, in the under-5 year age group in developing countries, rotaviruses are responsible for mild diarrhea in 110 million individuals, for moderately severe diarrhea in over 9 million, for severe diarrhea in over 8 million, and for the death of over 870,000 infants and young children.[10]

DEVELOPMENT OF ROTAVIRUS VACCINES BY THE "JENNERIAN" APPROACH

The development of a vaccine to prevent or modify rotavirus illness is one of the highest priorities in international health, with most efforts aimed at developing an orally administered vaccine effective against each of the four epidemiologically important serotypes, numbered 1-4.[5,10] Serotypes have been defined by VP7, one of the two outer capsid proteins that induces neutralizing antibodies; VP7 is coded for by RNA segment 7, 8, or 9.[5,11] VP4, the other outer capsid protein which also induces neutralizing antibodies, is encoded by RNA segment 4.[11] A serotyping system based on VP4 specificity of selected human rotaviruses has recently been described.[12] In the current report, certain strategies being pursued to achieve the goal of a rotavirus vaccine and the progress being made in field trials with some vaccine candidates will be summarized.

The most extensively evaluated strategy is that pioneered by Edward Jenner in 1798 for smallpox vaccination of humans in which an antigenically related, live, attenuated agent from a non-human host is used as the immunizing agent. The Jennerian approach was evaluated in early clinical trials with a bovine rotavirus strain NCDV (RIT 4237), (with VP7 serotype 6 specificity) by Vesikari et al. and later with bovine rotavirus strain WC3 (serotype 6, also) by Clark et al., with promising results initially in older infants and young children but with variable success in infants under 6 months of age.[5,13-19] We have pursued this strategy with another surrogate strain--simian rhesus rotavirus (RRV) strain MMU18006--that shares VP7-neutralization specificity with human rotaviruses belonging to serotype 3.[5,20-21] Besides the theoretical advantage of sharing serotype specificity with an epidemiologically important human rotavirus strain, the RRV strain grows to relatively high titer (10^6 PFU/ml) in a semi-continuous strain of fetal rhesus diploid lung cells.[22] This property was considered important because adventitious agents frequently contaminate primary monkey kidney cells, the latter being quite efficient for propagation of most rotaviruses.[5]

The orally administered RRV vaccine was shown (a) to be safe and antigenic in adults, (b) to induce an unacceptably high rate of reactions (fever or diarrhea or watery stools) with a 10^5 PFU dose in infants who were predominantly over 6 months of age, and (c) to induce a mild, transient febrile response with a 10^4 PFU dose in about one-third of 2- to 5-month-old infants in certain locations; it is likely that passively acquired maternal antibody in this younger age group attenuated the reactogenicity of the vaccine.[5,21,23-36]

These reactions with the 10^4 PFU dose in the target population were considered acceptable, and thus phase II double-blind field trials were initiated with collaborators worldwide. Seven RRV vaccine field trials involving over 1,000 infants and young children have been described by various collaborators in the United States and overseas.[21,27-34]

The protective efficacy of a single oral dose of the RRV vaccine was variable, ranging from nil to 85% against clinically significant rotavirus diarrhea.[21,27,29-36] A likely explanation for the variable efficacy became apparent after the rotavirus strains recovered from ill children in the various studies were serotyped. The predominant strains responsible for rotavirus diarrhea in the Venezuelan study (in which the 85% protection was observed) belonged to VP7 serotype 3, the same serotype as the vaccine.[32,33] However, in trials in which vaccine efficacy was nil, other serotypes predominated, suggesting that VP7 serotype-specific immunity against each of the four epidemiologically important serotypes was necessary for optimal protection.[34-36] In addition, serologic studies demonstrated that the RRV vaccine failed to induce heterotypic serum antibodies in young infants not primed by previous rotavirus infection.[37,38] Thus, the variable efficacy of the monovalent bovine or simian rotavirus vaccines might be explained as follows: infants primed by prior subclinical or clinical rotavirus infection develop a broadened (i.e., homotypic and heterotypic) antibody response after subsequent vaccination with a monovalent vaccine and thus become immune to homotypic and heterotypic strains; infants not primed by prior subclinical or clinical rotavirus infection develop a homotypic antibody response following vaccination and thus are immune only to that serotype.[38]

DEVELOPMENT OF ROTAVIRUS VACCINES BY THE MODIFIED "JENNERIAN" APPROACH

As a consequence, we altered our strategy and pursued a modified "Jennerian" approach by developing a quadrivalent vaccine in an attempt to achieve broader antigenic coverage and thus protect against each of the four epidemiologically important serotypes.[21,39] This vaccine included not only the monovalent RRV VP7 serotype 3 vaccine described above, but also three reassortant viruses generated by coinfection of cell cultures with RRV and a human rotavirus of VP7 serotype 1, 2, or 4 specificity under selective pressure of antibody against the RRV strain.[40,41] Each reassortant contains 10 RRV genes (including RRV gene 4 that encodes outer capsid protein VP4) and a single human rotavirus gene that encodes VP7 serotype 1 (DxRRV), 2 (DS-1xRRV), or 4 (ST-3xRRV) specificity.[40,41]

In phase I studies, individual reassortant vaccines were similar in reactogenicity and antigenicity to the 10^4 PFU RRV vaccine in the target population and were thus acceptable for further evaluation in phase II efficacy trials.[42-48]

Two phase II studies with human-rhesus rotavirus reassortant vaccines have been completed and analyzed for efficacy. In Finland, a single oral dose of 10^4 PFU of human-RRV reassortant vaccine DxRRV (VP7 serotype 1) or 10^5 PFU of DS-1xRRV (VP7 serotype 2) or placebo were evaluated in a double-blind trial in 359 infants 2 to 5 months of age studied for two rotavirus "seasons" following vaccination.[47] By IgA ELISA, 61% of the vaccinees who were given the VP7 serotype 1 reassortant vaccine and 75% who were given the VP7 serotype 2 reassortant vaccine developed a serologic response (i.e., a seroconversion) ($P<0.05$). The significant difference in the number of seroresponses may have resulted from (a) the administration of a higher dose of the serotype 2 reassortant vaccine, or (b) the finding that study individuals tested tended to have a higher prevalence of detectable preexisting serum neutralizing antibody to VP7 serotype 1 (Wa strain) than to VP7 serotype 2 (DS-1 strain), which may have inhibited the vaccine "takes."[47]

Protection against rotavirus diarrhea was evaluated in two successive rotavirus "seasons" by comparing the rate of rotavirus diarrhea in the vaccinees with that of the placebo group.[47] In the first season, only serotype 1 strains were detected in the study group whereas in the second season, although serotype 1 strains were predominant, serotype 4 strains were also found. The protective efficacy against serotype 1 rotavirus diarrhea of the serotype 1 reassortant vaccine was 67% in the first rotavirus season and negligible in the second season, whereas for the serotype 2 reassortant vaccine, it was 67% and 80% in the first and second seasons, respectively (Table 1). The protective efficacy in the second season against all rotavirus diarrhea regardless of serotype was negligible for the DxRRV vaccine and 67% for the DS-1xRRV vaccine (see Table 1 footnote**). If infants who did not seroconvert following vaccination are excluded from this analysis, overall vaccine efficacy was increased; the overall rate of protection against all rotavirus diarrhea among those who seroconverted, for both vaccines combined, was 92% and 59% for the first and second rotavirus seasons, respectively.[47] There is no clear explanation for the unexpected heterotypic protection induced by the serotype 2 reassortant vaccine. One hypothesis assumes that priming with a naturally occurring rotavirus strain(s) prior to vaccination led to a broadened antibody response after vaccination that resulted in protection against the heterotypic serotype 1 strain. With this assumption, the placebo group would also be primed but would not achieve a broadening of antibody response until a second natural infection (which could be clinical or subclinical) had occurred.

Table 1. Protective efficacy of a single dose of DxRRV reassortant (VP7 serotype 1) or DS-1xRRV (VP7 serotype 2) reassortant rotavirus (RV) vaccine against VP7 serotype 1 RV diarrhea during two RV seasons in infants inoculated at 2 to 5 months of age in the Finland study.

Inoculum (and number of infants inoculated)	First season		Second season		First and second seasons combined	
	Diarrhea episodes	Protective efficacy	Diarrhea episodes	Protective efficacy	Diarrhea episodes	Protective efficacy
DxRRV vaccine (120)	3[a]	67%	4	negligible	7[d]	50%
DS-1xRRV vaccine (119)	3[b]	67%	1[c]	80%	4[e]	71%
Placebo (120)	9[a,b]		5[c]		14[d,e]	
TOTAL	15*		10**		25	

*All strains recovered in the first season belonged to VP7 serotype 1.
**20 RV diarrhea episodes occurred in the second season: 10 serotype 1 (as shown); 7 serotype 4; 2 both 1 and 4; 1 both 1 and 2. In the second season, of the 20 RV diarrhea episodes, 8 were detected in the DxRRV vaccinees, 3 in the DS-1xRRV vaccinees, and 9 in the placebo group.
[a]$P=0.14$; [b]$P=0.14$; [c]$P=0.21$; [d]$P=0.17$; [e]$P=0.02$ (Fisher exact test, 2-tailed)
Data from ref. 47.

In Rochester, a single oral dose of the DxRRV (VP7 serotype 1) or RRV (VP7 serotype 3) vaccine or a placebo was evaluated in a double-blind trial in 223 infants 2 to 4 months of age.[48] The RRV and DxRRV vaccines were antigenic, inducing an IgA ELISA seroresponse in 69% and 47% of the infants, respectively, whereas none of the placebo recipients developed an IgA ELISA seroresponse. However, by neutralization assay (to RRV, DxRRV, or Wa strain) and/or IgA ELISA, 78% of the RRV vaccinees, 71% of the DxRRV vaccinees, and 12% of the placebo recipients developed a serologic response.

Efficacy of each vaccine against rotavirus diarrhea was evaluated over three rotavirus seasons; however, the analysis presented here will be limited to the first rotavirus season because the number of rotavirus diarrheal episodes in the second and third seasons was too few to determine vaccine efficacy (27 rotavirus strains were detected in the first season, 6 in the second season, and 6 in the third season). Rotavirus VP7 serotype 1 was prevalent throughout the three rotavirus seasons: in the first season, 21 of the 27 strains were serotype 1 (6 could not be typed); in the second season 5 of the 6 were serotype 1 (1 could not be typed); and in the third season, 5 of the 6 were serotype 1, and 1 reacted with reagents to both serotypes 1 and 2. The DxRRV and the RRV vaccine each induced protection against rotavirus diarrhea (Table 2). The protective efficacy of the DxRRV vaccine was 77% and that of the RRV vaccine 66%. Protection against non-rotavirus diarrhea was not observed.[48]

Table 2. Protective efficacy of a single dose of DxRRV reassortant (VP7 serotype 1) or RRV (VP7 serotype 3) vaccine against rotavirus (RV) gastrointestinal illness during the first RV season in infants inoculated at 2 to 4 months of age in the Rochester study.

Inoculum	Number of infants inoculated	Number of episodes of RV gastroenteritis* (percentage of those inoculated)	Protective efficacy
DxRRV vaccine	74	4 (5%)[a]	77%
RRV vaccine	76	6 (8%)[b]	66%
Placebo	73	17 (23%)[a,b]	----

*21 of the 27 strains were classified as VP7 serotype 1, and 6 were nontypeable.
[a]P=0.002; [b]P=0.012 (F.E.T., 2-tailed)
Adapted from ref. 48.

Unexpectedly, as in the Finnish trial described above, heterotypic protection was observed as the serotype 3 RRV vaccine protected against rotavirus diarrhea caused by the heterotypic serotype 1 strain.[48] This was in sharp contrast to a previous rotavirus vaccine trial in Rochester in which the RRV vaccine failed to induce protection against rotavirus diarrhea caused by serotype 1.[34] Again, one explanation for this anomaly is that the study group in the recent trial was primed by rotavirus infection before or during the vaccination period, which set the stage for a broadened serum antibody response following vaccination. Indeed, during the vaccination period, RRV vaccinees in the recent study had a significantly greater number of neutralizing antibody responses to human rotavirus Wa (VP7 serotype 1) than did the RRV vaccinees in the earlier trial (19.1% vs. 4.7%) (P=0.004), suggesting that the recent vaccinees may have been primed with serotype 1 infection at or around the period of vaccination.[48] In addition, during the same period, placebo recipients in the recent study developed a significantly greater number of neutralizing antibody responses to the Wa strain than did the placebo recipients in the earlier trial (9.6% vs. 0%) (P=0.006).

Because individual reassortant vaccines were similar to the extensively studied 10^4 PFU dose of RRV vaccine in reactogenicity and immunogenicity in 2- to 5-month-old infants, the RRV and the three human-RRV reassortant vaccines were combined into a single quadrivalent formulation comprised of RRV (VP7 serotype 3), DxRRV (VP7 serotype 1), DS-1xRRV (VP7 serotype 2), and ST-3xRRV (VP7 serotype 4). In initial studies carried out in Venezuela, the quadrivalent vaccine was similar in reactogenicity to that observed with 10^4 PFU of RRV vaccine in the target population, but the frequency of seroresponses to individual serotypes by neutralization was disappointing.[49,50]

Therefore, we adjusted the amount of antigen or the number of doses administered or the sequence of administration of various components in an attempt to reach a seroresponse rate of at least 50% by neutralization against each of the four serotypes represented in the vaccine without increasing the reactogenicity.[29,49-52] We found that by increasing the amount of virus in each component in the quadrivalent vaccine to 10^5 PFU (4 x 10^5 PFU) and by giving two doses, the reactogenicity was not increased and a "take rate" approaching, reaching, or exceeding the 50% level for each serotype could be achieved (Fig. 1).[51] In a final phase I study in this series, three doses of 10^6 PFU of each component did not appear to offer an advantage in antigenicity over three doses of the 4 x 10^5 PFU dose.[51] Therefore, we are planning a double-blind, placebo-controlled "catchment study" (i.e., passive surveillance of hospitalized study patients with diarrhea) in 3500 infants and young children in Caracas, Venezuela, in the spring of 1992. Three doses of the 4 x 10^5 PFU formulation of the quadrivalent vaccine will be given orally beginning at 8 to 10 weeks of age; the administration of each single dose of quadrivalent vaccine will be separated by an interval of 1 month. It should be noted that a three-dose (rather than two-dose) schedule is being adopted in this trial to conform with projected immunization schedules in this age group.

That the quadrivalent vaccine approach was promising was shown recently in the United States, where various pediatric centers under the auspices of Wyeth-Ayerst Research carried out a three-cell double-blinded efficacy trial in which three oral doses of a 4 x 10^4 PFU quadrivalent vaccine formulation (RRV, DxRRV, DS-1xRRV, ST-3xRRV) or three oral doses of 10^4 PFU of DxRRV reassortant or placebo were given to a total of approximately 900 infants and young children.[53] Preliminary analysis indicates that after the first rotavirus season, three doses of the quadrivalent vaccine or three doses of the DxRRV vaccine each had an efficacy rate of approximately 57% against predominantly VP7 serotype 1 rotavirus diarrhea. The second year of this trial is under analysis.

Finally, in an attempt to circumvent the reactogenicity of the quadrivalent vaccine and to continue to take advantage of the protective efficacy of the VP7 of the four epidemiologically important human rotavirus serotypes, previously prepared single gene substitution reassortant rotavirus strains with a single gene-encoding VP7 of human rotavirus serotype 1, 2, 3, or 4 and the remaining 10 genes from a bovine rotavirus UK strain are under evaluation as vaccine candidates.[40,41,54] This was prompted by the lack of reactogenicity in infants and young children of bovine rotavirus vaccine strain NCDV (RIT 4237) or WC3 by others.[13-19] The VP7 serotype 1 and serotype 2 reassortant components have been safety tested and have been undergoing clinical evaluation in adult volunteers and children and infants at Johns Hopkins University.[54] We plan to ascertain the reactogenicity and antigenicity of each of the four strains to determine (a) whether these human-bovine rotavirus reassortants are less reactogenic than the human-RRV reassortant vaccines and (b) whether they are as antigenic. In addition, because the VP4 of the rhesus (strain MMU 18006) and the bovine (strain UK) rotaviruses are distinct from each other, we are considering the administration of various combinations of the human-bovine and human-RRV reassortants together or sequentially.[40,55] This may overcome the possibility that RRV VP4 antibody induced by an initial dose of the RRV-based quadrivalent vaccine would inhibit the "take" rate of a second dose of the same vaccine.[51] It is of interest that in early studies in adult volunteers, we evaluated the UK strain as a vaccine candidate but discontinued further clinical studies with it when we decided to pursue the RRV MMU 18006 strain as the surrogate strain in lieu of the UK strain for the "Jennerian" approach to rotavirus vaccination as described above.[24,56-58] It is promising that two oral doses of a human-bovine rotavirus single gene substitution reassortant vaccine candidate (WI79-9) with a human rotavirus VP7 serotype 1 encoding gene and 10 genes of the bovine rotavirus WC3 strain was recently shown in a study in the United States by Clark et al. to be effective against predominantly serotype 1 rotavirus diarrhea following administration to 2- to 11-month-old infants.[59]

If the modified "Jennerian" approach should not prove to be successful, other strategies that consider the immunogenicity of both VP7 and VP4 of human rotaviruses will be examined, such as (a) further evaluation of human rotavirus strains, e.g., M37, recovered from asymptomatic human infection in neonates, (b) evaluation of cold-adapted strains of human rotaviruses, (c) development of reassortants between human rotavirus strains that have different VP4 and VP7 specificities, (d) exploration of recombinant DNA

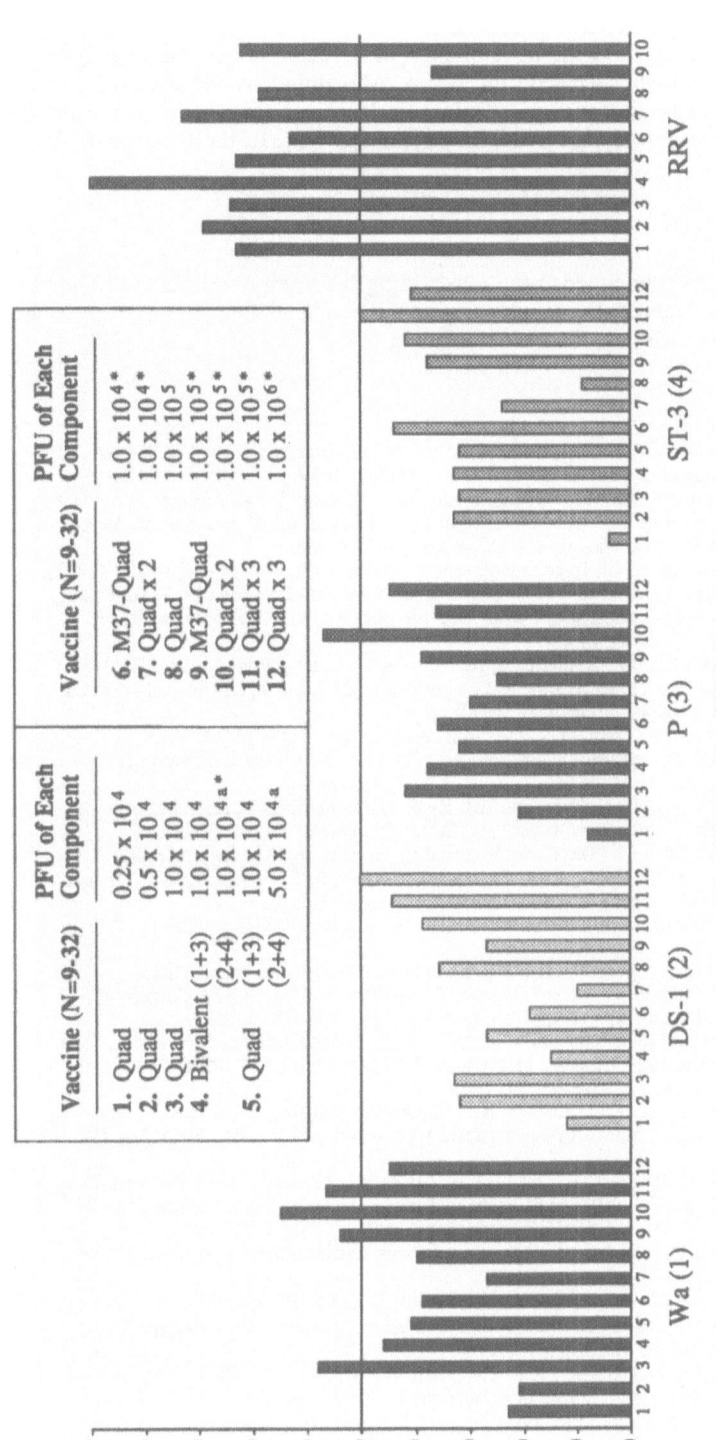

*One month interval (approx.) between each dose.
aTube neutralization vs. ST-3.
NOTE: Assays 11 and 12 represent preliminary partial data. (ref. 51)
Data from ref. 29, 49-51; categories 1-7 from ref. 52.

Figure 1. Progression of Quadrivalent (Quad) (VP7 Serotype 1, 2, 3, and 4) Rotavirus Vaccine Phase 1 Studies in Infants 10 to 20 Weeks of Age in Venezuela.

to deliver various rotavirus antigens, and (e) administration of rotavirus antigens to defined intestinal sites by microencapsulation procedures.[5,11,29,40,60-65]

Progress is being made in the development of a rotavirus vaccine to prevent the severe diarrheal episodes associated with this agent. Development of such a vaccine has received high priority in the quest for a "Children's Vaccine" that incorporates multiple antigens aimed at preventing important childhood illnesses--one of the goals of the World Summit for Children.

ACKNOWLEDGMENTS

We would like to acknowledge the outstanding laboratory assistance provided by Annie L. Pittman and Harvey D. James, Jr., and the outstanding editorial assistance provided by Amy H. Darwin and Todd J. Heishman.

REFERENCES

1. J.A. Walsh and K.S. Warren, Selective primary health care. An interim strategy for disease control in developing countries, *N. Engl. J. Med.* 301:967 (1979).
2. Institute of Medicine, Prospects for immunizing against rotavirus, in: "New Vaccine Development. Establishing Priorities. Diseases of Importance in the United States," Volume I, National Academy Press, Washington, DC, pp. 410-423 (1985).
3. J.D. Connor and E. Barrett-Connor, Infectious diarrheas, *Pediatr. Clin. North Am.* 14:197 (1967).
4. M.D. Yow, J.L. Melnick, R.J. Blattner, W.B. Stephenson, N.M. Robinson, and M.A. Burkhardt, The association of viruses and bacteria with infantile diarrhea, *Am. J. Epidemiol.* 92:33 (1970).
5. A.Z. Kapikian and R.M. Chanock, Rotaviruses, in: "Virology," Second Edition, B.N. Fields, D.M. Knipe, R.M. Chanock, M.S. Hirsch, J.L. Melnick, T.P. Monath, and B. Roizman, eds., Raven Press, New York, pp. 1353-1404 (1990).
6. A.Z. Kapikian, R.G. Wyatt, R. Dolin, T.S. Thornhill, A.R. Kalica, and R.M. Chanock, Visualization by immune electron microscopy of a 27nm particle associated with acute infectious nonbacterial gastroenteritis, *J Virol.* 10:1075 (1972).
7. R.F. Bishop, G.P. Davidson, I.H. Holmes, and B.J. Ruck, Virus particles in epithelial cells of duodenal mucosa from children with viral gastroenteritis, *Lancet* 2:1281 (1973).
8. A.Z. Kapikian and R.M. Chanock, The Norwalk group of viruses, in: "Virology," Second Edition, B.N. Fields, D.M. Knipe, R.M. Chanock, M.S. Hirsch, J.L. Melnick, T.P. Monath, and B. Roizman, eds., Raven Press, New York, pp. 671-693 (1990).
9. D.O. Matson and M.K. Estes, Impact of rotavirus infection at a large pediatric hospital, *J. Infect. Dis.* 162:598 (1990).
10. Institute of Medicine, The prospects for immunizing against rotavirus, in: "New Vaccine Development. Establishing Priorities. Diseases of Importance in Developing Countries," Volume II, National Academy Press, Washington, DC, pp. 308-318 (1986).
11. M.K. Estes, Rotaviruses and their replication, in: "Virology," Second Edition, B.N. Fields, D.M. Knipe, R.M. Chanock, M.S. Hirsch, J.L. Melnick, T.P. Monath, and B. Roizman, eds., Raven Press, New York, pp. 1329-1352 (1990).
12. M. Gorziglia, G. Larralde, A.Z. Kapikian, and R.M. Chanock, Antigenic relationships among human rotaviruses as determined by outer capsid protein VP4, *Proc. Natl. Acad. Sci. USA* 87:7155 (1990).
13. T. Vesikari, E. Isolauri, E. D'Hondt, A. Delem, F.E. Andre, and G. Zissis, Protection of infants against rotavirus diarrhoea by RIT4237 attenuated bovine rotavirus strain vaccine, *Lancet* 1:977 (1984).
14. T. Vesikari, Clinical and immunological studies of rotavirus vaccines, *Southeast Asian J. Trop. Med. Pub. Hlth.* 19:437 (1988).
15. H.F. Clark, R.E. Borian, L.M. Bell, K. Modesto, V. Gouvea, and S. Plotkin, Protective effect of WC3 vaccine against rotavirus diarrhea in infants during a predominantly serotype 1 rotavirus season, *J. Infect. Dis.* 158:570 (1988).
16. D.I. Bernstein, V.E. Smith, D.S. Sander, K.A. Pax, G.M. Schiff, and R.L. Ward, Evaluation of WC3 rotavirus vaccine and correlates of protection, *J. Infect. Dis.* 162:1055 (1990).
17. M.C. Georges-Courbet, J. Monges, M.R. Siopathis, J.B. Roungou, G. Gresenguet, L. Bellec, C. Lanckriet, M. Cadoz, L. Hessel, V. Gouvea, F. Clark, and A.J. Georges, Evaluation of the efficacy of a low passage bovine rotavirus (strain WC3) vaccine in children in Central Africa, *Res. Virol.* 142:405 (1991).
18. T. Vesikari, T. Ruuska, A. Delem, F. Andre, G.M. Beards, and T.H. Flewett, Efficacy of two doses of RIT4237 bovine rotavirus vaccine for prevention of rotavirus diarrhoea, *Acta. Pediatr. Scand.* 80:173 (1991).

19. R.S. Daum, B. Watson, H.F. Clark, and S.A. Plotkin, New developments in vaccines, *Adv. Pediatr. Inf. Dis.* 6:1 (1991).

20. A.Z. Kapikian, J. Flores, Y. Hoshino, R.I. Glass, K. Midthun, M. Gorziglia, and R.M. Chanock, Rotavirus: the major etiologic agent of severe infantile diarrhea may be controllable by a "Jennerian" approach to vaccination, *J. Infect. Dis.* 153:815 (1986).

21. A.Z. Kapikian, J. Flores, K. Midthun, Y. Hoshino, K.Y. Green, M. Gorziglia, K. Nishikawa, R.M. Chanock, L. Potash, and I. Perez-Schael, Strategies for the development of a rotavirus vaccine against infantile diarrhea with an update on clinical trials of rotavirus vaccines, *Adv. Exp. Biol. Med.* 257:67 (1989).

22. R.E. Wallace, P.J. Vasington, J.C. Petricciani, H.E. Hopps, D.E. Lorenz, and Z. Kadanka, Development of a diploid cell line from fetal rhesus monkey lung for vaccine production, *In Vitro.* 8:323 (1973).

23. A.Z. Kapikian, J. Flores, Y. Hoshino, K. Midthun, K.Y. Green, M. Gorziglia, R.M. Chanock, L. Potash, I. Perez-Schael, M. Gonzalez, T. Vesikari, L. Gothefors, G. Wadell, R.I. Glass, M.M. Levine, M.B. Rennels, G.A. Losonsky, C. Christy, R. Dolin, E.L. Anderson, R.B. Belshe, P.F. Wright, M. Santosham, N.A. Halsey, M.L. Clements, S.D. Sears, M.C. Steinhoff, and R.E. Black, Rationale for the development of a rotavirus vaccine for infants and young children, in: "Progress in Vaccinology," G.P. Talwar, ed., Springer-Verlag, New York, pp. 151-180 (1989).

24. A.Z. Kapikian, Y. Hoshino, J. Flores, K. Midthun, R.I. Glass, O. Nakagomi, T. Nakagomi, R.M. Chanock, L. Potash, M.M. Levine, R. Dolin, P.F. Wright, R.E. Belshe, E.L. Anderson, T. Vesikari, L. Gothefors, G. Wadell, and I. Perez-Schael, Alternative approaches to the development of a rotavirus vaccine, in: "Development of Vaccines and Drugs against Diarrhea. 11th Nobel Conf., Stockholm 1985," J. Holmgren, A. Lindberg, and R. Mollby, eds., Studentlitteratur, Lund, Sweden, pp. 192-214 (1986).

25. G.A. Losonsky, M.B. Rennels, A.Z. Kapikian, K. Midthun, P.J. Ferra, D.N. Fortier, K.M. Hoffman, A. Baig, M.M. Levine, Safety, infectivity, transmissibility and immunogenicity of rhesus rotavirus vaccine (MMU 18006) in infants, *Pediatr. Infect. Dis.* 5:25 (1986).

26. T. Vesikari, A.Z. Kapikian, A. Delem, and G. Zissis, A comparative trial of rhesus monkey (RRV-1) and bovine (RIT4237) oral rotavirus vaccines in young children, *J. Infect. Dis.* 153:832 (1986).

27. L. Gothefors, G. Wadell, P. Juto, K. Taniguchi, A.Z. Kapikian, and R.I. Glass, Prolonged efficacy of rhesus rotavirus vaccine in Swedish children, *J. Infect. Dis.* 159:753 (1989).

28. I. Perez-Schael, M. Gonzalez, N. Daoud, M. Perez, I. Soto, D. Garcia, G. Daoud, A.Z. Kapikian, and J. Flores, Reactogenicity and antigenicity of the rhesus rotavirus vaccine in Venezuelan children, *J. Infect. Dis.* 155:334 (1987).

29. J. Flores and A.Z. Kapikian, Vaccines against viral diarrhoea, *Ballieres Clin. Gastroenterol.* 4:675 (1990).

30. M.B. Rennels, G.A. Losonsky, M.M. Levine, A.Z. Kapikian, and the Clinical Study Group, Preliminary evaluation of the efficacy of rhesus rotavirus strain MMU18006 in young children, *Pediatr. Infect. Dis.* 5:587 (1986).

31. T. Vesikari, T. Rautanen, T. Varis, G.M. Beards, and A.Z. Kapikian, Clinical trial in children vaccinated between two and five months of age, *Am. J. Dis. Child.* 144:285 (1990).

32. J. Flores, I. Perez-Schael, M. Gonzalez, D. Garcia, M. Perez, N. Daoud, W. Cunto, and A.Z. Kapikian, Protection against severe rotavirus diarrhea by rhesus rotavirus vaccine in Venezuelan infants, *Lancet* 1:882 (1987).

33. I. Perez-Schael, D. Garcia, M. Gonzalez, N. Daoud, M. Perez, W. Cunto, A.Z. Kapikian, and J. Flores, Prospective study of diarrheal diseases in Venezuelan children to evaluate the efficacy of rhesus rotavirus vaccine, *J. Med. Virol.* 30:219 (1990).

34. C. Christy, H.P. Madore, M.E. Pichichero, C. Gala, P. Pincus, D. Vosefski, Y. Hoshino, A.Z. Kapikian, and R. Dolin, Field trial of rhesus rotavirus vaccine in infants, *Pediatr. Infect. Dis. J.* 7:645 (1988).

35. M. Santosham, G.W. Letson, M. Wolff, R. Reid, S. Gahagan, R. Adams, C. Callahan, R.B. Sack, and A.Z. Kapikian, A field study of the safety and efficacy of two candidate rotavirus vaccines in a native American population, *J. Infect. Dis.* 163:483 (1991).

36. M.B. Rennels, G.A. Losonsky, A.E. Young, C.L. Shindledecker, A.Z. Kapikian, M.M. Levine, and the Clinical Study Group, An efficacy trial of the rhesus rotavirus vaccine in Maryland, *Am. J. Dis. Child.* 144:601 (1990).

37. G.A. Losonsky, M.B. Rennels, Y. Lim, G. Krall, A.Z. Kapikian, and M.M. Levine, Systemic and mucosal immune responses to rhesus rotavirus vaccine MMU18006, *Pediatr. Infect. Dis. J.* 7:388 (1988).

38. K.Y. Green, K. Taniguchi, E.R. Mackow, and A.Z. Kapikian, Homotypic and heterotypic epitope-specific antibody responses in adult and infant rotavirus vaccinees: implication for vaccine development, *J. Infect. Dis.* 161:667 (1990).

39. A.Z. Kapikian, J. Flores, K. Midthun, Y. Hoshino, K.Y. Green, M. Gorziglia, K. Taniguchi, K. Nishikawa, R.M. Chanock, L. Potash, I. Perez-Schael, R. Dolin, C. Christy, M. Santosham, N.A. Halsey, M.L. Clements, S.D. Sears, R.E. Black, M.M. Levine, G.A. Losonsky, M.B.

Rennels, L. Gothefors, G. Wadell, R.I. Glass, T. Vesikari, E.L. Anderson, R.B. Belshe, P.F. Wright, and S. Urasawa, Development of a rotavirus vaccine by a "Jennerian" and a modified "Jennerian" approach, in: "Modern Approaches to New Vaccines Including Prevention of AIDS," R.M. Chanock, R.A. Lerner, F. Brown, and H. Ginsberg, eds., Cold Spring Harbor Laboratory, Cold Spring Harbor, New York, pp. 151-158 (1988).

40. K. Midthun, H.B. Greenberg, Y. Hoshino, A.Z. Kapikian, R.G. Wyatt, and R.M. Chanock. Reassortant rotaviruses as potential live rotavirus vaccine candidates, J. Virol. 53:949 (1985).

41. K. Midthun, Y. Hoshino, A.Z. Kapikian, and R.M. Chanock, Single gene substitution rotavirus reassortants containing the major neutralization protein (VP7) of human rotavirus serotype 4, J. Clin. Microbiol. 24:822 (1986).

42. N.A. Halsey, E.L. Anderson, S.D. Sears, M. Steinhoff, M. Wilson, R.B. Belshe, K. Midthun, A.Z. Kapikian, R.M. Chanock, R. Samoridin, B. Burns, and M.L. Clements, Human-rhesus reassortant rotavirus vaccines: safety and immunogenicity in adults, infants, and children, J. Infect. Dis. 158:1261 (1988).

43. J. Flores, I. Perez-Schael, M. Blanco, M. Vilar, D. Garcia, M. Perez, N. Daoud, K. Midthun, and A.Z. Kapikian, Reactogenicity and antigenicity of two human-rhesus rotavirus reassortant vaccine candidates of serotypes 1 and 2 in Venezuelan infants, J. Clin. Microbiol. 27:512 (1989).

44. T. Tajima, J. Thompson, P.F. Wright, Y. Kondo, S.J. Tollefson, J. King, and A.Z. Kapikian, Evaluation of a reassortant rhesus rotavirus vaccine in young children, Vaccine 8:70 (1990).

45. T. Vesikari, T. Varis, K. Green, J. Flores, and A.Z. Kapikian, Immunogenicity and safety of rhesus-human rotavirus reassortant vaccines with serotype 1 or 2 VP7 specificity, Vaccine 9:334 (1991).

46. P.F. Wright, J. King, K. Araki, Y. Kondo, J. Thompson, S.J. Tollefson, M. Kobayashi, and A.Z. Kapikian, Simultaneous administration of two human-rhesus rotavirus reassortant strains of VP7 serotype 1 and 2 specificity to infants and young children, J. Infect. Dis. 164:271 (1991).

47. T. Vesikari, T. Ruuska, K.Y. Green, J. Flores, and A.Z. Kapikian, Protective efficacy against serotype 1 rotavirus diarrhea by live oral rhesus-human reassortant rotavirus vaccines with human rotavirus VP7 serotype 1 or 2 specificity, Pediatr. Infect. Dis. J. in press (1992).

48. H.P. Madore, C. Christy, M. Pichichero, C. Long, P. Pincus, D. Vosefsky, A.Z. Kapikian, R. Dolin, and the Elmwood, Panorama, and Westfall Pediatric Groups, Field trial of rhesus rotavirus or human-rhesus reassortant vaccine of VP7 serotype 3 or 1 specificity in infants, J. Infect. Dis. in press (1992).

49. I. Perez-Schael, M. Blanco, M. Vilar, D. Garcia, L. White, R. Gonzalez, A.Z. Kapikian, and J. Flores, Clinical studies of a quadrivalent rotavirus vaccine in Venezuelan infants, J. Clin. Microbiol. 28:553 (1990).

50. J. Flores, I. Perez-Schael, M. Blanco, L. White, D. Garcia, M. Vilar, W. Cunto, R. Gonzalez, C. Urbina, J. Boher, M. Mendez, and A.Z. Kapikian, Comparison of reactogenicity and antigenicity of M37 rotavirus vaccine and rhesus-rotavirus-based quadrivalent vaccine, Lancet 336(2):330 (1990).

51. J. Flores, I. Perez-Schael, et al. (unpublished studies).

52. A.Z. Kapikian, J. Flores, T. Vesikari, T. Ruuska, H.P. Madore, K.Y. Green, M. Gorziglia, Y. Hoshino, R.M. Chanock, K. Midthun, and I. Perez-Schael, Recent advances in development of a rotavirus vaccine for prevention of severe diarrheal illness of infants and young children, Adv. Exp. Med. Biol. 310:255 (1991).

53. B.L. Davidson, E.T. Zito, E.G. Starr, J.R. Forro, and R.L. Ward (unpublished studies).

54. K. Midthun, et al. (unpublished studies).

55. Y. Hoshino, M.M. Sereno, K. Midthun, J. Flores, A.Z. Kapikian, and R.M. Chanock, Independent segregation of two antigenic specificities (VP7 and VP3) involved in neutralization of rotavirus infectivity, Proc. Natl. Acad. Sci. USA 82:8701 (1985).

56. R.G. Wyatt, A.Z. Kapikian, Y. Hoshino, J. Flores, K. Midthun, H.B. Greenberg, R.I. Glass, J. Askaa, M.M. Levine, R.E. Black, M.L. Clements, L. Potash, and W.T. London, Development of rotavirus vaccines, in: "Control and Eradication of Infectious Diseases an International Symposium, Pan American Health Organization Copublication Series No. 1, Washington, DC, pp. 17-28 (1985).

57. J.C. Bridger, and G.N. Woode, Neonatal calf diarrhoea: identification of a reovirus-like (rotavirus) agent by immunofluorescence and immune elctron microscopy, Br. Vet. J. 131:528 (1975).

58. G.L. Stuker, L. Oshiro, and N.L. Schmidt, Antigenic composition of two new rotaviruses from rhesus monkeys, J. Clin. Microbiol. 11:202 (1980).

59. H.F. Clark, F.E. Borian, and S.A. Plotkin, Immune protection of infants against rotavirus gastroenteritis by a serotype 1 reassortant of bovine rotavirus WC3, J. Infect. Dis. 161:1099 (1990).

60. J. Flores, I. Perez-Schael, M. Blanco, L. White, D. Garcia, M. Vilar, W. Cunto, R. Gonzalez, C. Urbina, J. Boher, M. Mendez, and A.Z. Kapikian, Comparison of reactogenicity and

68

antigenicity of M37 rotavirus vaccine and rhesus-rotavirus-based quadrivalent vaccine, *Lancet* 2:330 (1990).

61. K. Midthun, N.A. Halsey, M. Jett-Goheen, M.L. Clements, M. Steinhoff, J.C. King, R. Karron, M. Wilson, B. Burns, V. Perkis, R. Samoridin, and A.Z. Kapikian, Safety and immunogenicity of human rotavirus vaccine strain M37 in adults, children and infants, *J. Infect. Dis.* 164:792 (1991).

62. T. Vesikari, T. Ruuska, H.-P. Koivu, K.Y. Green, J. Flores, and A.Z. Kapikian, Evaluation of the M37 human rotavirus vaccine in 2- to 6-month-old infants, *Pediatr. Infect. Dis. J.* 10:912 (1991).

63. S. Matsuno, S. Murakami, M. Tagaki, M. Hayashi, S. Inouye, A. Hasegawa, and K. Fukai, Cold-adaptation of human rotavirus, *Virus. Res.* 7:273 (1987).

64. M.E. Andrew, D.B. Boyle, B.E. Coupar, P.L. Whitfeld, G.W. Both, and A.R. Bellamy, Vaccinia virus recombinants expressing the SA11 rotavirus VP7 glycoprotein gene induce serotype-specific neutralizing antibodies, *J. Virol.* 61:1054 (1987).

65. J.H. Eldridge, R.M. Gilley, J.K. Staas, Z. Modoveanu, J.A. Meulbroek, and T.R. Tice, Biodegradable microsphere: vaccine delivery system for oral immunization, *Curr. Topics Microbiol. Immunol.* 146:59 (1989).

... ...
... ...
... ...
... ...

INDUCTION OF MUCOSAL AND SERUM IMMUNE RESPONSES TO A SPECIFIC ANTIGEN OF PERIODONTAL BACTERIA

Shigeyuki Hamada, Tomohiko Ogawa, Hidetoshi Shimauchi
and Yutaka Kusumoto

Department of Oral Microbiology, Osaka University Faculty of Dentistry
Yamadaoka, Suita-Osaka, 565 Japan

INTRODUCTION

Protection against pathogenic microorganisms at mucosal surfaces is of critical importance because these surfaces are the sites where infectious agents first interact with the host. The route of immunization and subsequent priming of B cells in different tissues have been shown to lead to distinct humoral mucosal immunity in experimental animals and humans. IgA antibodies synthesized by plasma cells in mucosal secretory tissues play a role in host defense against mucosal infections[1-4]. Although both IgA and IgG are found in various external secretions, IgA has been of special concern since IgA antibodies correlate better with protection against mucosal infections and specific induction of IgA antibodies develops mainly through mucosal lymphoid tissues such as intestinal Peyer's patches. It is well known that oral administration of antigen induces preferentially antigen-specific IgA responses in external secretions including saliva. Therefore, oral immunization with appropriate antigen could provide an effective way to induce specific immune responses and prevent infections at mucosal surfaces[5-8].

Periodontitis is a chronic inflammatory disease characterized by gingival inflammation, progressive destruction of periodontal ligament fibers and alveolar bone loss. The infiltration of plasma cells and B and T lymphocytes may indicate that immune responses occur locally. In the advanced phase of periodontitis, plasma cells comprise most of the immune cells[9-11]. Brandtzaeg[12] reported that IgG-producing plasma cells are most predominant in the gingiva, with a ratio of IgG: IgA plasma cells of ca. 7 : 1.

Porphyromonas gingivalis, a strictly anaerobic black-pigmented bacterium, has been associated with the development of the adult type of periodontitis, while *Actinobacillus actinomycetemcomitans* is usually associated with the juvenile type of periodontitis[10,11]. Evidence indicates that elevated levels of serum and gingival crevicular fluid antibodies to whole cells of *P. gingivalis* are found in patients with adult periodontitis[13]. It was also observed that periodontitis lesions of patients exhibit higher numbers of plasma cells producing antibodies to *P. gingivalis* in gingiva[14]. However, no specific antigen epitopes of this pathogen were demonstrated in these reports. In this short review, we have summarized a series of studies on host immune responses to a purified protein antigen (*i.e.*, fimbriae) of *P. gingivalis* strain 381 administered orally or systemically to mice. This has allowed a direct comparison of local antibody responses in mucosal tissues with those found in serum and saliva.

PURIFICATION AND CHARACTERIZATION OF *P. gingivalis* FIMBRIAE

Whole cells of *P. gingivalis* possess a number of components that exhibit diverse effects on host lymphoreticular cells[15]. The outer membrane of this organism contains lipopolysaccharides (LPS) and lipoprotein that elicit profound B cell responses including those from C3H/HeJ mice, a low responder strain to LPS derived from the *Enterobacteriaceae*[16]. In addition, *P. gingivalis* has been shown to possess fimbrial appendages which emerge from the cell surface[17]. Although bacterial fimbriae have been suggested to facilitate attachment to the epithelial cells of mucosal tissues[18-20], the adhesive ability of *P. gingivalis* fimbriae has yet to be demonstrated experimentally.

It was demonstrated that *P. gingivalis* fimbriae can be isolated from whole cells and purified chromatographically[21]. SDS-PAGE of the purified fimbriae has revealed a distinct and single band of 41kDa for the fimbrial subunit, fimbrilin[22]. The gene encoding *P. gingivalis* 381 fimbrilin has been successfully cloned and sequenced by Dickinson et al.[23], and its molecular weight was calculated to be 35,294. Recently, a variety of immunological and biological properties of purified fimbriae of *P. gingivalis* were demonstrated as summarized in Table 1. It has been shown that the fimbrial protein exhibits strong immunogenic responses in various experimental animals and humans[21-31]. The fimbriae were used for experimental induction of humoral immune responses in mice.

Table 1. Some immunobiological activities of *P. gingialis* fimbriae.

Activity	Cells / Species	Reaction	References
Antigenicity	Rabbit / Human	Yes	21,24
	Mice / Human / Rabbit	Yes	25 -28
Hemagglutination	Erythrocyte (Sheep, Horse, Rabbit)	Yes	29
	Erythrocyte (Human, Rabbit, Sheep, Mice, Chicken, etc)	Yes	28
	Erythrocyte (Chicken)	No	21
Mitogenicity	Splenocyte (Mice)	Yes	28,30
	Thymocyte (Mice)	No	28
Polyclonal B cell activation	Splenocyte (Mice)	Yes	28
Thymocyte-activating factor production	Fibroblast (Human)	Yes	30
TNF-α / IL-6 production	Mononuclear leukocyte (Human)	Yes	28
IL-1 production	Monocyte / Macrophage (Mice)	Yes	31

Very recently, we have prepared synthetic peptides which mimic the segments of the fimbrilin. ELISA inhibition studies indicate that some peptides appear to be immunodominant epitopes. Other peptides induce tumor necrosis factor-α (TNF-α) and interleukin-6 (IL-6) in human mononuclear cell cultures, although these cytokine-inducing activities were weaker than those of native fimbriae[28]. These results support the concept that there are defined peptide segments within the fimbrial molecule that are responsible for immunomodulation.

HUMORAL IMMUNE RESPONSES TO *P. gingivalis* IN HUMANS

Antibody Responses in Sera

When the level of *P. gingivalis* fimbriae-specific antibodies was differentially assessed in sera of patients with periodontitis or gingivitis, higher titers of fimbriae-specific antibodies were observed in sera obtained from patients with adult periodontitis followed by rapidly progressive periodontitis and gingivitis[26]. Serum samples from juvenile periodontitis, however, contained only low levels of fimbriae-specific antibodies. It was also shown that sera from patients with adult periodontitis and gingivitis contained high levels of fimbriae-specific IgG antibodies (IgG3>>IgG1>IgG2>IgG4) with significant levels of fimbriae-specific IgA antibody responses (IgA1>IgA2)[25]. On the other hand, sera from normal control subjects were found to exhibit remarkably lower levels of fimbriae-specific antibodies[26]. Of interest were the findings that surgical treatment of inflamed gingiva resulted in marked reductions in fimbriae-specific IgG and IgA levels in sera of patients. Thus, a possible mechanism could be suggested for the recruitment of larger numbers of fimbriae-specific antibody-secreting cells at the local disease site in the gingiva[27] (also see below).

Cellular Responses in Inflamed Gingiva

In chronic inflammatory diseases such as rheumatoid arthritis, portions of immunoglobulins may be derived from the blood circulation[32]. However, significant amounts may be locally synthesized and released by plasma cells at the disease site. We have analyzed the distribution of antibody-secreting cells in gingiva at different stages of periodontal disease. Surgically removed gingival tissues were dissociated into single cell suspensions, and mononuclear cells rich in lymphoid cells and macrophages were obtained by the Ficoll-Hypaque method. Patients with adult periodontitis were found to contain increased numbers of antibody-secreting cells with a pattern of IgG>IgA>>IgM[33]. Solid-phase enzyme-linked immunospot (ELISPOT)[34] analysis revealed a progressive increase in spot-forming cell (SFC) numbers in gingiva of patients with slight, moderate, and advanced stages of periodontitis, and high numbers of SFC secreting *P. gingivalis* fimbriae-specific IgG were observed followed by IgA SFC in gingival mononuclear cells from these patients. In contrast, the mononuclear cells from non-inflamed gingiva contained very low numbers of SFC, which were mainly of the IgG isotype[25,27]. These findings strongly suggest that local immunobiological stimulation by *P. gingivalis* fimbriae at diseased gingival sites induced B cell responses. These antigen-specific B cells terminally differentiate into plasma cells producing either IgG or IgA antibodies, a process which is regulated by cytokines, mitogens and antigens in the local inflamed tissues[35].

IMMUNE RESPONSES TO *P. gingivalis* FIMBRIAE IN MICE

P. gingivalis fimbriae have been shown to exert strong immunogenic activity in experimental animals. High-titered polyclonal and monoclonal antibodies specific for fimbriae can be obtained from sera of hyperimmunized rabbits and mice, respectively; the antibodies were shown to be specific for fimbrial protein or fimbrilin of various strains of *P. gingivalis*[36]. Nevertheless, a need exists for a good animal model system to analyze the immunologic aspects of periodontitis, since it is of interest to know how mucosal and serum immune systems function, how immunocompetent cells are primed by a specific antigen of periodontopathic bacteria and how they differentiate into antibody-producing plasma cells, and how they transmigrate into lymphoid tissues or other tissues and organs.

We have recently shown that *P. gingivalis* fimbriae incorporated into liposomes markedly enhanced the levels of anti-fimbriae antibodies in serum, particularly of the IgG class, when oral primary and secondary immunizations were done in BALB/c mice (Fig. 1). It was shown that inclusion of adjuvant compounds such as lipophilic muramyldipeptide, MDP-Lys(L18), or a semi-synthetic compound, GM-53, into the liposome-fimbriae complex significantly enhanced the immune response. In this regard, GM-53, β-*N*-acetylglucosamyl-(1→4)-*N*-acetylmuramyl-L-alanyl-D-isoglutaminyl-(L)-*meso*-2, 6-diaminopimelic acid-(D)-amide-D-alanine[37] was found to be an effective adjuvant for this purpose. In these experiments, liposomes were prepared using lecithin and cholesterol dissolved in chloroform. This mixture was ultrasonicated to obtain homogeneous unilamellar liposome vesicles[38]. The subclass profile in the fimbriae-specific serum IgG responses was IgG1>>IgG2b>IgG2a>IgG3. In saliva, however, fimbriae-specific IgA antibodies were especially raised. In contrast, subcutaneous immunization with the same immunogen complex resulted in serum IgG responses and salivary IgG and IgA responses. The level of fimbriae-specific serum IgG was approximately 20-fold higher than that obtained after oral immunization[22]. It should be noted here that not only IgA, but also IgG, antibodies specific for fimbriae were induced in saliva at similar levels on a concentration basis after systemic immunization, and that serum IgG responses were induced by oral immunization with the fimbriae-adjuvant-liposome complex. In this regard, Kaijser[39] reported that serum IgG antibodies to *Escherichia coli* O4K12 were significantly raised following oral immunization of healthy adults with live cells of *E. coli*.

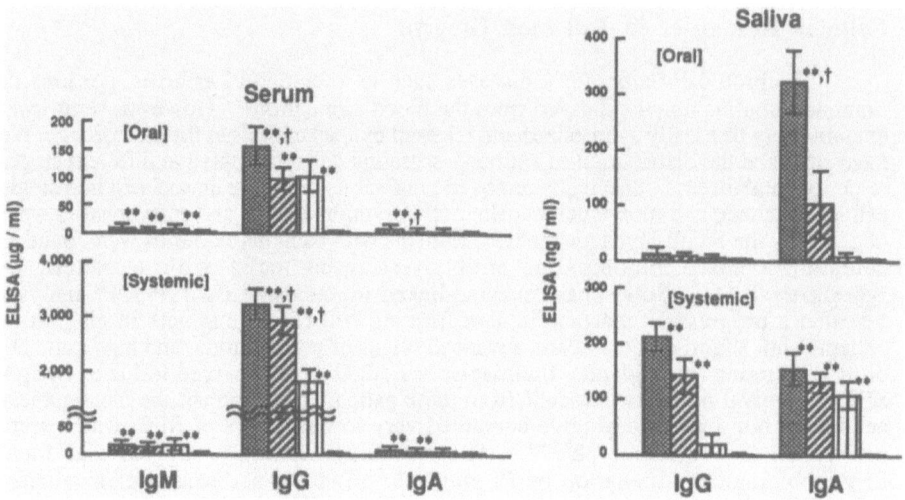

Figure 1. Serum and salivary antibody responses in the mice either orally or systemically immunized with *P. gingivalis* fimbriae and adjuvant in liposomes. Four groups of 8 to 12 BALB/c mice (male, 6-week-old) were immunized on days 0, 1, 27, and 28 by oral administration of fimbriae + GM-53, ▓; fimbriae + MDP-Lys(L18), ▨; fimbriae alone, ▯; liposomes alone (control), ▭. Another four groups of BALB/c mice were immunized by subcutaneous administration of liposomes containing fimbriae in the presence or absence of adjuvant on days 0 and 28. Anti-fimbriae antibody levels were determined by ELISA 5 days after the booster immunization. Values (mean ± SE) are expressed as μg/ml serum or ng/ml saliva. **Statistical difference from the value for the control (liposomes alone) group at $P < 0.01$; †, statistical difference from the value of the group given fimbriae alone at $P < 0.05$ (Ref. 22).

In the next series of studies, longitudinal changes in the fimbriae-specific antibody levels in BALB/c mice that had been immunized orally with *P. gingivalis* fimbriae were examined[40]. As depicted in Fig. 2, levels of serum anti-fimbriae IgG and salivary IgA antibodies were raised following primary immunization. Fimbriae-specific IgM and IgA responses in serum were also observed, but these were not as prominent in terms of antibody levels. In particular, the level of fimbriae-specific IgA became much higher following a second booster immunization. Our results also indicated that marked local IgA anamnestic responses occurred following the boosted immunization. In this regard, some investigators reported that an IgA anamnestic response was difficult to induce[41], while others demonstrated enhanced IgA responses in intestinal secretions following oral administration of live bacteria[42]. Our study[40] clearly showed that even purified protein antigen (*i.e.*, fimbriae) could induce immunocompetent cells in the small intestine of mice to synthesize IgA in secretions when the antigen was given in an entrapped form in liposomes with appropriate adjuvants such as lipophilic MDP compounds. Enhanced salivary IgA responses found in this study could be the outcome of stimulation of IgA-enhancing helper T cells and IgG/IgM suppressor T cells in the responsible mucosal tissues. Further, evidence indicates that the oral administration of antigens leads to the induction of antigen-specific precursors of IgA plasma cells in peripheral blood, and their presence precedes the appearance of IgA (and IgG) antibodies in external secretions[43,44].

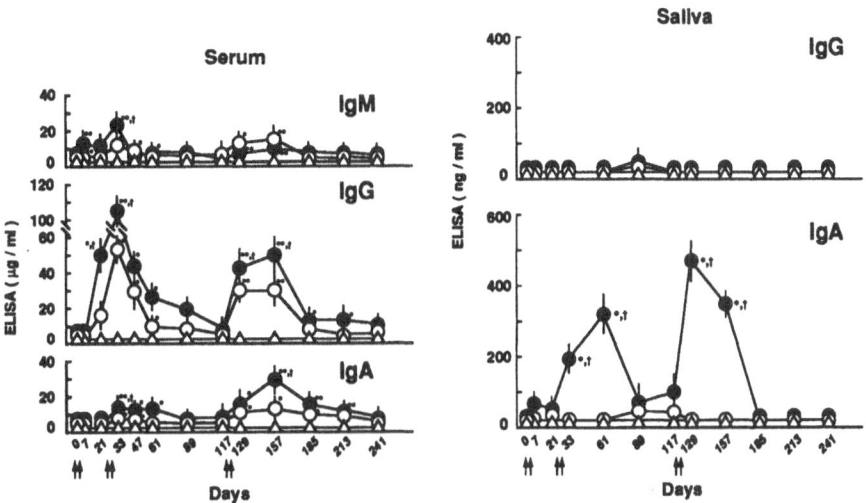

Figure 2. Temporal changes in the level of serum and salivary anti-fimbriae antibodies following oral immunization with *P. gingivalis* fimbriae in BALB/c mice. Mice were immunized orally on days 0, 1, 27, 28, 127, and 128 with 500 μg of fimbriae plus 500 μg of GM-53 (●), 500 μg of fimbriae alone (O), or liposomes alone (control) (Δ). Values (mean ± SE) are expressed as μg/ml serum or ng/ml saliva. *Statistical difference from the value for the control group, $P < 0.05$; †, statistical difference from the value of the group given fimbriae alone, $P < 0.05$ (Ref. 40).

GENETIC ASPECTS OF THE MOUSE RESPONSE TO *P. gingivalis* FIMBRIAE

It has been shown that the major histocompatibility complex (MHC) could influence the immune response to certain bacterial antigens and susceptibility to infection[45-47]. It would be of interest to determine whether the immune response to *P. gingivalis* administered orally as described above is controlled or dominated by the specific genetic background of mice (Fig. 3)[48]. It was shown that oral immunization of

BALB/c mice (H-2^d) with *P. gingivalis* fimbriae incorporated in the liposomes-GM-53 complex led to specific IgG production in serum as well as IgA production in saliva. When six inbred strains of mice were examined for their serum and saliva immunme responses, BALB/c and DBA/2 mice (H-2^d) were found to be high responders, CBA/J and C3H mice (H-2^k) were intermediate, C57BL/6 mice (H-2^b) were low responders. We next investigated the role of H-2 haplotypes which correspond to the three strains of mice using H-2 recombinants on the B10 background. B10.D2 (H-2^d) were shown to be high responders, while B10.BR (H-2^k) and C57BL/10 (B10, H-2^b) were intermediate and low rresponders to the fimbrial complex, respectively[48]. These results strongly suggest that the humoral immune responses in serum and mucosal secretions to the orally administered *P. gingivalis* fimbriae were dominated by the H-2^d haplotype. We also found that F1 hybrids between high and low strains reflected the phenotype of the low responder. In this regard, Haber and Grinnell[49] reported that the serum antibody response following immunization with type 1 and type 2 fimbriae of *Actinomyces viscosus* was genetically controlled; however, they used inbred mice that had diverse genetic backgrounds and the gene affecting the response was not specified.

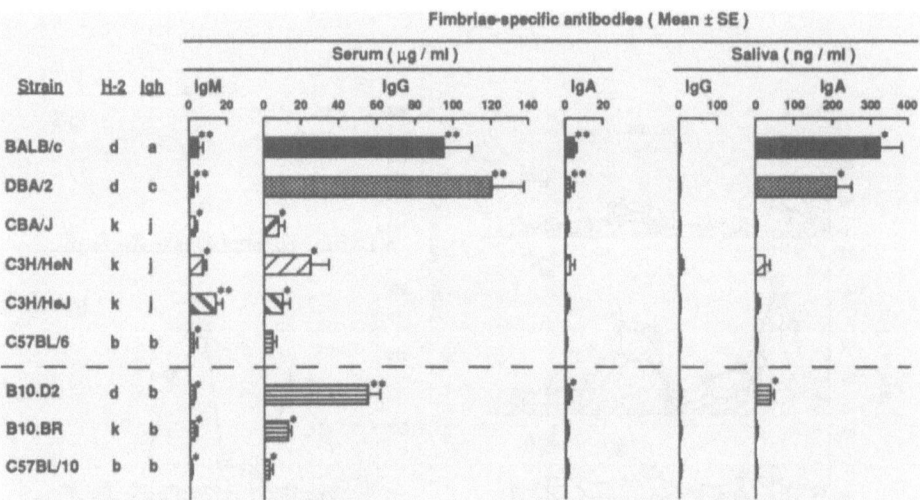

Figure 3. Serum and salivary antibody responses to *P. gingivalis* fimbriae administered orally with GM-53 in liposomes in inbred and congenic strains of mice. Mice (male, 6-week-old) were immunized on days 0, 1, 27 and 28 by oral administration of 500 µg fimbriae and GM-53 in liposomes. *, **Statistical differences from the value for the control group (liposomes alone, data not shown). $P <$ 0.05 and $P <$ 0.01, respectively. (Ref. 48).

Elson *et al.*[50] showed that cholera toxin administered orally generated high IgA antibodies in the intestinal secretions and IgG antibodies in serum of H-2^b and H-2^q mice, while H-2^k and H-2^d mice induced only marginal levels of anti-cholera toxin IgA in the secretions. These results indicate that MHC Ir genes may control the secretory IgA response. Except for the haplotype involved in the response, Elson's results are basically in agreement with our findings in terms of the MHC gene regulation of mucosal and systemic immune responses to orally administered antigen. Further, it should be noted here that MDP adjuvanticity was also genetically affected in mice; H-2^d and H-2^k haplotypes were immunopotentiated strongly, while H-2^b responded only weakly. Thus, combinations of adjuvant and antigen may modulate host immune responses[51].

MIGRATION OF ANTIGEN-SPECIFIC PLASMA CELLS FOLLOWING IMMUNIZATION

Following immunization, a series of cellular interactions and proliferation, differentiation and distribution of activated B cells occur in the responding host. Current evidence suggests that a combination of signals involving T cells, B cells, and macrophages are required for these B cell responses. Various types of T cell populations have been identified even within the T helper series. They may participate in B cell activation and differentiation at different stages with the aid of specific recognition of antigen epitope given through the cellular interaction and the influence of B cell growth and differentiation factors[52]. The anatomical site of cellular interactions *in vivo* occurs in the cortex of the lymph node and the white pulp of the spleen. Although B and T lymphocytes develop independently and may be anatomically separated, they often appear together following antigen stimulation[53]. In view of these findings and the cellular events in periodontal lesions, it would be of interest to determine how B cells: 1) primed with antigen(s) of specific periodontopathic bacteria, 2) differentiate into mature plasma cells producing specific antibody, and, 3) migrate to and inhabit the specific inflammatory sites as well as lymphoid tissues in the host.

Analyses were done to determine the temporal changes in the induction and distribution of antigen-specific antibody-producing cells in various lymphoid tissues and in the blood in BALB/c mice that had been immunized orally or systemically with *P. gingivalis* fimbriae[54]. As shown in Figures 4 and 5, groups of mice were immunized orally by intubation of the liposomes-fimbriae-GM-53 complex on days 0, 1, 27 and 28, while other groups of mice were subcutaneously injected with the immunogen complex on days 0 and 28. As described earlier, the orally immunized mice exhibited markedly enhanced fimbriae-specific IgG and IgA responses in serum, while those immunized systemically developed low levels of IgM followed by very high levels of IgG antibodies specific for fimbriae in serum (Fig. 5).

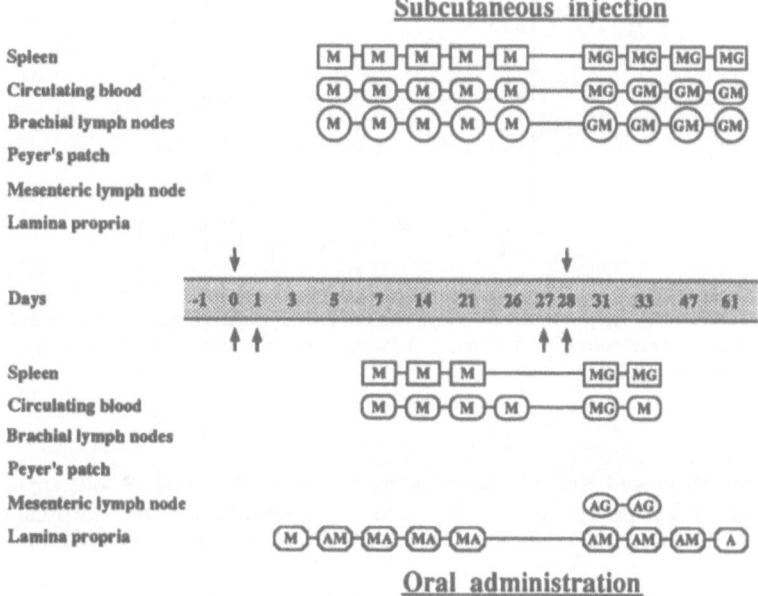

Figure 4. Emergence of *P. gingivalis* fimbriae-specific spot-forming cells (SFC) in various lymphoid tissues and circulating blood of BALB/c mice immunized subcutaneously on days 0 and 28 with the fimbriae-GM-53-FIA or orally on days 0, 1, 27 and 28 with the fimbriae-GM-53-liposomes. A, G or M indicates the presence of IgA, IgG or IgM SFC at the indicated experimental day by the ELISPOT assay (modified from Ref. 54).

ELISPOT assays indicated that significant numbers of fimbriae-specific IgA SFC were first seen in lamina propria and later in mesenteric lymph nodes and salivary glands of orally immunized mice. In contrast, fimbriae-specific IgM and IgG SFC were found in the spleen and in the blood. On the other hand, antigen-specific IgM SFC were seen earlier in brachial lymph nodes, blood and spleen, followed by IgG SFC in these tissues. It was noted that very low numbers of fimbriae-specific IgA SFC were detected in the systemically immunized mice. These results clearly indicate that both oral and systemic immunization induce distinct and compartmentalized antigen-specific antibody responses in lymphoid cells (Fig. 4). Of interest was the finding that antigen-specific SFC were not seen in Peyer's patches throughout the experimental period regardless of the immunization route. Involvement of Peyer's patch lymphocytes in antibody production has been controversial. Some investigators have suggested that intestinal lamina propria cells but not Peyer's patch cells produce

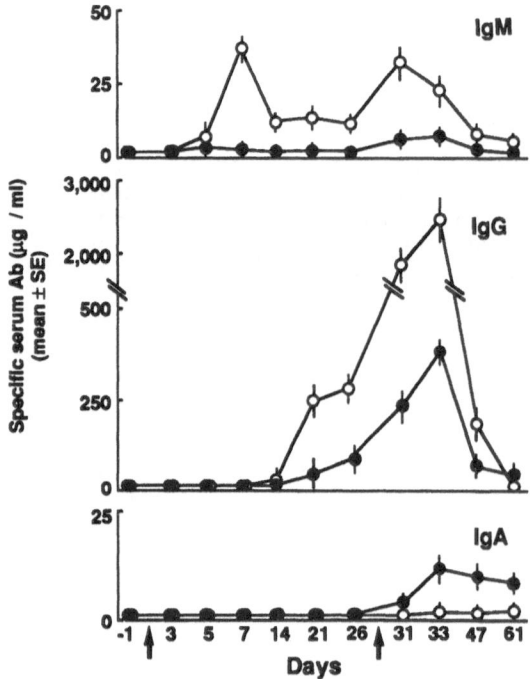

Figure 5. Sequence of appearance of *P. gingivalis* fimbriae-specific antibodies in serum of BALB/c mice immunized subcutaneously (O) or orally (●) with the fimbrial immunogen as described in the legend of Figure 4 (Ref. 54).

antigen-specific antibody[55,56], while others claim that Peyer's patch cells are critically important for IgA production[57,58]. It should be mentioned further that the total number of antibody-secreting cells was increased following the administration of the immunogen complex either orally or systemically including in Peyer's patches. It appears that a combination of fimbriae, GM-53 and liposomes nonspecifically stimulates other B cells in addition to antigen-specific clones to become antibody-producing cells[54]. It should be noted that purified *P. gingivalis* fimbriae exhibit significant *in vitro* polyclonal B cell activation[28]. This property may contribute to the increased appearance of divergent specificities of antibodies in chronically inflamed gingiva[59].

SUMMARY

The dynamics of the host immune response to periodontal bacteria not only may be informative from the standpoint of specific mucosal protection to these pathogens, but also may reveal the capacity of the mucosal immune response to provide protection of the host. To this end, we have examined the immune response to chromatographically purified fimbriae of *P. gingivalis* administered orally or systemically with liposomes and adjuvant in BALB/c mice, high responders to this antigen.

Oral administration of *P. gingivalis* fimbriae clearly enhanced the fimbriae-specific salivary IgA response. ELISPOT analysis revealed that significant numbers of fimbriae-specific IgA SFC were seen in lamina propria and mesenteric lymph nodes but not in Peyer's patches of mice immunized orally. In contrast, antigen-specific IgM and IgG SFC were seen mainly in the circulating blood mononuclear cells. On the other hand, subcutaneous injection of fimbriae with GM-53 also raised the fimbriae-specific IgG followed by IgM and IgA responses in serum, and both IgA and IgG responses in saliva. Oral immunization was less effective than subcutaneous injection in terms of the serum antibody response. However, the salivary antibody level of mice injected subcutaneously was similar to that of mice immunized orally. In the subcutaneously immunized mice, fimbriae-specific SFC were detected in the spleen, blood, and brachial lymph nodes by ELISPOT assay. Fimbriae-specific IgM SFC appeared earlier and antigen-specific IgG SFC were seen later. These results show that the combined use of fimbriae together with the adjuvant results in sharply increased IgA responses in saliva and IgG responses in serum. In summary, it is clear that the nature of the host's antibody response in serum and mucosal secretions is distinct, and depends on the route of antigen administration, the use of adjuvant and/or liposomes, and the temporal phase of the humoral immune response following various immunization regimes.

ACKNOWLEDGMENT

The authors wish to thank Drs. Jerry R. McGhee and Hiroshi Kiyono for many helpful discussions and suggestions.

REFERENCES

1. J. Bienenstock and A.D. Befus, Mucosal immunology, *Immunology* 41:249 (1980).
2. P.C. McNabb, and T.B. Tomasi TB, Host defense mechanisms at mucosal surfaces, *Annu. Rev. Microbiol.* 35:477 (1981).
3. J. Mestecky and J.R. McGhee, Immunoglobulin A (IgA): molecular and cellular interactions involved in IgA biosynthesis and immune response, *Adv. Immunol.* 40:153 (1987).
4. T.B. Tomasi, Jr., Mechanisms of immune regulation at mucosal surfaces, *Rev. Infect. Dis.* 5:S784 (1983).
5. K.-C. Bergmann and R.H. Waldman, Stimulation of secretory antibody following oral administration of antigen, *Rev. Infect. Dis.* 10:939 (1988).
6. J.R. McGhee, J. Mestecky, C.O. Elson, and H. Kiyono, Regulation of IgA synthesis and immune response by T cells and interleukins, *J. Clin. Immunol.* 9:175 (1989).
7. A.M. Mowat and A.M. Donachie, ISCOMS-a novel strategy for mucosal immunization?, *Immunol. Today* 12:383 (1991).
8. M.W. Russell and J. Mestecky, Induction of the mucosal immune response, *Rev. Infect. Dis.* 10:S440 (1988).
9. R.C. Page and H.E. Schroeder. "Periodontitis in Man and Other Animals," Karger, Basel (1982).
10. S. Hamada, S.C. Holt, and J.R. McGhee, eds., "Periodontal Disease: Pathogens and Host Immune Responses," Quintessence, Tokyo (1991).
11. J. Slots and M.A. Listgarten, *Bacteroides gingivalis, Bacteroides intermedius* and *Actinobacillus actinomycetemcomitans* in human periodontal diseases, *J. Clin. Periodontol.* 15:85 (1988).
12. P. Brandtzaeg, Immunology of flammatory periodontal lesions, *Int. Dent. J.* 23:438 (1973).

13. J.M.A. Wilton, N.W. Johnson, M.A. Curtis, I.R. Gillett, R.J. Carman, J.L.M. Bampton, G.S. Griffiths, and J.A.C. Sterne, Specific antibody responses to subgingival plaque bacteria as aids to the diagnosis and prognosis of destructive periodontitis, *J. Clin. Periodontol.* 18:1 (1991).

14. S.E. Schonfeld and J.M. Kagan, Specificity of gingival plasma cells for bacterial somatic antigens, *J. Periodontal Res.* 17:60 (1982).

15. D. Mayrand and S.C. Holt, Biology of asaccharolytic black-pigmented *Bacteroides* species, *Microbiol. Rev.* 52:134 (1988).

16. S. Hamada, H. Takada, T. Ogawa, T. Fujiwara, and J. Mihara, Lipopolysaccharides of oral anaerobes associated with chronic inflammation: chemical and immunomodulating properties, *Intern. Rev. Immunol.* 6:247 (1990).

17. P.S. Hardley and L.S. Tipler, An electron microscope survey of the surface structures and hydrophobicity of oral and non-oral species of the bacterial genus *Bacteroides*, *Arch. Oral Biol.* 31:325 (1986).

18. S. Clegg and G.F. Gerlach, Enterobacterial fimbriae, *J. Bacteriol.* 169:934 (1987).

19. K.A. Krogfelt, Bacterial adhesion: genetics, biogenesis, and role in pathogenesis of fimbrial adhesins of *Escherichia coli*, *Rev. Infect. Dis.* 13:721 (1991).

20. S.E. Mergenhagen, A.L. Sandberg, B.M. Chassy, M.J. Brennan, M.K. Yeung, J.A. Donkersloot , and J.O. Cisar, Molecular basis of bacterial adhesion in the oral cavity, *Rev. Infect. Dis.* 9:S467 (1987).

21. F. Yoshimura, T. Takahashi, Y. Nodasaka, and T. Suzuki, Purification and characterization of a novel type of fimbriae from the oral anaerobe *Bacteroides gingivalis*, *J. Bacteriol.* 160:949 (1984).

22. T. Ogawa, H. Shimauchi, and S. Hamada, Mucosal and systemic immune responses in BALB/c mice to *Bacteroides gingivalis* fimbriae administered orally, *Infect. Immun.* 57:3466 (1989).

23. D.P. Dickinson, M.A. Kubiniec, F. Yoshimura, and R.J. Genco, Molecular cloning and sequencing of the gene encoding the fimbrial subunit protein of *Bacteroides gingivalis*, *J. Bacteriol.* 170:1658 (1988).

24. F. Yoshimura, T. Sugano, M. Kawanami, H. Kato, and T. Suzuki, Detection of specific antibodies against fimbriae and membrane proteins from the oral anaerobe *Bacteroides gingivalis* in patients with periodontal diseases, *Microbiol. Immunol.* 31:935 (1987).

25. T. Ogawa, A. Tarkowski, M.L. McGhee, Z. Moldoveanu, J. Mestecky, H.Z. Hirsch, W.J. Koopman, S. Hamada, J.R. McGhee, and H. Kiyono, Analysis of human IgG and IgA subclass antibody-secreting cells from localized chronic inflammatory tissue, *J. Immunol.* 142:1150 (1989).

26. T. Ogawa, Y. Kusumoto, S. Hamada, J.R. McGhee, and H. Kiyono, *Bacteroides gingivalis*-specific serum IgG and IgA subclass antibodies in periodontal diseases, *Clin. Exp. Immunol.* 82:318 (1990).

27. T. Ogawa, Y. Kono, M.L. McGhee, J.R. McGhee, J.E. Roberts, S. Hamada, and H. Kiyono, *Porphyromonas gingivalis*-specific serum IgG and IgA antibodies originate from immunoglobulin-secreting cells in inflamed gingiva, *Clin. Exp. Immunol.* 83:237 (1991).

28. T. Ogawa, Y. Kusumoto, H. Uchida, S. Nagashima, H. Ogo, and S. Hamada, Immunobiological activities of synthetic peptide segments of fimbrial protein from *Porphyromonas gingivalis*, *Biochem. Biophys. Res. Commun.* 180:1335 (1991).

29. K. Okuda, J. Slots, and R.J. Genco, *Bacteroides gingivalis, Bacteroides asaccharolyticus* and *Bacteroides melaninogenicus* subspecies: Cell surface morphology and adherence to erythrocytes and human buccal epithelial cells, *Curr. Microbiol.* 6:7 (1981).

30. S. Hanazawa, K. Hirose, Y. Ohmori, S. Amano, and S. Kitano, *Bacteroides gingivalis* fimbriae stimulate production of thymocyte-activating factor by human gingival fibroblasts, *Infect. Immun.* 56:272 (1988).

31. S. Hanazawa, Y. Murakami, K. Hirose, S. Amano, Y. Ohmori, H. Higuchi, and S. Kitano, *Bacteroides (Porphyromonas) gingivalis* fimbriae activate mouse peritoneal macrophages and induce gene expression and production of interleukin-1, *Infect. Immun.* 59:1972 (1991).

32. R.M. Wernick, P.E. Lipsky, E. Marban-Arcos, J.J. Maliakkal, D.E. Edelbaum, and M. Ziff, IgG and IgM rheumatoid factor synthesis in rheumatoid synovial membrane cultures, *Arthritis Rheum.* 28:742 (1985).

33. M.L. McGhee, T. Ogawa, A.M. Pitts, Z. Moldoveanu, J. Mestecky, J.R. McGhee, and H. Kiyono, Cellular analysis of functional mononuclear cells from chronically inflamed gingival tissue, *Regional Immunol.* 2:103 (1989).

34. C.C. Czerkinsky, L. Nilsson, H. Nygren, O. Ouchterlony, and A. Tarkowski, A solid-phase enzyme-linked immunospot (ELISPOT) assay for enumeration of specific antibody-secreting cells, *J. Immunol. Methods* 65:109 (1983).

35. Y. Kono, K.W. Beagley, K. Fujihashi, J.R. McGhee, T. Taga, T. Hirano, T. Kishimoto, and H. Kiyono, Cytokine regulation of localized inflammation: Induction of activated B cells and IL-6-mediated polyclonal IgG and IgA synthesis in inflamed human gingiva, *J. Immunol.* 146:1812 (1991).

36. T. Ogawa, T. Mukai, K. Yasuda, H. Shimauchi, Y. Toda, and S. Hamada, Distribution and immunochemical specificities of fimbriae of *Porphyromonas gingivalis* and related bacterial species, *Oral Microbiol. Immunol.* 6:332 (1991).

37. R. Furuta, S. Kawata, S. Naruto, A. Minami, and S. Kotani, Synthesis and biological activities of *N*-acetylglucosaminyl-β-1-4-*N*-acetylmuramyl tri- and tetrapeptide derivatives, *Agric. Biol. Chem.* 50:2561 (1986).

38. K. Inoue, Permeability properties of liposomes prepared from dipalmitoyllecithin, dimyristoyllecithin, egg lecithin, rat liver lecithin and beef brain sphingomyelin, *Biochim. Biophys. Acta.* 339:390 (1974).

39. B. Kaijser, Peroral immunization of healthy adults with live *Escherichia coli* O4K12 Bacteria, *Int. Allergy Appl. Immun.* 70:164 (1983).

40. T. Ogawa, H. Shimauchi, Y. Kusumoto, and S. Hamada, Humoral immune response to *Bacteroides gingivalis* fimbrial antigen in mice, *Immunology* 69:8 (1990).

41. T.B. Tomasi, and H.M. Grey, Structure and function of immunoglobulin A, *in*: "*Progress Allergy*," P. Kallos, B.H. Waksman, and A.D. Weck, eds., Karger, Basel (1972).

42. D.F. Keren, S.E. Kern, D.H. Bauer, P.J. Scott, and P. Porter, Direct demonstration in intestinal secretions of an IgA memory response to orally administered *Shigella flexneri* antigens, *J. Immunol.* 128:475 (1982).

43. C. Czerkinsky, S.J. Prince, S.M. Michalek, S. Jackson, M.W. Russell, Z. Moldoveanu, J.R. McGhee, and J. Mestecky, IgA antibody-producing cells in peripheral blood after antigen ingestion: evidence for a common mucosal immune system in humans, *Proc. Natl. Acad. Sci. USA* 84:2449 (1987).

44. C. Czerkinsky, M.W. Russell, N, Lycke, M. Lindblad, and J. Holmgren, Oral administration of a streptococcal antigen coupled to cholera toxin B subunit evoke strong antibody responses in salivary glands and extramucosal tissues, *Infect. Immun.* 57:1072 (1989).

45. T. Sasazuki, HLA: genetic control of immune response and disease susceptibility, *Jpn. J. Human Genet.* 27:81 (1982).

46. R.R.P. de Vries, Regulation of T cell responsiveness against mycobacterial antigens by HLA class 2 immune response genes, *Rev. Infect. Dis.* 11:S400 (1989).

47. C.E. Hormaeche, K.A. Harrington, and H.S. Joysey, Natural resistance to Salmonellae in mice: Control by genes within the major histocompatibility complex, *J. Infect. Dis.* 152:1050 (1985).

48. H. Shimauchi, T. Ogawa, and S. Hamada, Immune response gene regulation of the humoral immune response to *Porphyromonas gingivalis* fimbriae in mice, *Immunology* 74:362 (1991).

49. J. Haber and C. Grinnell, Analysis of the serum antibody responses to type 1 and type 2 fimbriae in mice immunized with *Actinomyces viscosus* T14V, *J. Periodontal Res.* 24: 81 (1989).

50. C.O. Elson and W. Ealding, Ir gene control of the murine secretory IgA response to cholera toxin, *Eur. J. Immunol.* 17:425 (1987).

51. M.J. Staruch and D.D. Wood, Genetic influences on the adjuvanticity of muramyl dipeptide *in vivo*, *J. Immunol.* 128:155 (1982).

52. E.S. Vitetta, R. Fernandez-Botran, C.D. Myers, and V.M. Sanders, Cellular interactions in the humoral immune response, *Adv. Immunol.* 45:1 (1989).

53. W. van Ewijk, Immunohistology of lymphoid organs, *Curr. Opinions Immunol.* 1:954 (1989).

54. T. Ogawa, Y. Kusumoto, H. Kiyono, J.R. McGhee, and S. Hamada, Distinct transmigration pathways for antigen-specific B cells following oral or systemic immunization with *Porphyromonas gingivalis* fimbriae, *Intern. Immunol.* 4:In press (1992).

55. J. Bienenstock and J. Dolezel, Peyer's patches: lack of specific antibody-containing cells after oral and parenteral immunization, *J. Immunol.* 106:938 (1971).

56. R.V. Heatley, J.M. Stark, P. Horsewood, E. Bandouvas, F. Cole, and J. Bienenstock, The effects of surgical removal of Peyer's patches in the rat on systemic antibody responses to intestinal antigen, *Immunology* 44:543 (1981).

57. G.A. Enders, S. Ballhaus, and W. Brendel, The influence of Peyer's patches on the organ-specific distribution of IgA plasma cells, *Immunology* 63:411 (1988).

58. R. Clancy and A. Pucci, Sensitisation of gut-associated lymphoid tissue during oral immunisation, *AJEBAK.* 56:337 (1978).

59. S.M. Mallison III, A.K. Szakal, R.R. Ranney, and J.G. Tew, Antibody synthesis specific for nonoral antigens in inflamed gingiva, *Infect. Immun.* 56:823 (1988).

IgA1 PROTEASES AND HOST–PARASITE RELATIONSHIPS

IN THE ORAL CAVITY

Mogens Kilian,[1] Jesper Reinholdt,[2] Knud Poulsen,[1] and Hans Lomholt[1]

[1]Institute of Medical Microbiology
[2]Department of Oral Biology
Faculty of Health Sciences
University of Aarhus
DK–8000 Aarhus C, Denmark

INTRODUCTION

Beginning at birth the oral microflora develops under the influence of the humoral immune system. The newborn receives S–IgA antibodies to oral bacteria including *Streptococcus mutans* in colostrum and milk from the mother.[1] In addition, maternal IgG antibodies to oral bacteria are present in serum of the newborn.[2] The young infant's own response to antigens of oral streptococci is detectable early after birth. While salivary levels of total S–IgA do not reach adult levels until the age of 1 to 2 years,[3] specific S–IgA antibodies to oral streptococcal antigens may be detected in saliva as early as 3 weeks after birth although the spectrum of antigenic determinants detected may not be comparable to that of S–IgA antibodies found in adult saliva.[4] In serum, IgG and IgA antibodies to oral bacteria appear within the first years of life and may reach the oral cavity during tooth eruption and later with crevicular fluid. However, serum antibodies to streptococcal glucosyltransferases (GTF) remain at a relatively low level for several years.[2]

Several *in vivo* and *in vitro* studies have demonstrated that S–IgA antibodies in secretions efficiently protect mucosal surfaces against both viral and bacterial pathogens. The molecular mechanisms include the ability of S–IgA to interfere with the adhesion of microorganisms to surfaces by blocking their adhesins, by rendering microorganisms hydrophilic, and by agglutinating them to enhance their elimination with the fluids that bathe the surfaces. Like antibodies of other isotypes, S–IgA antibodies are capable of neutralizing viruses and microbial enzymes and toxins. In contrast to other isotypes, IgA has anti–phlogistic properties that are important for maintaining the structural integrity and the barrier function of mucosal tissues (for review see [5]). However, there is no convincing evidence that naturally occurring S–IgA antibodies interfere with the life–long persistence and accumulation of bacteria on tooth surfaces. Only after passive or active vaccination resulting in significantly increased levels of antibodies in saliva is it possible to demonstrate an effect on specific members of the dental plaque flora, as it has been done with *Streptococcus mutans*.[6,7] The same situation applies to members of the subgingival microflora. Although specific antibodies to putative periodontal pathogens can be detected in serum and saliva of periodontitis patients[8,9] there is yet no evidence that such antibodies eventually result in the elimination of the bacteria from the flora.

Several factors may explain this apparent functional incompetence of the immune system: 1) a relatively low responsiveness of the immune system to members of the commensal microflora; 2) the ability of commensal oral bacteria to evade an immune response by molecular mimicry including their ability to mask their own surface antigens by binding host proteins from the secretions; 3) the production of proteases capable of inactivating antibodies, such as IgA1 proteases; 4) antigenic variation within the species; and 5) various immune–modulating and –suppressive factors and toxins. However, only a few of these putative mechanisms have been evaluated in any detail. This review deals with IgA1 proteases and their possible influence on the host–parasite relationship in the oral cavity.

IgA1 PROTEASES

Several members of the oral microflora secrete proteases that are able to cleave human IgA1 and S–IgA1, in spite of the relative resistance of the latter to proteolytic attack. The first example of an IgA1 protease was demonstrated in *Streptococcus sanguis*.[10] Subsequent studies revealed similar enzymes in other members of the oral microflora: *Streptococcus oralis* ("*mitior*"),[11] some strains of *Streptococcus mitis* biovar 1[12], and in oral *Capnocytophaga* and *Prevotella* ("*Bacteroides*") species.[13,14] The property is shared by the three leading causes of bacterial meningitis, *Neisseria meningitidis*,[15] *Haemophilus influenzae*, and *Streptococcus pneumoniae*,[16] and by important pathogens of the genital tract, *Neisseria gonorrhoeae*[15] and *Ureaplasma urealyticum*.[17,18]

Common to the IgA1 proteases from all these bacteria is their substrate specificity. They all cleave exclusively one of the Pro–Ser or Pro–Thr peptide bonds within the duplicated

Table 1. Characteristics of bacterial IgA1 proteases.

Proteinase type/species	Enzyme specificity*
Serine proteinases	
H. influenzae	Pro–Ser (231–232) Type 1 or
H. aegyptius	Pro–Thr (235–236) Type 2
N. meningitidis	Pro–Ser (237–238) Type 1 or
N. gonorrhoeae	Pro–Thr (235–236) Type 2
Metallo–proteinases	
S. pneumoniae	Pro–Thr (227–228)
S. sanguis	Pro–Thr (227–228)
S. mitis	Pro–Thr (227–228)
Cysteine–proteinases	
Prevotella species	Pro–Ser (223–224)
Yet unclassified	
Capnocytophaga species	Pro–Ser (223–224)
Ureaplasma urealyticum	Pro–Thr (235–236)

*Amino acid residues between which the peptide bond is cleaved.

heavily glycosylated octapeptide Thr–Pro–Pro–Thr–Pro–Ser–Pro–Ser present in the hinge region of the human α1 chain (Table 1). The phylogenetically older IgA2 is resistant to cleavage due to a lack of the susceptible sequence in the α2 chain.[19] A similar amino acid sequence is present in the precursor of the IgA1 proteases from the pathogenic neisseriae and *H. influenzae* and is cleaved by the protease itself during its maturation and eventual release to the extracellular space.[20,21] Apart from this, there is as yet no conclusive evidence that the bacterial IgA1 proteases have substrates *in vivo* other than human IgA1.

Some proteins including the lymphocyte CD8 molecule and granulocyte–macrophage-colony stimulating factor (GM–CSF) show amino acid sequence homology to the sequence cleaved by IgA1 proteases[22] but there is no evidence yet that cleavage of these proteins takes place in their natural configuration. Cleavage of human IgA1 appears to depend on structural properties of the hinge region as removal of its O–glycosidically linked carbohydrate side chains significantly reduces the susceptibility to IgA1 proteases of oral *S. sanguis*.[12]

Recently, Shoberg and Mulks[23] presented data suggesting that IgA1 proteases play a role in the modulation of outer membrane proteins in gonococci. In this context it is of potential great interest that sequence analyses of several cell wall proteins from oral streptococci reveal stretches with amino acid sequences homologous to that of the hinge region of human IgA1.[24]

Characterization of IgA1 proteases by sequence analysis of cloned genes and studies employing inhibitors of various proteinase types have revealed that the IgA1 proteases, in spite of their virtual functional identity, have developed through at least three independent evolutionary lines.[25-28] Thus, at least three of the major types of proteinases are represented among bacterial IgA1 proteases (Table 1).

Biological Significance of IgA1 Proteases

Several findings prove that IgA1 proteases are active *in vivo*. Thus, characteristic fragments of IgA1 and S–IgA1 have been demonstrated in vaginal secretions from women with gonococcal infection, in cerebrospinal fluid of meningitis patients, in nasopharyngeal secretions of children with a history of atopic diseases, and in human dental plaque (for review see [29]). Monomeric Fab fragments of IgA can be detected on bacteria in dental plaque[30] in accordance with the finding that Fab fragments released from IgA1 after cleavage with IgA1 proteases retain their antigen–binding properties.[31] Evidence for *in vivo* activity of *Prevotella* and *Capnocytophaga* IgA1 proteases in periodontitis patients was recently obtained though the demonstration of serum antibodies against a neoepitope on Fab_α fragments exposed only after cleavage induced by these particular proteases.[32] Cleavage of S–IgA1 is usually extensive and not restricted to antibodies against the IgA1 protease–producing microorganism.

Fab fragments released after IgA1 protease cleavage do no harm to the microorganisms to which they bind as the anti-microbial functions of S–IgA and IgA depend on the Fc part of the molecule[5]. Direct support for this conclusion comes from *in vitro* studies showing that IgA1 protease may negate the ability of S–IgA to inhibit binding of oral streptococci to saliva–coated hydroxyapatite,[33] and from the *in vivo* observation that Fab fragments of specific antibacterial antibodies, in contrast to intact antibodies, are incapable of interfering with the colonization of bacteria in the gut and human oral cavity.[7,34] Fab_α fragments binding to surface epitopes of bacteria are not only functionally incompetent, they may, in addition, block access of intact antibodies and thereby protect the bacteria. In support of this concept, we have shown that Fab_α fragments of IgA1 antibodies, like intact IgA, are capable of inhibiting the complement–activating activity of coexisting IgG antibodies with the same or related antigen specificity.[35] We have hypothesized that this may be an important mechanism in the pathogenesis of invasive infections due to IgA1 protease–producing bacteria.[36] Thus, it is possible that IgA1 protease produced by subgingival bacteria, in this way, protect members of the subgingival microflora associated with periodontal disease against the immune system. In addition, this mechanism may facilitate the penetration of intact bacteria or bacterial products including outer membrane vesicles through the crevicular epithelium.

Thus, as a result of their unique cleavage specificity, the IgA1 proteases differ in several ways from non–specific proteases produced by many other members of the subgingival plaque flora. They are not inhibited by physiological proteinase inhibitors such as α2–macroglobulin, and are not likely to serve any nutritional purpose. In contrast, they appear to represent an intriguing way of turning a specific immune response into an advantage for the microorganism.

Streptococci constitute the predominant part (~ 80%) of the bacteria that initiate supragingival plaque formation. Of these streptococci 86% (median value; range 48–95%) are IgA1 protease–producing *S. sanguis, S. oralis,* and *S. mitis* biovar 1. The property is distinctly associated with the initial colonizers as only 17% (median, value; range 2–59%) of streptococci in mature dental plaque produce IgA1 protease.[37,38] It is conceivable that IgA1 protease makes some streptococcal species able to evade the colonization–inhibiting effect of S–IgA. However, cleavage of S–IgA1 (and ensuing coating with Fab$_\alpha$ fragments) does not appear essential for their adherence as the pattern of colonization of cleaned tooth surfaces in selectively IgA–deficient, and IgM–compensating, subjects is identical to that seen in normal subjects (Reinholdt and Kilian, in preparation).

The ability of IgA1 proteases to interfere with the protective mechanisms of IgA *in vivo* is likely to depend on the subclass distribution of IgA antibodies in secretions and in serum and on the ability of the immune system to respond with protease–neutralizing antibodies.

IgA Subclasses

The two IgA subclasses differ in their relative proportions in serum and secretions. While IgA1 is vastly predominant in serum (~ 90%), mucosal secretions show distinct differences. In the nasopharynx S–IgA1 accounts for more than 90% of the total IgA, whereas in the gut IgA2 appears to be slightly predominant as reflected by the distribution of IgA1– and IgA2–producing plasma cells in the respective mucosal tissues.[39,40] The average proportion of S–IgA1 in saliva is approximately 65–70%. However, studies by Smith et al.[3] revealed significant individual differences among infants. In fact, their data suggest that infants may be divided into two distinct groups, one with 80–100% IgA1, and another with 30–55% IgA1 in saliva. Similar individual differences are observed in the subclass distribution of S–IgA antibodies to particular antigens including oral streptococci.[41,42] It is an intriguing possibility that such differences may be associated with differences in plaque–forming potential and in plaque–associated disease activity.

Neutralizing Antibodies to IgA1 Proteases

Antibodies capable of inhibiting the activity of IgA1 proteases are present in human colostrum and in serum of patients convalescing from infections with IgA1 protease–producing bacteria.[42,43] *In vitro* neutralization of IgA1 proteases of the pathogenic neisseriae and *H. influenzae* with antibodies raised in rabbits have disclosed an interesting high degree of antigenic variation. This variation appears to be a result of horizontal gene transfer and homologous recombination between IgA1 protease genes within and between species, resulting in a mosaic–like gene structure in these bacteria.[44–47]

It is conceivable that these variations increase the immune–escape potential of the IgA1 proteases. However, while numerous antigenic forms of IgA1 proteases exist in the population of *N. meningitidis, N. gonorrhoeae,* and *H. influenzae,* only one or two variants have been detected in each of the IgA1 protease–producing oral species.[12,47] It is not yet known if this reflects a high degree of conservation of the IgA1 protease genes among oral bacteria as only one example of the *S. sanguis iga* gene has been sequenced so far.[28] The lack of antigenic variation among *S. sanguis* IgA1 proteases may be due to lack of an immune selection pressure. In support of this hypothesis, Gilbert et al.[42] detected very low levels of neutralizing antibodies to the IgA1 protease of *S. sanguis* in colostrum and serum. However, ongoing studies in our laboratories suggest significant individual differences.

CONCLUSION

As outlined above, previous research has not only clarified important molecular and genetic aspects of IgA1 protease activity; it has also shown that IgA1 protease–induced cleavage may nullify the inhibitory effect of S–IgA antibodies on bacterial colonization. However, direct evidence for a role of IgA1 proteases in the interaction of pathogenic or potentially pathogenic indigenous organism with the host immune system is still lacking. In accordance with the notion that IgA1 proteases act by negating the effect of IgA1 antibodies, the potential significance of the protease activity at mucosal surfaces can be expected to depend on the level of S–IgA antibodies with specificity for the relevant antigen. Thus, the role of IgA1 protease in the colonization of oral surfaces may be limited due to relatively low responsiveness of the secretory immune system to the commensal microflora. On the other hand, it seems plausible that the IgA1 protease produced by the same oral flora can interfere with the function of protective salivary S–IgA antibodies that may be induced by immunization. This possibility deserves attention in the context of prospective oral immunization against oral pathogens.

REFERENCES

1. R.R. Arnold, J. Mestecky, and J.R. McGhee, Naturally occurring secretory immunoglobulin A antibodies to *Streptococcus mutans* in human colostrum and saliva, *Infect. Immun.* 14:335–362 (1976).
2. Z. Luo, D.J. Smith, M.A. Taubman, and W.F. King, Cross–sectional analysis of serum antibody to oral streptococcal antigens in children, *J. Dent. Res.* 67:554–560 (1988).
3. D.J. Smith, W.F. King, and M.A. Taubman, Isotype, subclass and molecular size of immunoglobulins in salivas from young infants, *Clin. Exp. Immunol.* 76:97–102 (1989).
4. D.J. Smith, W.F. King, and M.A. Taubman. Salivary IgA antibody to oral streptococcal antigens in predentate infants, *Oral. Microbiol. Immunol.* 5:57–62 (1990).
5. M. Kilian, J. Mestecky, and M.W. Russell, Defence mechanisms involving Fc–dependent functions of immunoglobulin A and their subversion by bacterial immunoglobulin A proteases, *Microbiol. Rev.* 52:296–303 (1988).
6. B. Cohen, S.L. Peach, and R.R.B. Russell, Immunization against dental caries, *in* "Medical Microbiology, vol. 2: Immunization Against Bacterial Disease," C.S.F. Easmon and J. Jeljaszewicz, ed., Academic Press, London, pp. 255–294 (1983).
7. J.K.–C. Ma, M. Hunjan, R. Smith, C. Kelly, and T. Lehner, An investigation into the mechanism of protection by local passive immunization with monoclonal antibodies against *Streptococcus mutans, Infect. Immun.* 58:3407–3414 (1990).
8. J.M.A. Wilton, N.W. Johnson, M.A. Curtis, I.R. Gillett, R.J. Carman, J.L.M. Bampton, G.S. Griffiths, and J.A.C. Sterne, Specific antibody responses to subgingival plaque bacteria as aids to the diagnosis and prognosis of destructive periodontitis, *J. Clin. Periodontol.* 18:1–15 (1991).

9. J.M.A. Wilton, J.M. Slaney, J.A.C. Sterne, D. Beighton, and N.W. Johnson, Salivary IgA antibodies against bacteria incriminated as periodontal pathogens in Kenyan adolescents: correlation with disease status and demonstration of antibody specificity, *Microb. Ecol. Hlth. Dis.* 4:293–301 (1991).

10. A.G. Plaut, R. Wistar, and J.D. Capra, Differential susceptibility of human IgA immunoglobulins to streptococcal IgA protease, *J. Clin. Invest.* 54:1295–1300 (1974).

11. M. Kilian and K. Holmgren, Ecology and nature of immunoglobulin A1 protease producing streptococci in the human oral cavity and pharynx, *Infect. Immun.* 31:868–873 (1981).

12. J. Reinholdt, M. Tomana, S.B. Mortensen, and M. Kilian, Molecular aspects of immunoglobulin A1 degradation by oral streptococci, *Infect. Immun.* 58:1186–1194 (1990).

13. M. Kilian, Degradation of immunoglobulin A1, A2, and G by suspected principal periodontal pathogens, *Infect. Immun.* 34:757–765 (1981).

14. E.V.G. Frandsen, E. Theilade, B. Ellegaard, and M. Kilian, Proportions and identity of IgA1 degrading bacteria in periodontal pockets from patients with juvenile and rapidly progressive periodontitis, *J. Periodont. Res.* 21:613–623 (1986).

15. A.G. Plaut, R.J. Genco, and T.B. Tomasi, Isolation of an enzyme from *Streptococcus sanguis* which specifically cleaves IgA, *J. Immunol.* 113:289–191 (1974).

16. M. Kilian, J. Mestecky, and R.L. Schrohenloher, Pathogenic species of the genus *Haemophilus* and *Streptococcus pneumoniae* produce immunoglobulin A1 protease, *Infect. Immun.* 26:143–149 (1979).

17. J. A. Robertson, M.E. Stemler, and G.W. Stemke, Immunoglobulin A protease activity of *Ureaplasma urealyticum*, *J. Clin. Microbiol.* 19:255–258 (1984).

18. M. Kilian, M.B. Brown, T.A. Brown, E.A. Freundt, and G.H. Cassell, Immunoglobulin A1 protease activity in strains of *Ureaplasma urealyticum*, *Acta Pathol. Microbiol. Scand.* Section B. 92:61–64 (1984).

19. A.G. Plaut, R. Wistar, and J.D. Capra, Differential susceptibility of human IgA immunoglobulins to streptococcal IgA protease, *J. Clin. Invest.* 54:1295–1300 (1974).

20. J. Pohlner, R. Halter, K. Beyreuther, and T.F. Meyer, Gene structure and extracellular secretion of *Neisseria gonorrhoeae* IgA protease, *Nature* 335:458–462 (1987).

21. K. Poulsen, J. Brandt, J.P. Hjorth, H.C. Thøgersen, and M. Kilian, Cloning and sequencing of the immunoglobulin A1 protease gene (*iga*) of *Haemophilus influenzae* serotype b, *Infect. Immun.* 57:3097–3105 (1989).

22. J. Pohlner, R. Halter, and T.F. Meyer, *Neisseria gonorrhoeae* IgA protease. Secretion and implications for pathogenesis, *Antonie van Leeuwenhoek* 53:479–484 (1989).

23. R.J. Shoberg and M.H. Mulks, Proteolysis of bacterial membrane proteins by *Neisseria gonorrhoeae* type 2 immunoglobulin A1 protease, *Infect. Immun.* 29:2535–2541 (1991).

24. I. Takahashi, N. Okahashi, C. Sasakawa, M. Yoshikawa, S. Hamada, and T. Koga, Homology between surface protein antigen genes of *Streptococcus sobrinus* and *Streptococcus mutans*, *FEBS Lett.* 249:383–388 (1989).

25. W.W. Bachovchin, A.G. Plaut, G.R. Flentke, M. Lynch, and C.A. Kettner, Inhibition of IgA1 proteases from *Neisseria gonorrhoeae* and *Haemophilus influenzae* by peptide prolyl boronic acids, *J. Biol. Chem.* 265:3738–3743 (1990).

26. M. Kilian, J. Mestecky, R. Kulhavy, M. Tomana, and W.T. Butler, IgA1 proteases from *Haemophilus influenzae*, *Streptococcus pneumoniae*, *Neisseria meningitidis*, and *Streptococcus sanguis*: comparative immunochemical studies, *J. Immunol.* 124:2596–2600 (1980).

27. S.B. Mortensen and M. Kilian, Purification and characterization of an immunoglobulin A1 protease from *Bacteroides melaninogenicus*, *Infect. Immun.* 45:550–557 (1984).

28. J.V. Gilbert, A.G. Plaut, and A. Wright, Analysis of the immunoglobulin A protease gene of *Streptotococcus sanguis*, *Infect. Immun.* 59:7–17 (1991).

29. M. Kilian and J. Reinholdt, Interference with IgA defence mechanisms by extracellular bacterial enzymes, in: "Medical Microbiology," C.S.F. Easmon and J. Jeljaszewicz, ed., Academic Press, London, pp. 173–208 (1986).

30. T. Ahl and J. Reinholdt, Detection of immunoglobulin A1 protease–induced Fab$_\alpha$ fragments on dental plaque bacteria, *Infect. Immun.* 59:563–569 (1991).

31. B. Mansa and M. Kilian, Retained antigen–binding activity of Fab$_\alpha$ fragments of human monoclonal immunoglobulin A1 (IgA1) cleaved by IgA1 protease, *Infect. Immun.* 52:171–174 (1986).

32. E.V.G. Frandsen, J. Reinholdt, and M. Kilian, Immunoglobulin A1 (IgA1) proteases from *Prevotella* (*Bacteroides*) and *Capnocytophaga* species in relation to periodontal diseases. *J. Periodont. Res.* 26:297–299 (1991).

33. J. Reinholdt and M. Kilian, Interference of IgA protease with the effect of secretory IgA on adherence of oral streptococci to saliva–coated hydroxyapatite, *J. Dent. Res.* 66:492–497 (1987).

34. E.J. Steele, W. Chaicumpa, and D. Rowley, Further evidence for cross–linking as a protective factor in experimental cholera: properties of antibody fragments, *J. Infect. Dis.* 132:175–180 (1975).

35. M.W. Russell, J. Reinholdt and M. Kilian, Anti–inflammatory activity of human IgA antibodies and their Fab$_\alpha$ fragments: inhibition of IgG–mediated complement activation, *Eur. J. Immunol.* 19:2243–2249 (1989).

36. M. Kilian and J. Reinholdt, A hypothetical model for the development of invasive infection due to IgA1 protease–producing bacteria, *Adv.Exp.Med.Biol.* 216B:1261–1269 (1987).

37. B. Nyvad and M. Kilian, Microbiology of the early colonization of human enamel and root surfaces in vivo, *Scand. J. Dent. Res.* 95:369–380 (1987).

38. E.V.G. Frandsen, V. Pedrazolli, and M. Kilian, Ecology of viridans streptococci in the oral cavity and pharynx, *Oral Microbiol. Immunol.* 6:129–133 (1991).

39. K.P. Kett, P. Brandtzaeg, J. Radl, and J.J. Haaijman, Different subclass distribution of IgA–producing cells in human lymphoid organs and various secretory tissues, *J. Immunol.* 136:3631–3635 (1986).

40. J. Mestecky, C. Lue, A. Tarkowski, I. Ladjeva, J.H. Peterman, Z. Moldoveanu, M.W. Russell, T.A. Brown, J. Radl, J.J. Haaijman, H. Kiyono and J.R. McGhee, Comparative studies of the biological properties of human IgA subclasses, *Protid. Biol. Fluids.* 36:173–182 (1989).

41. T.A. Brown and J. Mestecky, Immunoglobulin A subclass distribution of naturally occurring salivary antibodies to microbial antigens, *Infect. Immun.* 49:459–462 (1985).

42. T. Ahl and J. Reinholdt, Subclass distribution of salivary immunoglobulin A antibodies to oral streptococci, *Infect. Immun.* 59:3619–3625 (1991).

43. J.V. Gilbert, A.G. Plaut, B. Longmaid, and M.E. Lamm. Inhibition of microbial IgA1 proteases by secretory IgA and serum, *Mol. Immunol.* 20:1039–1049 (1983).

44. K. Kobayashi, Y. Fujiyama, K. Hagiwara, and H. Kondoh, Resistance of normal IgA serum and secretory IgA to bacterial IgA proteases: Evidence for the presence of enzyme–neutralizing antibodies in both serum and secretory IgA, and also in serum IgG, *Microbiol. Immunol.* 31:1097–1106 (1987).

45. R. Halter, J. Pohlner, and T.F. Meyer, Mosaic–like organization of IgA protease genes in *Neisseria gonorrhoeae* generated by horizontal genetic exchange *in vivo*, *EMBO J.* 8:2737–2744 (1989).

46. K. Poulsen, J. Reinholdt and M. Kilian, A comparative genetic study of serologically distinct *Haemophilus influenzae* type 1 immunoglobulin A1 proteases, *J. Bacteriol.* 174:2913–2921 (1992).

47. H. Lomholt, K. Poulsen, D. Caugant, and M. Kilian, Molecular polymorphism and epidemiology of immunoglobulin A1 proteases from *Neisseria meningitidis*, *Proc. Natl. Acad. Sci. USA* 89:2120–2124 (1992).

48. E.V.G. Frandsen, J. Reinholdt and M. Kilian, Enzymatic and antigenic characterization of immunoglobulin A1 proteases from *Bacteroides* and *Capnocytophaga* spp, *Infect. Immun.* 55:631–638 (1987).

TRANSPORT OF IgA IMMUNE COMPLEXES ACROSS EPITHELIAL MEMBRANES: NEW CONCEPTS IN MUCOSAL IMMUNITY

Michael E. Lamm, Janet K. Robinson, and Charlotte S. Kaetzel

Institute of Pathology
Case Western Reserve University
Cleveland, OH 44106

INTRODUCTION

The traditional concept of Immunoglobulin A's (IgA's) role in mucosal immunity is to act as a lumenal barrier to prevent exogenous substances from attaching to or penetrating a mucosal lining (e.g., in the oropharynx, respiratory, intestinal or genital tracts).[1,2] Based on this concept, there is much current interest in developing vaccines designed to stimulate the secretion of IgA antibodies into particular exocrine sites, as dictated by the route of pathogenesis of infectious agents that afflict mucous membranes or that enter the body through a particular mucosal portal of entry.[3]

Traditional concepts of mucosal immunity, however, have largely neglected the possibility, more likely the probability, that significant amounts of immune complexes containing IgA antibodies are being formed on an ongoing basis within the various mucous membranes. For example, in the intestinal tract, macromolecular antigens or fragments thereof derived from food and the intestinal microflora have an opportunity to penetrate to some extent through or between mucosal epithelial cells to enter the lamina propria. Infections of mucous membranes offer another opportunity for antigens to be released into the mucous membrane. If by these or other means significant amounts of antigen can find their way into the mucous membrane, and if there is a concurrent local immune response dominated, as is usual in mucosal sites, by the IgA isotype, the stage would be set for significant amounts of IgA immune complexes to form within the lamina propria.

How IgA immune complexes within the lamina propria would be handled by the host is not precisely known. Some complexes could enter the local vascular drainage and reach the blood circulation, where they would be exposed to the mononuclear phagocyte system. In some animal species, like the rat, circulating IgA immune complexes can be cleared efficiently through the liver to be excreted into the bile, a pathway that does not appear to be significant in humans.[4] Consistent with this line of thinking, investigations of how the body handles IgA immune complexes have frequently employed model systems based on parenteral injection of IgA immune complexes. This question is important because the mononuclear phagocyte system is saturable and excess circulating immune complexes have the potential to cause disease directly or, with certain antigens, perhaps to lead to autoimmune reactions.

On the other hand, we believe that introduction of IgA immune complexes parenterally may not optimally model the most important means by which the body deals with IgA immune complexes formed within mucous membranes. Instead, we have postulated that significant amounts of IgA immune complexes are regularly being formed within mucous membranes, including healthy subjects, and that such immune complexes are quickly and directly excreted through adjacent lining epithelial cells into the lumen by the same route that free dimeric IgA normally enters secretions: binding to the extracellular domain (secretory component) of the polymeric immunoglobulin (poly Ig) receptor on the basolateral surface of the epithelial cell, followed by transcytosis and release from the apical surface.[5] Such an excretory route for immune complexes would make good use of the relatively low phlogistic potential of IgA (i.e., immune complexes containing IgA antibodies might form and be excreted without eliciting the degree of local inflammation that IgG, IgM or IgE antibodies might provoke), at the same time preventing the entry of antigen and immune complexes into the circulation, where they would have greater access to the more phlogistic IgG antibodies. An excretory function for IgA could thus limit the systemic load of circulating immune complexes and their attendant inflammatory sequelae.

EXPERIMENTAL TRANSPORT OF IgA IMMUNE COMPLEXES

We decided to begin our studies on transport of IgA immune complexes across epithelial cells with an *in vitro* system. The system depends on the availability of cell lines that both express poly Ig receptor and can grow in polarized monolayers. In such monolayers the individual cells are differentiated, exhibiting polarity; furthermore, each cell is attached to its neighbors by tight junctions so that free diffusion across the monolayer is prevented. If such monolayers are grown on a permeable surface separating an upper and lower compartment, molecules can pass from one compartment to the other only by passing transcellularly. The system we have employed uses dog kidney epithelial cells (MDCK) that were transfected with cDNA encoding the rabbit poly Ig receptor and expressing the receptor on the basolateral surface.[6] The cells are allowed to grow into monolayers on permeable nitrocellulose filters.

The IgA antibodies used are rat monoclonals with specificity for the dinitrophenyl (DNP) hapten (derived from the LO-DNP-64 hybridoma and kindly provided by Dr. J.-P. Vaerman). The antigen is DNP-coupled bovine serum albumin (DNP-BSA). For purposes of analysis, IgA and BSA can be radioiodinated or coupled to biotin, which allows binding to a streptavidin-Sepharose solid phase.

Experiments performed to date have yielded the following results:

- Transport of soluble IgA immune complexes (IgA anti-DNP antibodies and DNP-BSA) depends, as expected, on having oligomeric IgA that can bind to epithelial cell surface poly Ig receptor.

- Immune complexes in which the IgA is exclusively monomeric are not transported. Thus, simple aggregation of monomeric IgA, as occurs in an immune complex, is not sufficient to provide a site or conformation capable of binding to poly Ig receptor. Such binding requires J chain-containing oligomeric IgA.

- Mixed immune complexes containing both monomeric and dimeric IgA in the same complex over a range of antigen/antibody ratios are transported.

IgA immune complexes are not appreciably degraded intracellularly during transport and appear to follow the same route through epithelial cells and be treated in the same manner as free dimeric IgA.

DISCUSSION AND CONCLUSION

We believe our results suggest new ways of looking at the role of IgA in mucosal immunity and host defense. In the first place, since mixed immune complexes containing both monomeric and dimeric IgA antibody can be transported across epithelial cells, the data support a role for monomeric IgA antibody in mucosal immunity, an idea that is at variance from the traditional view that because monomeric IgA cannot bind to poly Ig receptor, it has no such role to play. According to our results, however, if a molecule of monomeric IgA antibody can participate in the same immune complex with dimeric IgA, the entire complex, including the monomeric IgA, can be transported through epithelial cells. A role for monomeric IgA in mucosal immunity is in keeping with the concept that individual plasma cells synthesize both monomeric and dimeric IgA.[7] If so, according to our view, the entire secretory product of mucosal plasma cells could be directed toward mucosal defense.

Second, if, as seems likely to us, the lamina propria of a mucous membrane is indeed a site where IgA antibodies have ample opportunity to meet antigens, then IgA immune complexes should be forming regularly. If so, direct elimination through the adjacent epithelium would afford a much more efficient route than disposal following entry into the circulation. For this excretory function, the relatively low inflammatory potential of IgA would be particularly advantageous. The net effect for host defense would be two mucosal barriers to the entry of antigens or immune complexes into the systemic circulation, with intra-lumenal IgA providing the other barrier.

In summary, the traditional view of mucosal IgA function is "secretory": free IgA antibodies are secreted into the lumen in order to function there in host defense. In addition to this classical role, we are now proposing that mucosal IgA also has an "excretory" function, serving as a vehicle for the excretion of antigens from the body proper (Table 1).

Table 1. The "secretory" and "excretory" functions of mucosal IgA.[*]

Pathway	Lamina propria	Epithelium	Lumen
Secretory	IgA	IgA	IgA+Ag=IgAIC
Excretory	IgA+Ag=IgAIC	IgAIC	IgAIC

[*] In the secretory pathway, IgA antibody produced by mucosal plasma cells passes through the epithelium to meet antigen in the lumen, resulting in the formation of IgA immune complexes, which prevent attachment and entry of foreign molecules. In the excretory pathway, after secretion from mucosal plasma cells, IgA antibody meets antigen to form IgA immune complexes within the lamina propria; these immune complexes are then transported through the epithelium into the lumen.

ACKNOWLEDGMENT

This work was supported by NIH grants AI-26449, CA-51998 and HL-37117.

REFERENCES

1. M.E. Lamm, Cellular aspects of immunoglobulin A., *Adv. Immunol.* 22:223 (1976).
2. J. Mestecky and J.R. McGhee, Immunoglobulin A (IgA): Molecular and cellular interactions involved in IgA biosynthesis and immune response, *Adv. Immunol.* 40:153 (1987).
3. J.R. McGhee and J. Mestecky, In defense of mucosal surfaces, Development of novel vaccines for IgA responses protective at the portals of entry of microbiol pathogens, *Infect. Dis. Clin. N. America* 4:315 (1990).
4. B.J. Underdown and J.M. Schiff, Immunoglobulin A: strategic defense initiative at the mucosal surface, *Ann. Rev. Immunol.* 4:389 (1986).
5. C.S. Kaetzel, J.K. Robinson, K.R. Chintalacharuvu, J.-P. Vaerman, and M.E. Lamm, The polymeric immunoglobulin receptor (secretory component) mediates transport of immune complexes across epithelial cells: A local defense function for IgA, *Proc. Natl. Acad. Sci. USA* 88:8796 (1991).
6. K.E. Mostov and D.L. Deitcher, Polymeric immunoglobulin receptor expressed in MDCK cells transcytoses IgA, *Cell* 46:613 (1986).
7. A. Tarkowski, Z. Moldoveanu, W.J. Koopman, J. Radl, J.J. Haaijman, and J. Mestecky, Cellular origins of human polymeric and monomeric IgA: enumeration of single cells secreting polymeric IgA1 and IgA2 in peripheral blood, bone marrow, spleen, gingiva and synovial tissue, *Clin. Exp. Immunol.* 85:341 (1991).

EFFECT OF MUCOSAL MICROENVIRONMENT ON IMMUNE RESPONSE TO VIRUSES

Rebecca Abraham and Pearay L. Ogra

Department of Pediatrics
University of Texas Medical Branch
Galveston, TX 77550

INTRODUCTION

Mucosal tissue constitutes the primary portal of entry for many mammalian pathogens or orally introduced antigens. The immunological events initiated by such host-pathogen interactions are diverse and appear to be profoundly influenced by the physiologic and pathologic state of the mucosal microenvironment. It is not surprising that mucosal surfaces are equipped with an elaborate system of defense including the mucosa-associated lymphoid tissue (MALT).[1] The defense system in each mucosal area is in constant communication with other mucosal areas and thus the existence of a common mucosal system has been elaborated by a number of prominent workers.[2-4] In addition, a number of nonimmune mucosal microenvironmental factors also appear to govern the outcome of immune responses to mucosal-introduced antigens.

MUCOSAL MICROENVIRONMENT

The gastrointestinal tract comprises a large area of mucosal tissue. The microflora of the gastrointestinal tract constitutes an important but often forgotten aspect of the host's environment as they influence a wide variety of processes in the healthy and diseased individual. The gastrointestinal tract of the neonate is sterile at birth and the enteric flora is derived largely from the environment. The establishment of intestinal flora follows a definite time sequence related to the suckling period.[5] Within 3 to 4 weeks after birth, the flora characteristics are well established. An increasing interest in the indigenous intestinal flora has been generated in the last few years as they appear to interact with the immunologic system in promoting host resistance to enteric diseases.[6,7]

The ability of the mucosal tissue to absorb intact macromolecules, including proteins, can be best demonstrated in the neonatal rat intestine particularly during the suckling period.[8] In the human infant passive immunity is acquired "in utero" and a minimal amount of intact immunoglobulins is absorbed by the suckling infant.[9] However, in humans such absorption is markedly elevated during recovery from diarrhea and other mucosal injuries.[10]

Genetically Engineered Vaccines, Edited by
J.E. Ciardi *et al.*, Plenum Press, New York, 1992

The proteolytic activity in the small intestine also appears to determine the outcome of immune responses to mucosal antigens. Intraluminal proteases such as trypsin and chymotrypsin could play a role in antigen processing and also influence the nature of the immune responses elicited by exposing hidden or otherwise inaccessible antigenic determinants.[11]

PATHOLOGIC CHANGES IN THE MICROENVIRONMENT

Changes in the Microbial Flora

A variety of factors such as infectious agents, antibiotics and drugs can produce changes in the intestinal microflora. Recent studies conducted in our laboratory have tried to determine a possible relationship between breast feeding, intestinal microflora and the natural acquisition of rotavirus infection.[12] The possibility that bacterial colonization induced by breast feeding might modify diarrheal disease was originally reported by Mata et al.[6] Previous studies have also shown that breast feeding is associated with significant reduction in bacteria-associated enteritis and to a lesser extent acute respiratory infections.[13] Present studies revealed that breast-fed infants maintained a predominant growth of bifidobacteria, whereas bottle-fed infants showed a higher diversity of enteric flora. The number of rotavirus particles shed was inversely proportional to levels of bifidobacteria. There was no significant difference in infection rates between breast-fed and bottle-fed groups of infants. However, the clinical manifestation of the disease was milder in breast-fed infants.[12] The above observations strongly suggest that alterations in gut microflora, in combination with specific antibody or dietary factors present in milk, may play a role in reducing the morbidity observed in breast-fed infants. Another interesting possibility is that bifidobacteria may interact with secretory IgA and protect the intestinal villous epithelium from attachment by viruses or viral antigens in the luminal environment. The study therefore suggests a functional role for bifidobacteria as a potential marker of mucosal events.

Immune Deficiency

Under normal circumstances antigens crossing the mucosal epithelium elicit a local immune response by the production of dimeric IgA from primed IgA producing plasma cells in the lamina propria. Specific IgA[14] is transported across the epithelial surface by secretory component. On the mucosal surface, IgA antibody acts to modulate the further uptake of antigen by interfering with antigen attachment and uptake at the epithelial surface.[15] It has been postulated that immunoglobulins suppress the transport of antigens by complexing with them in the lumen, thereby interfering with absorption.[16] The concentrations of IgA in saliva, stool and serum of newborn animals and humans is low, and it has been hypothesized that this transient deficiency may in part account for the increased transport of macromolecules in newborn animals.[17] Immune deficiency appears to alter the mucosal microenvironment, resulting in the uptake of macromolecules and antigens. This hypothesis has been supported by studies in patients with selective IgA deficiency where an increase in circulating immune complexes and precipitating antibodies to bovine milk proteins was demonstrated.[18]

Malnutrition

Nutritional deficiencies have been reported to affect immunological response and increase susceptibility to infections. The combination of malnutrition and diarrheal disease has been shown to be the most important cause of death in infants, particularly in developing countries.[19]   Acute enteric infections aggravate and increase nutritional deficiencies. Malnutrition, on the other hand, increases the attack rate and severity of diarrhea with prolonged duration of symptoms.[20]

Rotavirus exclusively infects the terminally differentiated villus enterocytes in the small intestine. The mucosal damage includes villus atrophy and necrosis of villus epithelial cells, followed by replacement with immature cells and altered absorptive capacities.[21]   Severe malnutrition also causes small-intestinal atrophy and altered intestinal mucosal defense.[22]   It has therefore been suggested that the intestine during malnutrition and infection may become more permeable to macromolecules with increased passage of environmental antigenic materials into the systemic circulation, resulting in adverse effects.[17]   The murine model has been extensively used in our laboratory to study the pathogenesis of rotavirus infection, particularly under conditions of malnutrition.[23,24]

Malnutrition was induced by expanding the litter size within 12 to 16 hr after birth from 6-8 pups to 18-20 pups.   By 4 days of age there was a significant decrease in weight and albumin in the malnourished litters as compared with the normally nourished.   At 6 days of age malnourished and control mice were fed mouse rotavirus (MRV) and the kinetics of infection followed.   Infection in malnourished animals was characterized by lowered minimal infectious dose, shorter incubation period and earlier, virus replication as evidenced by increased number of infected enterocytes.   Peak shedding of virus in the feces occurred significantly earlier, and the malnourished animal also showed more severe and prolonged disease.[23]   It is not very clear whether these alterations in the course of rotavirus infection in malnutrition are due to impaired immune response or due to changes in the gut morphology.   Studies in adult mice have demonstrated that malnutrition does not affect the state of refractoriness to symptomatic rotavirus infection.[24]   Other studies.[25] have also demonstrated that the young of severely protein malnourished mothers are more severely affected when infected with rotavirus.   It has also been postulated that thinning of the jejunal membrane occurs during malnutrition.[26]   The present studies indicate that such thinning could result in more rapid penetration of the virus and its replication.[24]

More recent experiments have been undertaken in our laboratory to study the effects of malnutrition in BALB/c mice infected with rhesus rotavirus (RRV).[27]   Malnutrition appeared to have a significant effect on the extramucosal spread of RRV.   Clinical symptoms were more severe and of longer duration in the malnourished group.   The mortality rates were 9 percent in the malnourished group and 0 percent in the control group.   Alterations in the intestinal transport mechanisms might underlie the severity of rotavirus infection and the extent of extra mucosal spread of RRV.

EFFECT OF MUCOSAL MICROENVIRONMENT ON HOST PATHOGEN INTERACTIONS

Interactions Involving the Host

Antigenic Processing and Uptake.   The primary interaction between the immuno-competent cells of the host and the mucosal-introduced antigens takes place in the Peyer's patches.   This results in the initial activation of and commitment of IgA precursor B cells.   The antigenic sensitized precursor cells undergo mitotic changes, and the resulting B lymphoblasts bearing IgA on the cell surface migrate to regional lymph nodes and eventually to the systemic circulation via the thoracic duct and other lymphatics. The migration process of the antigen sensitized precursor cells terminates with their homing to the lamina propria of the gut and possibly the submucosal regions of the bronchial tissue. At these sites the cells initiate local production of IgA antibody as mature plasma cells. It has also been reported that during the course of their circulation through the blood stream, the bronchial or intestinal antigen sensitized precursor cells also seed the stroma of the mammary glands, female genital tract and the mucosal tissues of the pharynx and buccal cavity.[28]

Serum and Secretory Antibody Responses. The primary replication of either live or orally introduced vaccine virus occurs in the pharyngeal or intestinal mucosa with subsequent replication in the tonsils and Peyer's patches. Considerable insight into the mechanism of mucosal immunity has been obtained by extensive studies carried out in our laboratory on poliovirus. It has been demonstrated that the induction of virus-specific antibody is determined by the type of antigen (i.e., live virulent, attenuated or inactivated). Similar serum levels of IgM and IgG antibodies are present after immunization of humans with either oral live attenuated poliovaccine (OPV-Sabin) or parenteral inactivated vaccine IPV-Salk. IgA is detected in the serum later and often remains at low levels.[29] However, a significant secretory antibody response in the alimentary tract and nasopharynx occurred only after oral immunization. The viral specific S-IgA had neutralizing activity and resulted in decreased implantations and shorter periods of virus excretion on reinfection challenge. Immunization with inactivated poliovaccine resulted in some degree of local immunity, but the level of protection was less than in children immunized with OPV or after natural infection.[29] On the other hand, the new enhanced potency IPV (IPV-EP) was able to elicit ELISA antibody and neutralizing secretory antibody activity in nasopharyngeal samples which may limit pharyngeal spread of poliovirus.[30] However the neutralizing antibody activity in nasopharyngeal samples after OPV vaccination was higher in frequency and titer as compared to (IPV-EP) vaccination.[30]

The antigenicity and immunogenicity of poliovirus virion proteins have also been extensively studied in our laboratory. It has been reported that VP1 is the important capsid protein which mediates induction of neutralizing antibody responses to poliovirus types 1,2 and 3 respectively.[31,32] The antibody responses in the serum and secretory IgA responses in nasopharyngeal samples to different virion proteins after immunization with OPV, IPV, or a combination of both, have been studied in our laboratory recently. Infants administered oral and inactivated vaccine developed similar S-IgA responses to VP1 and VP2 in the nasopharyngeal samples. However, the VP3 specific S-IgA response was good (75%) in subjects immunized with OPV as compared to the IPV group (10-15%). These observations interestingly suggested that in addition to VP1, neutralization sites associated with VP3 may also be involved in eliciting protective antibody response after administration of the oral vaccine.[11]

Immune Exclusion and Macromolecular Uptake. The role of mucosal damage in the uptake of antigens has been investigated in a number of experimental models involving both the respiratory.[33] and gastric mucosa.[34] Possible mechanisms for these alterations include increased permeability of the mucosal epithelial barrier during viral infections, which facilitate access of antigens to the antibody forming cells. Extensive studies have been carried out in our laboratory to understand the mechanism of macromolecular uptake concomitant with viral infection.[35,36]

Since respiratory syncytial virus (RSV) is the most common respiratory pathogen among neonates, the kinetics of antigen uptake were studied using the mouse model.[35] Groups of BALB/c mice were sham-infected or inoculated intranasally (IN) with live RSV. From day 4 to 8 after infection the animals were exposed IN to ovalbumin (OVA) with or without alum adjuvant. At different intervals, levels of OVA concentration in serum, IgG anti-OVA antibody activity in serum and IgA anti-OVA antibody activity in bronchial washing were determined employing the ELISA technique. The IgE anti-OVA antibody titers in serum and bronchial washing were assessed by passive cutaneous anaphylaxis (PCA). RSV-infected animals developed significantly higher OVA-specific antibody titers of IgG isotype in serum and IgA isotype in bronchial washings than uninfected controls. Alum enhanced the immune response less markedly but significantly in uninfected mice. An IgE antibody response to OVA in serum was demonstrated in 50 percent of RSV-infected mice immunized with OVA and serum. All uninfected animals and RSV-infected animals immunized with OVA alone failed to demonstrate a detectable IgE response (Table 1).

Table 1. Serum OVA concentration[a] measured by ELISA in RSV-infected and noninfected controls.

Experimental Group	OD ± SEM at 1/200 dilution	p value versus Group B
(A) RSV infection, OVA/IN[b]	0.644 ± 0.103	$p < 0.001$
(B) Sham-infected, OVA/IN	0.179 ± 0.035	–
(C) RSV-infected, OVA/alum/IN	0.571 ± 0.119	$p < 0.001$
(D) Sham-infected, OVA-alum/IN	0.243 ± 0.048	$p < 0.01$

[a] Representative data observed on samples taken 2 hr after IN administration of 20 μg OVA with or without alum adjuvant on day 6 after RSV inoculation or Sham inoculation.

[b] IN = intranasal

The increased serum IgG and IgE antibody responses to OVA in RSV infected and adjuvant-treated animals could be due to the large amount of antigen entering the circulation. Stimulation of the lymphoid tissue of the respiratory tract seems to play a role in potentiating the systemic immune response. The increased IgA antibody response could be due to increased access of antigens to bronchus-associated lymphoid tissue. Viral agents and adjuvants may release lymphokines and other mediators which may alter antigen processing and the number and function of B lymphocytes and regulatory T cells. This may partly explain the altered IgE response. RSV infection and adjuvant may both act in enhancing the immune response to OVA.

The effects of malnutrition and rotavirus infection on intestinal barrier function have been analyzed in our laboratory.[36] Intestinal absorption of ovalbumin OVA was examined in malnourished and normally nourished mice after infection with rotavirus. The malnourished infants exhibited more severe symptoms and increased number of rotavirus containing enterocytes in intestinal sections as compared to rotavirus-infected animals. The uptake of ovalbumin was rapid and showed significantly higher serum levels as compared to well-nourished or uninfected controls (Table 2).

Table 2. Serum concentration of ovalbumin (ng/ml) after oral administration to normally nourished and malnourished mice.

Time (min)	Normally nourished Serum concentration Mean ± s.e.m.		Malnourished Serum concentration Mean ± s.e.m.		p value*
0	0		0		--
5	7.8	± 1.58	36.3	± 11.63	< 0.01
15	12.6	± 5.76	29.0	± 3.10	< 0.001
30	15.7	± 3.43	29.1	± 8.15	< 0.005
45	13.2	± 6.66	21.8	± 3.85	< 0.05
60	19.6	± 5.14	46.7	± 8.22	< 0.001
90	11.9	± 2.54	27.6	± 4.82	< 0.001
180	7.6	± 2.51	29.5	± 9.17	< 0.005
360	8.4	± 3.7	20.9	± 3.07	< 0.005

* Statistical analysis by Student's t-test.

The data therefore suggest that both malnutrition and acute infection results in increased intestinal absorption of dietary macromolecules. Under conditions of malnutrition, however, the serum and tissue levels of ovalbumin were significantly high, suggesting that under conditions of protein calorie malnutrition the intestine may become more permeable to the absorption of dietary and other environmental antigens. Similar data from other animal models have also shown increased uptake of protein during starvation[37] or in children during malnutrition.[38] Malnutrition causes atrophy of the intestine leading to increased antigen uptake. The combination of rotavirus infection and protein deficiency seemed to intensify mucosal atrophy and disruption of microvilli.

Regulation of T Cell Function. The concept of humoral immune response at mucosal sites is now well established. Activated T cells are also believed to follow the path from Peyer's patches and preferentially home to the intestinal lamina propria.[39] Macrophages and mast cells participating along with lymphocytes are capable of responding to invading foreign antigens. Cell-mediated cytolysis of virus-infected cells has been reported to occur through mechanisms involving cytotoxic T cells and antibody-dependent cell-mediated cytotoxicity.

Kumagai et al.[40] have demonstrated the presence of natural killer cell activity in mucosal lymphocytes in response to experimental infection with several respiratory viruses. The role of cellular immune response in protection and recovery from rotavirus infection has, however, not been extensively investigated. Immunocompromised infants with rotavirus infection have been shown to produce chronic infection and intermittent diarrhea lasting from 6 weeks to 2 years.[41] The observations of Kumagai et al.[40] suggested the possible role of the immune system in recovery from the disease. Studies in our laboratory have been undertaken to examine the relative role of T and B cell immune responses in rotavirus infection. For this study three groups of mice were selected: (1) mice with isolated T cell deficiency (congenitally athymic nude mice), (2) mice with combined B and T cell deficiency (severe combined immunodeficiency, SCID) and (3) immunologically normal seronegative (control) suckling mice.[42,43] The three groups of infant mice were orally inoculated with mouse rotavirus and assessed for diarrhea and virus shedding. During the first 5 days post-inoculation all mice displayed 10-20 percent of antigen-positive enterocytes. By 12 days only 1-2 percent of enterocytes in nude and control mice were found to be positive and by 18 days no infected villus cells could be detected. In SCID mice however, 25 percent of enterocytes contained antigen at 7 days post-inoculation. The antigen-positive enterocytes increased to 35 percent by day 10 and dropped to 10 percent by day 22. The level of infected cells remained at 1 to 3 percent between 56 to 80 days post-inoculation.[44] Persistent fecal virus shedding was characteristic of SCID mice. The nude and control mice experienced diarrhea for 8.4 days compared to 15.5 days in SCID mice. Our findings therefore suggest that antibody response is critical for recovery from viral infection. Cell-mediated reactions at various mucosal sites could be important in recovery from certain infections once the mucosal epithelium is colonized.

Interactions Involving the Pathogen

Exposure of Inaccessible or Hidden Antigenic Determinants. The naturally acquired wild type virus or the oral vaccine virus must interact with proteolytic enzymes and other environmental components as it replicates in the gut. Such interaction may result in fragmentation of the foreign antigen by cellular proteases, and this is important for antigen presentation, processing and uptake. It has been experimentally shown that such cleavage can alter the antigenic nature of the virus.[45,46] It has been shown in the case of polioviruses that the serum antibody induced by oral administration of live poliovaccine (OPV) neutralizes the infectivity of trypsin or intestinal fluid treated poliovirus type 3 more effectively than that induced by parenteral immunization with inactivated poliovaccine alone (IPV).[45,46] Naturally acquired wild virus or immunization with OPV would therefore present different poliovirus antigenic determinants to the host immune system as compared to parenteral immunization with IPV. The development of secretory antibody responses to poliovirus virion proteins against intact or trypsin-treated poliovirus type 3 have been studied in the nasopharyngeal samples of four groups of infants immunized with either OPV, IPV-EP or a combined vaccination of IPV-EP followed by OPV [11]

Table 3. Effect of trypsin treatment of poliovirus type 3 on the detection of neutralizing antibody activity and secretory ELISA IgA antibody activity in nasopharyngeal secretions collected after immunizations with OPV[(a)] or EIPV.[(b)]

Vaccination Group (n)	Mean neutralizing antibody to poliovirus type 3 (log^2)		Mean secretory ELISA antibody titer poliovirus type 3	
	intact	cleaved	intact	cleaved
1. OPV, OPV, OPV (17)	3.2	4.6†	7.2	8.4
2. EIPV, EIPV, EIPV (23)	2.0	0.8†	4.5	0.6*
3. EIPV, OPV, OPV (20)	1.6	3.1†	6.2	7.5†
4. EIPV, EIPV, OPV	2.4	3.4	5.1	6.3

Modified from Zhaori et al.[11] The data are based on nasopharyngeal secretion samples collected after the third vaccine dose.

[(a)] OPV = live attenuated oral poliovirus vaccine.

[(b)] IPV-EP = enhanced potency inactivated poliovirus vaccine.

† p value < 0.05
* p value < 0.001

(Table 3). ELISA S-IgA and, less frequently, neutralizing antibody activity were detected in the nasopharyngeal secretions in IPV-EP- and IPV-Salk-vaccinated subjects when tested against the whole virus as compared to a decline in secretory antibody activity when tested against cleaved virus. These observations suggest that antigenic sites are exposed by trypsin cleavage in case of live oral vaccine which in turn elicits a mucosal secretory antibody response. Such antigenic determinants may not be available to the mucosal tissue with inactivated vaccine administered parenterally.

Induction of Mutations and Genetic Construct. A major concern with a number of viruses is the formation of neurovirulent mutants (revertant virus) as they replicate in the gut. This has been particularly demonstrated in the case of oral poliovaccine. The live attenuated poliovirus vaccine in current use is derived from strains developed by multiple passage of wild type viruses in nonhuman primates and in tissue culture. It has been reported that Sabin type 2 and 3 vaccine viruses may revert to neurovirulent phenotypes.[47] The types and potential sites of mutations associated with virulence are complex. Elegant studies by Evans and coworkers.[48] have shown that these mutations may include single base substitutions at either the 5' or 3' noncoding regions of the viral genome or changes in the amino acid sequences of capsid proteins.[49] However, the frequency of such reversion after different vaccination schedules has not been well defined. Recent studies[50] carried out in our laboratory have demonstrated that fecal shedding of nonvaccine-type poliovirus revertants is not uncommon after immunization with OPV, regardless of prior immunization status. The nature of the virus shed in four groups of infants immunized with OPV, EIPV, or a combined vaccination of OPV and EIPV, has been analyzed (Table 4). In group 1 receiving only one OPV dose, 2 out of 3 (67%) isolates were nonrevertants. In group 2 receiving two doses of OPV, 2 out of 3 (67%) isolates were found to be revertants. In groups 3 and 4 receiving doses of EIPV followed by OPV over 90 percent of isolates were found to be revertants. The frequency of reversion for different serotypes was examined. All type 2 poliovirus isolates tested were revertant types irrespective of the immunization group. Significantly, all poliovirus type 3 revertants were observed in subjects previously immunized with EIPV.

Table 4. Distribution of revertant and nonrevertant vaccine poliovirus isolated from feces after different schedules of vaccination.

Prior Immunization Status (n)	Fecal Virus Shedding[a]							
	Revertants serotype				Non revertants serotype			
	1	2	3	%	1	2	3	%
1. None	1	1	0	33.3	2	0	2	66.7
2. OPV × 1 (3)	1	1	0	66.7	1	0	0	33.3
3. EIPV × 1 (7)	3	3	1	100.0	0	0	0	0.0
4. EIPV × 2 (5)	0	1	3	80.0	1	0	0	20.0

Modified from Friehorst et al.[35]

[a] Fecal virus shedding 30-60 days after last OPV dose.

Nasopharyngeal antibody response was observed more significantly in subjects shedding revertants in the feces, irrespective of prior immunization with OPV or EIPV. It is, therefore, possible that formation of wild virulent strains during replication of OPV in the gut may provide a more potent stimulus for induction of mucosal immune response than attenuated strains of vaccine viruses.

CONCLUSION

It is evident from the observations summarized above that the mucosal microenvironment serves as an efficient barrier between the external and internal milieu. Exposure of mucosal surfaces to a variety of infectious agents or viral vaccines stimulates both the gut (GALT) and bronchus- associated lymphoid tissue (BALT). The outcome of such host-pathogen interactions is complex and is determined by both the nature of the mucosal environment and the effect of the environment on the host and pathogen. The primary effect of viral or bacterial infection is the appearance of antigen specific S-IgA which confers varying degrees of protection against subsequent reinfection challenges. In addition, cell-mediated immune responses can also act along with specific antibody at mucosal surfaces to provide local protection.

The immune response evoked at mucosal surfaces is directed against different viral proteins and antigenic determinants and is superior to the response induced after parenteral non-mucosal exposure to the virus. Oral poliovaccines have been shown to be superior to inactivated poliovaccine in inducing nasopharyngeal secretory immune responses as antigenic sites are exposed by intestinal trypsin cleavage during the replication of OPV in the gut. A combination of OPV and EIPV has been found to be effective in inducing systemic and secretory antibody response and in reducing the incidence of vaccine-associated poliomyelitis. It, however, affords no protection against generation of non-attenuated virulent revertants in vaccinees.

The murine model has been successfully used to study the kinetics of viral infection in relation to mucosal barrier function. It has been demonstrated that the mucosal tissue becomes more permeable to specific proteins and antigens under conditions of acute viral infection with rotavirus or respiratory syncytial virus. The mechanisms underlying these alterations are not known. Possible explanations may include the development of mucosal damage during acute viral infection, alterations in antigen processing, changes in mucosal permeability and receptor-mediated pinocytosis. Virus specific secretory antibody present in mucosal tissue is capable of modifying viral infection and completely preventing it. The immune system of the inoculated or infected animal is, however, important in recovery from infection as shown from the results of SCID and nude mice.

ACKNOWLEDGEMENT

The authors wish to thank Sandra Fuentes for her excellent secretarial assistance in the preparation of the manuscript.

REFERENCES

1. L.A. Hanson and P. Brandtzaeg, The mucosal defense system, in "Immunological Disorders in Infants and Children", E.R. Stiehm, ed., W.B. Saunders, Philadelphia (1989).

2. C. Czerkinsky, S.J. Prince, S.M. Michalek, S. Jackson, M.W. Russell, Z. Moldoveanu, J.R. McGhee, and J. Mestecky, IgA antibody producing cells in peripheral blood after antigen ingestion: Evidence for a common mucosal immune system in humans, *Proc. Natl. Acad. Sci. USA.* 84:2449 (1987).

3. M.R. McDermott and J. Bienenstock, Evidence for a common mucosal immunologiz system: I. Migration of B immunoblasts into intestinal, respiratory, and genital tissues, *J. Immunol.* 122: 1892 (1979).

4. P. Weisz-Carrington, M.E. Roux, M. McWilliams, J.M. Phillips- Quagliata,and M.E. Lamm, Organ and isotype distribution of plasma cells producing specific antibody after oral immunization: Evidence for a generalized secretory immune system, *J. Immunol.* 123: 1705 (1979).

5. P.L. Stark and A. Lee, The microbial ecology of the large bowel of breast-fed and bottle-fed infants during the first year of life, *J. Med. Microbiol.* 15: 189 (1982).

6. L.J. Mata, J.J. Urrutia, and A. Lechtig, Infection and nutrition of hildren of a low socioeconomic rural community, *Am. J. Clin. Nutr.* 24: 249 (1971).

7. J.L. Rasic and J.A. Kurmann, Bifidobacteria and their role, *Experientia Suppl.* 39: 1, (1983).

8. S.J. Henning, Otogeny of enzymes in the small intestine. *Ann. Rev. Physiol.* 47: 231 (1985).

9. W.A. Walker, Absorption of protein and protein fragments in the developing intestine: Role in immunologic/allergic reactions, *Pediatrics.* 75 (suppl): 167 (1985).

10. E. Lebenthal and Y.K. Leung, Feeding the premature and compromised infant: Gastrointestinal considerations, *Pediatric Clin North America* 35: 215, (1988).

11. G. Zhaori, M. Sun, H.S. Faden, and P.L. Ogra, Nasopharyngeal secretory antibody response to poliovirus Type 3 virion proteins exhibit different specificities after immunization with live or inactivated poliovirus vaccines, *J. Infect. Dis.* 159: 1018 (1989).

12. L.C. Duffy, M. Riepenhoff-Talty, T.E. Byers, L.J. La Scolea, M. A. Zielezny, D.M. Dryja, and P.L. Ogra, Modulation of rotavirus enteritis during breast feeding: Implications on alternation in the intestinal bacterial flora. *Am. J. Dis. Child.* 40: 1164 (1986).

13. P.L. Ogra and D.H. Dayton, "Immunology of Breast Milk", Raven Press, New York (1979).

14. P.L. Ogra, Otogeny of the local immune system, *Pediatrics* 64: 765 (1979).

15. C.O. Elson, M.F. Kagnoff, C. Fiocchi, A.D. Befus, S. Targan, Intestinal immunity and inflammation: Recent progress, *Gastroenterology.* 91: 746 (1986).

16. W.A. Walker, and K.J. Isselbacher, Uptake and transport of macromolecules by the intestine: Possible role in clinical disorders, *Gastroenterology.* 67: 531 (1974).

17. B. Haneberg, and D. Aarskog, Human faecal immunoglobulins in healthy infants and children, and in some with diseases affecting the intestinal tract of the immune system, *Clin. Exp. Immunology.* 22: 210 (1975).

18. C. Cunningham-Rundles, W.E. Brandeis, R.A. Good, and N.K. Day, Bovine antigens and the formation of circulating immune complexes in selective immunoglobulin A deficiency, *J Clin Invest* 64: 272 (1979).

19. G. Cukor and N.R. Blacklow, Human viral gastroenteritis, *Microbiol Rev.* 48:157 (1984).

20. R.E. Black, K.H. Brown, and S. Becker, Malnutrition is determining factor in diarrheal duration, but not incidence, among young children in a longitudual study in rural Bangladesh, *Am. J. Clin. Nutr.* 39:87 (1984).

21. G.P. Davidson and G.L. Barnes, Structural and functional abnormalities of the small intestine in infants and young children with rotavirus enteritis, *Acta. Pediatr. Scand.* 68: 181 (1979).

22. F.E. Viteri, and R.E. Schneider, Gastrointestinal alterations in protein-calorie malnutrition, *Med. Clin. North. Am.* 58:1487 (1974).

23. E. Offor, M. Reipenhoff-Talty, and P.L. Ogra, Effect of malnutrition on rotavirus infection in suckling mice: Kinetics of early infection, *Proc. Soc. Exp. Biol. Med.* 178:85 (1985).

24. M. Riepenhoff-Talty, E. Offor, K. Klossner, E. Kowalski, P.J. Carmody, and P.L. Ogra, Effect of age and malnutrition on rotavirus infection in mice, *Pediatr. Res.* 19:1250 (1985).

25. R.L. Noble, R.W. Sidwell, A.W. Mahoney, B.B. Barnett, and R.S. Spendlove, Influence of malnutrition and alterations in dietary proteins on murine rotaviral disease, *Proc. Soc. Exp. Biol. Med.* 173:417 (1983).

26. J. Takano, Intestinal changes in the protein-deficient rats, *Exp. Mol. Pathol.* 3:224 (1964).

27. I. Uhnoo, M. Riepenhoff-Talty, P. Chegas, J.E. Fisher, H.B. Greenberg, and P.L.Ogra, Effect of malnutrition on extraintestinal spread of rotavirus and development of hepatitis in mice, *Nutrition Research* 10:1419 (1990).

28. P.L. Ogra, Mucosal immune response to poliovirus vaccines in childhood, *Rev. Infect. Dis.* 6 (Suppl 2), 361 (1984).

29. P.L. Ogra, M. Fishaut, and M.R. Gallagher, Viral vaccination via the mucosal route, *Rev. Infect. Dis.* 2:352 (1980).

30. G. Zhaori, M. Sun, and P.L. Ogra, Characteristics of the immune response to poliovirus virion polypeptides after immunization with live or inactivated polio vaccines, *J. Infect. Dis.* 158: 160 (1988).

31. E.A. Emini, B.A. Jameson, A.J. Lewis, G.R. Larsen, and E. Wimmer, Poliovirus neutralization epitopes: analysis and localization with neutralizing monoclonal antibodies, *J. Virol.* 43:997 (1982).

32. D.M. Evans, P.D. Minor, G.S. Schild, and J.W. Almond, Critical role of an eight-amino acid sequence of VP1 in neutralization of poliovirus type 3, *Nature* (London) 304:459 (1983).

33. R.E. Gordon, B.W. Case, and J. Kleinerman, Acute NO_2 effects on penetration and transport of horseradish peroxidase in hamster respiratory epithelium, *Amer. Rev. Respir. Dis.* 128:528 (1983).

34. D. Rothman, M.C. Latham, and W.A. Walker, Transport of macromolecules in malnourished animals. I. Evidence for increased uptake of intestinal antigens, *Nutr. Res.* 2:467 (1982).

35. J. Freihorst, P.A. Piedra, Y. Okamoto, and P.L. Ogra, Effects of respiratory syncytial virus infection on the uptake of and immune response to other inhaled antigens, *Proc. Soc. Exp. Biol. Med.* 188:191 (1988).

36. I.S. Uhnoo, J. Freihorst, M. Riepenhoff-Talty, J.E. Fisher, and P.L. Ogra, Effect of rotavirus infection and malnutrition on uptake of a dietary antigen in the intestine, *Ped. Res.* 27:153 (1990).

37. D. Rothman, J.N. Udall, K.Y. Pang, S.E. Kirkham, and W.A. Walker, The effect of short-term starvation on mucosal barrier function in the newborn rabbit, *Pediatr. Res.* 19:727 (1985).

38. M. Heyman, G. Boudraa, S. Sarrut, M. Giraud, L. Evans, M. Touhami, and J.F. Desjeux, Macromolecular transport in jejunal mucosa of children with severe malnutrition, A quantitative study, J. Pediatr. *Gastorenterol. Nutr.* 3:357 (1984).

39. S.R. Targan, M.F. Kagnoff, M.D. Brogan, F. Shanahan, Immunologic mechanisms in intestinal diseases, *Ann. Intern. Med.* 106:853 (1987).

40. T. Kumagai, D.T. Wong, and P.L. Ogra, Development of cell-mediated cytotoxic activity in the respiratory tract after experimental infection with respiratory syncytial virus, *Clin. Exp. Immunol.* 61:351 (1985).

41. G.A. Losonsky, J.P. Johnson, J.A. Winkelstein, and R.H. Yolken, Oral administration of human serum immunoglobulin in immunodeficient gpatients with viral gastroenteritis. *J. Clin. Invest.* 76:2362 (1985).

42. M. Riepenhoff-Talty, T. Dharakul, E. Kowalski, S. Michalak, and P.L. Ogra, Persistent rotavirus infection in mice with severe combine immunodeficiency, *J. Virol.* 61:3345 (1987).

43. M. Riepenhoff-Talty, T. Dharakul, E. Kowalski, D. Sterman, and P.L.Ogra, Rotavirus infection in mice: pathogenesis and immunity in: "Recent Advances in Mucosal Immunology" J. Mestecky, J.R. McGhee, J. Bienenstock, P.L. Ogra ed, Plenum Press, New York. 1987.

44. I. Uhnoo, T. Dharakul, M. Riepenhoff-Talty, and P.L. Ogra, Immunological aspects of interaction between rotavirus and the intestine in infancy, *Immunol. Cell. Biol.* 66:135 (1988).

45. M. Roivainen and T. Hovi, Intestinal trypsin can significantly modify antigenic properties of polioviruses: Implications for the use of inactivated poliovirus vaccine. *J. Virol.* 61:3749 (1987).

46. M. Roivainen and T. Hovi, Cleavage of VP1 and modification of antigenic site 1 of Type 2 polioviruses by intestinal trypsin, *J. Virol.* 62:3536 (1988).

47. K.L. Burke, G. Dunn, M. Ferguson, P.D. Minor, and J.W. Almond, Antigen chimeras of poliovirus as potential new vaccines, *Nature* (London) 332:81 (1988).

48. D.M.A Evans, P.D. Minor, G.S. Schild, and J.W. Almond, Critical role of an eight-amino acid sequence of VP1 in neutralization of poliovirus type 3, *Nature* (London) 304:459 (1983).

49. D.M.A. Evans, G. Dunn, P.D. Minor, G.C. Schild, A.J. Cann, G. Stanway, J.W. Almond, K. Currey, and J.V. Maizel, Jr., Increased neurovirulence associated with a single nucleotide change in noncoding region of the Sabin Type 3 poliovaccine genome, *Nature* (London). 314:548 (1985).

50. P.L. Ogra, H.S. Faden, R. Abraham, L.C. Duffy, M. Sun, and P.D.Minor, Effect of prior immunity on the shedding of virulent revertant virus in feces after oral immunization with live attenuated poliovirus vaccine, *J. Infect. Dis.* 164:191 (1991).

INDUCTION OF T HELPER CELLS AND CYTOKINES FOR MUCOSAL IgA

RESPONSES

Jiangchun Xu-Amano, Kenneth W. Beagley, Junichi Mega[1], Kohtaro Fujihashi, Hiroshi Kiyono and Jerry R. McGhee

The Immunobiology Vaccine Center and the Departments of Microbiology, Oral Biology, Medicine, University of Alabama at Birmingham, Birmingham, AL 35294

[1]The Department of Crown and Bridge, School of Dentistry at Matsudo, Nihon University, Matsudo, Chiba, Japan

INTRODUCTION

The immune system of mucosal tissues can be divided into two interrelated compartments according to their functions in mucosal immune responses: 1) the sites where immune responses are induced, including the gut-associated and bronchus-associated lymphoreticular tissues (GALT and BALT); and 2) the larger mucosal areas where effector functions (e.g., local IgA responses and T cell-mediated responses occur including the lamina propria (LP) of the gastrointestinal (GI) tract and exocrine tissues such as the salivary glands. Distinct features of mucosal effector tissues (e.g., LP or salivary glands) include high frequencies of T and B cells, including plasma cells, most of which produce IgA, and the presence of an epithelial cell barrier (see Mega et al., this volume).[1] The epithelial cells which cover mucosal effector sites also produce the polymeric immunoglobulin receptor, secretory component (SC).[1] The polymeric (usually dimeric) IgA is synthesized locally in lamina propria or glandular acinar regions, and the IgA-producing plasma cells in these effector sites represent > 80% of the total plasma cells present. The locally produced polymeric IgA is transported via SC across the epithelium and released at apical surfaces as secretory IgA (S-IgA) antibodies.

The GALT are comprised of Peyer's patches (PP), appendix and smaller lymphoid follicles, and all are discrete aggregates of lymphoid tissues that represent major IgA inductive sites in mammals.[1] PP are unique in that a high proportion of B cells express surface IgA (sIgA+) and are thus committed to this isotype. PP contain all necessary immunocompetent cells required for IgA responses, and it has been shown that PP cell cultures from mice orally primed with sheep erythrocytes (SRBC) support IgM, IgG, and largely IgA anti-

SRBC antibody responses.[2,3] Major questions remain as to how the immune cells (i.e., B and T cells) and antigen-presenting cells interact in the inductive sites to ultimately provide for the immune responses which occur in distant effector tissues such as the salivary glands and why these responses are mainly restricted to the IgA isotype.

It is well established that CD4[+] Th cell clones can be divided into two distinct types, Th1 and Th2, based upon their profiles of cytokine production.[4-6] Murine Th1 cell clones produce IL-2, IFN-γ and lymphotoxin (TNF-β) upon antigen or mitogen stimulation, while established Th2 cell clones preferentially synthesize IL-4, IL-5, IL-6 and IL-10.[4-6] Both types of Th cell clones produce IL-3, GM-CSF and other cytokines. The mechanisms which regulate the development of Th cells into Th1- and Th2-type cells have not been clearly identified. Furthermore, the cytokines produced by the two types of Th cell clones regulate B cell responses differently. Cytokines produced by Th2-type cells (e.g., IL-5 and IL-6) have been shown to play an important role in the regulation of IgA-committed B cell responses.

To better understand the mechanisms involved in mucosal immune responses, we have briefly summarized our recent studies on the induction and regulatory functions of Th1 and Th2 cells in IgA-inductive sites when compared with T cells obtained from systemic lymphoid tissues. These studies have also included analysis of B cell subsets which respond to IL-5 and IL-6 for IgA synthesis. Our recent studies have shown that CD4[+] Th cells are required for mucosal IgA responses *in vivo* and, collectively, these studies suggest that careful analysis of mucosal vaccine delivery systems will require an understanding of the role of Th cell subsets and their derived cytokines for the induction of optimal immune protection in mucosal effector tissues.

SELECTIVE INDUCTION OF Th CELL SUBSETS IN MURINE PP BY ORAL IMMUNIZATION

Past work has shown that Th1 and Th2 clones regulate immune responses in an isotype-specific manner. For example, allogeneic or antigen stimulated Th1 cell clones predominantly support IgG2a synthesis in B cell cultures, an effect which is mainly due to the production of IFN-γ.[7] Further, this IgG2a response was enhanced in a synergistic manner by IL-2.[8,9] On the other hand, Th2 cells and derived cytokines supported IgG1, IgE and IgA B cell responses.[7,10-13] Th2 cell-derived and recombinant IL-4 have been shown to preferentially induce IgG1 and IgE synthesis in LPS-stimulated B cell cultures.[15,16] Studies have also shown that IL-5 enhanced IgA secretion in both LPS-treated splenic B cells[17-20] and in non-stimulated PP B cell cultures.[21] IL-6 was shown to induce terminal differentiation of activated human B cells to Ig-secreting plasma cells.[22] Recombinant IL-6 induced murine PP sIgA[+] B cells to become IgA-secreting plasma cells,[23] and other studies showed that the addition of this cytokine to human appendix B cell cultures resulted in the synthesis of both IgA1 and IgA2 subclasses.[24]

It is now well established that Th1 and Th2 cells can down-regulate each other by the production of specific cytokines. For example, IFN-γ produced by Th1 cells was shown to inhibit the induction of Th2 cells,[25,26] while IL-10, secreted by Th2 cells, suppressed the functions of Th1 cells via accessory cells.[27] Thus, the frequency of Th1 or Th2 cells in IgA-inductive sites (e.g., the PP) may influence the isotype specific responses that occur in mucosal tissues. Although the cytokine response patterns are distinct for long-term Th cell clones, the requirements for the induction and development of Th1 and Th2 cells are not yet clear. Furthermore, only limited information is available for the regulatory functions of Th1 and Th2 cells at IgA-inductive (e.g., PP) and effector (e.g., LP or salivary glands) sites. It is tempting to propose that PP may be sites where higher frequencies of Th2-type cells are induced and subsequently migrate to mucosal effector sites. Their subsequent activation would induce them to produce Th2-type cytokines (e.g., IL-5 and IL-6) for mucosal IgA responses. Cognate interactions between Th2-sIgA$^+$ B cells and secreted IL-5 and IL-6 may in turn induce the differentiation of PP sIgA$^+$ B cells to become IgA-secreting plasma cells.

Our recent studies have shown that CD4$^+$ T cells freshly isolated from PP of normal mice contained low numbers of IFN-γ and IL-5 spot forming cells (SFC) with an approximate equal frequency of IFN-γ and IL-5 producing T cells.[28] On the other hand, T cells from IgA effector regions, such as the LP lymphocytes (LPL) revealed large numbers of both IL-5 and IFN-γ SFC with an overall higher frequency of IL-5-producing T cells.[28] The appearance of equal frequencies of Th1- and Th2-type T cells in PP may seem to refute the idea that PP contain higher frequencies of Th2-type cells for regulation of IgA responses. Nevertheless, the above studies were performed in normal, naive mice whose PP had not received deliberate antigen stimulation. In this regard, it has been shown that appropriate oral immunization of mice with sheep erythrocytes (SRBC) induced CD4$^+$ Th cells in PP.[2,3] Furthermore, Th cell clones from this tissue preferentially induced sIgA$^+$ B cells to terminally differentiate into IgA-producing plasma cells.[29]

It was, therefore, of importance to study the nature of these CD4$^+$ T cells, especially the frequency of Th1- and Th2-type cells in PP following oral immunization with a T cell-dependent (TD) antigen. For these studies, mice were orally immunized with SRBC for 3 consecutive days, and as controls, other groups of mice were systemically immunized with SRBC by the intraperitoneal (I.P.) route.[30] One week following immunization, mice were sacrificed and T cell subsets from both PP and spleen (SP) assessed. The PP and SP CD3$^+$ and CD3$^+$, CD4$^+$ T cell subsets were isolated and cultured with SRBC, feeder cells and IL-2 and were tested at various intervals (day 0, 1, 3 and 6) for numbers of T cells producing either IFN-γ or IL-5 by use of an enzyme linked immunospot (ELISPOT) procedure to detect these two cytokines. Cultures of T cells from PP or SP of mice given SRBC by the oral route had a high frequency of IL-5 SFC, with lower numbers of IFN-γ SFC.[30] However, cultures of CD3$^+$ T cells and CD3$^+$, CD4$^+$ Th cells from SP of I.P. immunized mice exhibited predominantly IFN-γ SFC, with smaller but significant numbers of IL-5 SFC (Table 1).[30] These results clearly showed that Th2-type (IL-5 SFC) responses are preferentially induced in PP following oral administration with the TD antigen, SRBC.[30]

Table 1. Frequencies of IFN-γ (Th1) and IL-5 (Th2) SFC in PP and SP of mice given SRBC by either oral or I.P. routes.

Cytokines Tested	Days in Culture	Numbers of SFC /10^6 Cells and Route Of Antigen Delivery		I.P. Immunization[a]
		Oral Immunization		
		PP CD4$^+$	SP CD4$^+$	SP CD4$^+$
IFN-γ (Th1-type)	0	501 ± 129	100 ± 10	2,250 ± 296
	1	1,210 ± 174	2,075 ± 255	3,375 ± 111
	3	2,050 ± 544	4,500 ± 723	9,310 ± 512
	6	5,375 ± 674	5,150 ± 646	13,700 ± 1,088
IL-5 (Th2-type)	0	1,750 ± 233	400 ± 20	900 ± 191
	1	2,850 ± 171	1,450 ± 340	1,410 ± 205
	3	7,275 ± 632	3,400 ± 1,227	3,850 ± 386
	6	19,200 ± 616	7,250 ± 561	6,250 ± 178

[a] Separate analysis of PP T cell cultures from mice immunized with SRBC by the I.P. route gave less than 200 SFC /10^6 cells at all time points.

Further support for distinct Th1-type and Th2-type responses following systemic (I.P.) or oral immunization, respectively, was obtained by use of IFN-γ- and IL-5-specific cDNA probes and mRNA analysis. In these studies, appropriate aliquots of CD4$^+$ Th cells were obtained from cultures at the indicated times and the RNA was extracted and assessed by dot-blot analysis. The intensity of the blot was expressed as a relative gray scale of 0-200 (Table 2). In these studies, higher levels of IFN-γ mRNA were obtained from CD4$^+$ Th cells from I.P. immunized mice, while higher IL-5-specific mRNA was associated with PP and SP CD4$^+$ Th cells from mice orally immunized with SRBC (Table 2). Further, the respective mRNA levels increased with time in culture, clearly indicating outgrowth of Th1-type and Th2-type cells in the presence of antigen in I.P.-immunized or orally immunized mice, respectively. These studies provide direct proof that antigen-specific Th2 cells are preferentially induced in PP of mice given SRBC by the oral route.[30]

We have analyzed the frequencies of Th1 and Th2 clones with specificity for antigen in mice immunized by either the systemic or oral routes. Of interest was the finding that 74% of Th cell clones isolated from PP of mice given SRBC orally were IL-5 producers (Th2-type), while two-thirds of Th cell clones from SP of I.P. immunized mice were IFN-γ producers (Th1-type) (Table 3). These results now confirm, by clonal analysis, that oral administration of SRBC preferentially induces Th2 cell responses in GALT. Thus, it is likely that the induction of Th2 cells producing cytokines that favor IgA responses contributes to the preponderance of this isotype in mucosal sites.

Table 2. Analysis of IFN-γ (Th1) and IL-5 (Th2)-specific mRNA in CD4$^+$ T cell cultures of mice given SRBC by either the oral or I.P. routes.

Cytokines Tested	Days In Culture	mRNA Intensity as Relative Gray Scale (0-200)		
		Oral Immunization		I.P. Immunization
		PP CD4$^+$	SP CD4$^+$	SP CD4$^+$
IFN-γ (Th1-type)	0	20.6	9.7	35.8
	1	26.7	9.0	38.0
	3	41.8	16.8	47.2
	6	47.5	28.3	113.8
IL-5 (Th2-type)	0	78.8	60.4	26.9
	1	100.6	80.2	56.6
	3	124.1	102.5	42.7
	6	150.2	122.7	76.4

Table 3. Frequencies of Th1 and Th2 cell clones from SP or PP of mice immunized by either the I.P. or oral routes.

Source of CD4$^+$ Th Clones	Cytokine Tested	Frequency of IFN-γ- or IL-5-Producing Clones (Percent)
PP (oral)	IFN-γ	18
	IL-5	74
SP (I.P.)	IFN-γ	67
	IL-5	30

REGULATORY FUNCTIONS OF Th2-DERIVED CYTOKINES ON PP B CELL SUBSETS FOR INDUCTION OF IgA SYNTHESIS

The above studies have clearly shown that oral immunization with a TD antigen preferentially induced antigen-specific Th2-type cells in PP. The regulatory functions of Th2 cell-derived cytokines for IgA responses are also under investigation. Murine PP are organized secondary lymphoid tissues which architecturally consist of three main regions: 1) the dome area underneath an epithelium where specialized epithelial microfold cells (M cells) actively endocytose antigens from the lumen and deliver them into the underlying dome region; 2) lymphoid follicles or B cell zones, each with one or two germinal centers (GC), which contain cells undergoing active B cell proliferation and which avidly bind the lectin *Arachis hypogea* (peanut agglutinin; PNA); and 3) a parafollicular or thymus-dependent area (TDA) containing all major CD3-TCR+ T cell subsets.[1]

PP B cells respond to an array of cytokines which regulate their activation, isotype switching, proliferation, and differentiation (reviewed in 1). Among these cytokines, transforming growth factor-beta (TGF-β) has been shown to induce heavy-chain class switching to the IgA isotype in mitogen-activated splenic B cell cultures from both mice[31-33] and humans.[34,35] Further, the Th2 cytokines IL-4, IL-5 and IL-6 can act sequentially on B cells in at least three major and distinct stages: activation of quiescent B cells, proliferation, and terminal differentiation into plasma cells. Previous work from our group showed that murine recombinant IL-6 induced murine PP sIgA+ B cells to become IgA-secreting plasma cells.[23]

In our most recent studies, we have analyzed the kinetics of B cell responses to IL-6 and their dependence on *de novo* DNA and RNA synthesis for increased IgA secretion.[36] Addition of rIL-6 to cultures of PP B cells induced a six- to eightfold increase in IgA secretion in 7-day cultures; more than 80% of the IgA present was produced within the first 72 hours of culture.[36] Further, rIL-6 induced four- to fivefold increases in IgA SFC within 24 hours, with significant increases observed as early as 4-8 hours of culture. These results strongly suggest that the increase in IgA secretion was the result of IL-6-induced B cell differentiation.[36] This assumption was correct since culture of B cells with IL-6 resulted in no net increase in [^{3}H]-thymidine incorporation when compared with untreated controls. Further, treatment with the DNA synthesis inhibitor mitomycin C had no effect on the IL-6-induced increase in SFC numbers, while the RNA synthesis inhibitor actinomycin D totally abolished the increased IgA secretion.[36] These results clearly indicated that IL-6 induces PP B cells to terminally differentiate into IgA plasma cells.

The majority of sIgA+ B cells in the PP occur in the GC. Separation of PP B cells into PNA^{Hi} (GC) and PNA^{Lo} (non-GC) subpopulations prior to culture with IL-6 showed that only the PNA^{Lo} B cells transcribe increased levels of α mRNA and secrete high levels of IgA in response to this cytokine (Table 4). These results suggest that the PNA^{Lo}, sIgA+ B cells are sensitive to Th2-derived cytokines and may represent the immediate precursors of IgA plasma cells found in the lamina propria of the GI tract.[36] Furthermore, our results indicate that the GC B cells, which are enriched for sIgA+ B cells, are less sensitive to cytokines which induce terminal differentiation. It is tempting to speculate that GC are major sites for $\mu \rightarrow \alpha$ switches which occur prior to responses to antigen. It would follow that post-switched sIgA$^+$ B cells become responsive to antigen and Th2 cell help prior to migration to mucosal effector sites.

Table 4. Effect of IL-6 on IgA Synthesis by PP B Cells Separated into PNA^{Hi} and PNA^{Lo} Subsets.

| Experiment | PNA Subset | IL-6 In Culture (50 U/ml) | Secreted Immunoglobulin (ng/ml) | | | α mRNA Transcript |
			IgA	IgM	IgG	
One	PNA^{Hi}	-	56 ± 12	<4	22 ± 3	-
		+	225 ± 36	<4	85 ± 5	±
	PNA^{Lo}	-	366 ± 18	7 ± 3	37 ± 3	±
		+	3,064 ± 686	48 ± 7	188 ± 30	+++
Two	PNA^{Hi}	-	183 ± 37	90 ± 84	22 ± 6	-
		+	1,385 ± 278	73 ± 55	86 ± 9	±
	PNA^{Lo}	-	638 ± 55	165 ± 80	23 ± 3	+
		+	7,889 ± 900	214 ± 84	118 ± 10	+++

IMPORTANCE OF CD4$^+$ Th CELLS IN THE REGULATION OF MUCOSAL IgA RESPONSES

The induction and regulatory functions of Th2- type cells and their secreted cytokines in murine PP have already been discussed. Several past studies have shown that CD4$^+$ Th cells support IgA responses *in vitro*. However, few studies have evaluated the effects of *in vivo* suppression of CD4$^+$ Th cells on mucosal cell functions including IgA responses. Early studies had shown that neonatally thymectomized rabbits or rats produced few or no IgA antibodies.[37,38] Athymic (nude) mice displayed low levels of serum IgA (and IgG subclasses) and low IgA plasma cell numbers in mucosal effector sites.[39,40] These results indirectly suggested the importance of thymus-derived T cells in the regulation of IgA responses, but could not address the importance of T cell subsets including CD4$^+$ Th cells for induction of IgA plasma cells.

Our recent studies have now shown that CD4$^+$ T cells can be depleted in both inductive (PP) and IgA effector (LP) sites by weekly injections of anti-L3T4 (CD4) monoclonal antibody (mAb) to young BALB/c mice. The numbers of IgA plasma cells were dramatically decreased in small intestine and the LPL of anti-L3T4 mAb-treated mice showed an 80% reduction in the numbers of IgA SFC.[41] Further, the depletion of CD4$^+$ Th cells in PP resulted in the reduction of overall size of the PP as well as markedly reduced GC in this tissue. The numbers of spontaneous Ig-producing cells, including IgM, IgG and IgA, were dramatically suppressed. However, the relative frequency of sIgA+ B cells in this tissue did not change. These results suggest that the loss of Ig-secreting cells in PP and the LPL is due to the absence of CD4+ T cell help which would result in a net loss of Th2 cells and their derived cytokines for the induction of IgA plasma cell responses.[41]

Taken together, the results briefly summarized here point to a central role for CD4+ Th cells in IgA responses. In this regard, the PP are important reservoirs for B cells that commit to IgA and also represent sites where Th2 cells are preferentially induced. An environment enriched for Th2 cells and derived cytokines such as IL-5 and IL-6 offers optimal support for the preferential induction of IgA plasma cell responses in mucosal effector tissues.

SUMMARY

CD4+ Th cells and their derived cytokines play an important role in the regulation of IgA responses in the mucosal immune system. Th1 and Th2 cells induce different Ig isotype and IgG-subclass responses. Further, cytokines produced by Th2-type cells (e.g., IL-5 and IL-6) have been shown to induce PP sIgA+ B cells to secrete IgA. Our studies have now shown that oral immunization with SRBC selectively induces Th2-type cells in PP while systemic (I.P.) immunization with SRBC predominantly induces Th1-type cells. It is tempting to suggest that Th2 cells which produce IL-5 and IL-6 tend to be predominant in mucosal effector regions, such as the salivary glands and LP tissues and account for the predominant IgA responses which characterize these tissues. The PP contain B cell subsets which respond to IL-5 and IL-6, and these are largely restricted to the PNA^{Lo} non-GC (memory) sIgA+ B cells. The importance of CD4+ Th cells in the regulation of IgA responses has also been shown by the depletion of CD4+ Th cells in anti-L3T4 (CD4)-treated mice. Loss of CD4+ Th cells from mucosal tissues resulted in dramatically decreased numbers of IgA plasma cells in the small intestine and led to a reduction in IgA SFC in isolated LP cells. The overall size of PP was reduced and the GCs were absent; however, the relative frequency of sIgA+ B cells in PP did not change, possibly suggesting that CD4+ Th cells do not influence switches to IgA.

ACKNOWLEDGMENTS

The results summarized in this review were supported by U.S. Public Health Service grants DE 04217, AI 18958, DE 09837, AI 19674, DK 44240, DE 08182, DE 08228 and NIAID contract AI 15128. Hiroshi Kiyono is the recipient of NIH-NIDR RCDA DE 00237. We thank Ms. Debra Clisby for the preparation of this paper.

REFERENCES

1. J.R. McGhee, J. Mestecky, C.O. Elson, and H. Kiyono, Regulation of IgA synthesis and immune response by T cells and interleukins, *J. Clin. Immunol.* 9:175 (1989).
2. H. Kiyono, J.R. McGhee, M.J. Wannemuehler, M.V. Frangakis, D.M. Spalding, S.M. Michalek, and W.J. Koopman, *In vitro* immune responses to a T cell-dependent antigen by cultures of disassociated murine Peyer's patch, *Proc. Natl. Acad. Sci. U.S.A.* 79:596 (1982).
3. H. Kiyono, L.M. Mosteller, J.H. Eldridge, S.M. Michalek, and J.R. McGhee, IgA responses in *xid* mice: oral antigen primes Peyer's patch cells for *in vitro* immune responses and secretory antibody production, *J. Immunol.* 131:2616 (1983).
4. T.R. Mosmann, H. Cherwinski, M.W. Bond, M.A. Giedlin, and R.L Coffman, Two types of murine helper T cell clone. 1. Definition according to profiles of lymphokine activities and secreted proteins, *J. Immunol.* 136:2348 (1986).
5. T.R. Mosmann, and R.L. Coffman, Two types of mouse T helper T cell clone: implications for immune regulation, *Immunol. Today* 3:223 (1987).
6. T.R. Mosmann, and R.L. Coffman, Th1 and Th2 cells: different patterns of lymphokine secretion lead to different functional properties, *Annu. Rev. Immunol.* 7:145 (1989).
7. R.L. Coffman, B.W. Seymour, D.A. Lebman, D.D. Hiraki, J.A. Christiansen, B. Shrader, H.M. Cherwinski, H.F.J. Savelkoul, F.D. Finkelman, M.W. Bond, and T.R. Mosmann, The role of helper T cell products in mouse B cell differentiation and isotype regulation, *Immunol. Rev.* 102:5 (1988).
8. H.J. Leibson, M. Gefter, A. Zlotnik, P. Marrack, and J.W. Kappler, Role of γ-interferon in antibody-producing responses, *Nature* 309:799 (1984).
9. C.M. Snapper, C. Peschel, and W.E. Paul, IFN-γ stimulates IgG2a secretion by murine B cells stimulated with bacterial lipopolysaccharide, *J. Immunol.* 140:2121 (1988).
10. T.L. Stevens, A. Bossie, V.M. Sanders, R. Fernandez-Botran, R.L. Coffman, T.R. Mosmann, and E.S. Vitetta, Regulation of antibody isotype secretion by subsets of antigen-specific helper T cells, *Nature* 334:255 (1988).
11. R.L. Coffman, and J. Carty, A T cell activity that enhances polyclonal IgE production and its inhibition by interferon-γ, *J. Immunol.* 136:949 (1986).
12. D.A. Lebman, and R.L. Coffman, Interleukin 4 causes isotype switching to IgE in T cell-stimulated clonal B cell cultures, *J. Exp. Med.* 168:853 (1988).
13. J.E. Layton, E.S. Vitetta, J.W. Uhr, and P.H. Krammer, Clonal analysis of B cells induced to secrete IgG by T cell-derived lymphokine(s), *J. Exp. Med.* 160:1850 (1984).
14. S. Lutzker, P. Rothman, R. Pollock, R. Coffman, and F.W. Alt, Mitogen- and IL-4-regulated expression of germ-line Igδ2b transcripts: evidence for directed heavy chain class switching, *Cell* 53:177 (1988).

15. R.L. Coffman, J. Ohara, M.W. Bond, J. Carty, A. Zlotnik, and W.E. Paul, B cell stimulatory factor-1 enhances the IgE response of lipopolysaccharide-activated B cells, *J. Immunol.* 136:4538 (1986).

16. W.E. Paul, Interleukin 4/B cell stimulatory factor 1: one lymphokine, many functions, *FASEB J.* 1:456 (1987).

17. P.D. Murray, D.T. McKenzie, S.L. Swain, and M.F. Kagnoff, Interleukin 5 and interleukin 4 produced by Peyer's patch T cells selectively enhance immunoglobulin A expression, *J. Immunol.* 139:2669 (1987).

18. R.L. Coffman, B. Shrader, J. Carty, T.R. Mosmann, and M.W. Bond, A mouse T cell product that preferentially enhances IgA production. I. Biologic characterization, *J. Immunol.* 139:3685 (1987).

19. G.R. Harriman, D.Y. Kunimoto, J.F. Elliott, V. Paetkau, and W. Strober, The role of IL-5 in IgA B cell differentiation, *J. Immunol.* 140:3033 (1988).

20. R. Matsumoto, M. Matsumoto, S. Mita, Y. Hitoshi, M. Ando, S. Araki, N. Yamaguchi, A. Tominaga, and K. Takatsu, Interleukin-5 induces maturation but not class switching of surface IgA-positive B cells into IgA-secreting cells, *Immunol.* 66:32 (1989).

21. K.W. Beagley, J.H. Eldridge, H. Kiyono, M.P. Everson, W.J. Koopman, T. Honjo, and J.R. McGhee, Recombinant murine IL-5 induces high rate IgA synthesis in cycling IgA-positive Peyer's patch B cells, *J. Immunol.* 141:2035 (1988).

22. A. Muraguchi, T. Hirano, B. Tang, T. Matsuda, Y. Horii, K. Nakajima, T. Kishimoto, The essential role of B cell stimulatory factor 2 (BSF-2/IL-6) for the terminal differentiation of B cells, *J. Exp. Med.* 167:332 (1988).

23. K.W. Beagley, J.H. Eldridge, F. Lee, H. Kiyono, M.P. Everson, W.J. Koopman, T. Hirano, T. Kishimoto, and J.R. McGhee, Interleukins and IgA synthesis. Human and murine interleukin 6 induce high rate IgA secretion in IgA-committed B cells, *J. Exp. Med.* 169:2133 (1989).

24. K. Fujihashi, J.R. McGhee, C. Lue, K.W. Beagley, T. Taga, T. Hirano, T. Kishimoto, J. Mestecky, and H. Kiyono, Human appendix B cells naturally express receptors for and respond to interleukin 6 with selective IgA1 and IgA2 synthesis, *J. Clin. Invest.* 88:248 (1991).

25. R. Fernandez-Botran, V.M. Sanders, T.R. Mosmann, and E.S. Vitetta, Lymphokine-mediated regulation of the proliferative response of clones of T helper 1 and T helper 2 cells, *J. Exp. Med.* 168:543 (1988).

26. T.F. Gajewski, and F.W. Fitch, Anti-proliferative effect of IFN-γ in immune regulation. I. IFN-γ inhibits the proliferation of Th2 but not Th1 murine helper T lymphocyte clones, *J. Immunol.* 140:4245 (1988).

27. D.F. Fiorentino, M.W. Bond, and T.R. Mosmann, Two types of mouse T helper cell. IV. Th2 clones secrete a factor that inhibits cytokine production by Th1 clones, *J. Exp. Med.* 170:2081 (1989).

28. T. Taguchi, J.R. McGhee, R.L. Coffman, K.W. Beagley, J.H. Eldridge, K. Takatsu, and H. Kiyono, Analysis of Th1 and Th2 cells in murine gut-associated tissues: Frequencies of CD4+ and CD8+ T cells that secrete IFN-γ and IL-5, *J. Immunol.* 145:68 (1990).

29. H. Kiyono, M.D. Cooper, J.F. Kearney, L.M. Mosteller, S.M. Michalek, W.J. Koopman, and J.R. McGhee, Isotype-specificity of helper T cell clones. Peyer's patch Th cells preferentially collaborate with mature IgA B cells for IgA responses, *J. Exp. Med.* 159:798 (1984).

30. J. Xu-Amano, W.K. Aicher, T. Taguchi, H. Kiyono, and J.R. McGhee, Selective induction of Th2 cells in murine Peyer's patches by oral immunization, *Internat. Immunol.* (in press) (1992).

31. R.L. Coffman, D.A. Lebman, and B. Shrader, Transforming growth factor β specifically enhances IgA production by lipopolysaccharide-stimulated murine B lymphocytes, *J. Exp. Med.* 170:1039 (1989).

32. E. Sonoda, R. Matsumoto, Y. Hitoshi, T. Ishii, M. Sugimoto, S. Araki, A. Tominaga, N. Yamaguchi, and K. Takatsu, Transforming growth factor β induces IgA production and acts additively with interleukin 5 for IgA production, *J. Exp. Med.* 170:1415 (1989).

33. D.L. Lebman, D.Y. Nomura, R.L. Coffman, and F.D. Lee, Molecular characterization of germ-line immunoglobulin A produced during growth factor type-β-induced isotype switching, *Proc. Natl. Acad. Sci. U.S.A.* 87:3962 (1990).

34. K.B. Islam, L. Nilsson, P. Sideras, L. Hammarström, and C.I.E. Smith, TGF-β1 induces germ-line transcripts of both IgA subclasses in human B lymphocytes, *Internat. Immunol.* 3:1099 (1991).

35. L. Nilsson, K.B. Islam, O. Olafsson, I. Zalcoborg, C. Samakovlis, L. Hammarström, C.I.E. Smith, and P. Sideras, Structure of TGF-β1-induced human immunoglobulin Cα1 and Cα2 germ-line transcripts, *Internat. Immunol.* 3:1107 (1991).

36. K.W. Beagley, J.H. Eldridge, W.K. Aicher, J. Mestecky, S. DiFabio, H. Kiyono, and J.R. McGhee, Peyer's patch B cells with memory cell characteristics undergo terminal differentiation within 24 hours in response to interleukin-6, *Cytokine* 3:107 (1991).

37. J.D. Clough, J.H. Mims, and W. Strober, Deficient IgA antibody responses to arsanilic acid bovine serum albumin (BSA) in neonatally thymectomized rabbits, *J. Immunol.* 106:1624 (1971).

38. J.L. Ebersole, M.A. Taubman, and D.J. Smith, Thymic control of secretory antibody responses in the rat, *J. Immunol.* 123:19 (1979).

39. H. Pritchard, J. Riddaway, and H.S. Micklem, Immune responses in congenitally thymus-less mice. II. Quantiative studies of serum immmunoglobulins, the antibody response to sheep erythrocytes, and the effect of thymus allografting, *Clin. Exp. Immunol.* 13:125 (1973).

40. D. Guy-Grand, C. Griscelli, and P. Vassalli, Peyer's patches, gut IgA plasma cells and thymic function: Study in nude mice bearing thymic grafts, *J. Immunol.* 115:361 (1975).

41. J. Mega, M.G. Bruce, K.W. Beagley, J.R. McGhee, T. Taguchi, A.M. Pitts, M.L. McGhee, R.P. Bucy, J.H. Eldridge, J. Mestecky, and H. Kiyono, Regulation of mucosal responses by CD4+ T lymphocytes: effects of anti-L3T4 treatment on the gastrointestinal immune system, *Internat. Immunol.* 3:793 (1991).

CYTOKINE PRODUCTION AND T CELL RECEPTOR EXPRESSION BY SALIVARY GLAND T CELLS AND INTRAEPITHELIAL T LYMPHOCYTES FOR THE REGULATION OF THE IgA RESPONSE

Junichi Mega[1,3], Kohtaro Fujihashi[1], Masafumi Yamamoto[1], Jerry R. McGhee[2], Masatomo Hirasawa[1,4], and Hiroshi Kiyono[1]

Departments of Oral Biology[1] and Microbiology,[2] Research Center in Oral Biology, and The Immunobiology Vaccine Center, The University of Alabama at Birmingham Birmingham, AL

The Departments of Crown and Bridge[3] and Microbiology[4], School of Dentistry at Matsudo Nihon University Matsudo Chiba, Japan

INTRODUCTION

For optimal induction of the mucosal IgA response, one must always consider usage of an important aspect of mucosal immunity, the Common Mucosal Immune System. This concept was originally postulated by Heremans and Bazin[1] and experimentally suggested by Craig and Cebra[2] in 1971. In the latter study,[2] it was shown that the transfer of lymphoid cells isolated from rabbit Peyer's patches and appendix (IgA inductive tissue) to irradiated allogenic recipients resulted in the repopulation of donor IgA plasma cells in the intestinal lamina propria of the recipient. On the other hand, lymphocytes from systemic tissues did not migrate to these mucosal-associated effector tissues. Since then numerous studies have provided strong evidence for the existence of the Common Mucosal Immune System by using both experimental animal models and human studies (reviewed in 3-5). Mucosal delivery of antigens (e.g., oral immunization and intranasal administration) induces antigen-specific IgA responses in the remote mucosal areas (IgA effector sites) via the Common Mucosal Immune System (see Mestecky and McGhee in this volume).

The Common Mucosal Immune System consists of three major compartments which include: IgA inductive sites, a cell circulatory migration pathway and IgA effector sites. The gut-associated lymphoreticular tissues (GALT) including Peyer's patches (PP) and appendix have been extensively studied as IgA inductive tissues.[3,5] For example, PP contain all the necessary immunocompetent cells such as CD4[+], CD8[-] T helper (Th) cells, CD4[-], CD8[+] cytotoxic T lymphocytes (T$_{CTL}$) and T suppressor (Ts) cells, CD4[-], CD8[-] T cells which can augment Th cell function, high frequencies of IgA committed B cells and accessory cells (e.g., MØ and dendritic cells) for the induction and regulation of antigen-specific mucosal immune responses.[6] For the homing of antigen sensitized-surface IgA-positive (sIgA[+]) B cells from the GALT to IgA effector sites, mesenteric lymph nodes, thoracic duct and the peripheral blood circulation represent a circulatory migration pathway for cells in the Common Mucosal Immune System.

IgA effector sites comprise tissues and glands which form the external secretions bathing the mucous membranes of the body, and include the oral region and the intestinal, upper respiratory and genitourinary tracts as well as their associated glandular tissues (e.g., salivary glands). Oral immunization with whole *Streptococcus mutans*, its derived purified protein or carbohydrate antigens induced salivary antigen-specific S-IgA responses.[7,8] In addition, oral immunization of *S. mutans* protein antigen I/II conjugated to cholera toxin subunit B resulted in the induction of antigen-specific IgA-producing cells in salivary glands.[9,10] Immunization of mice with *Porphyromonas gingivalis* fimbriae antigen with muramyl dipeptide via the oral route resulted in salivary fimbriae-specific IgA antibodies[11] and antigen-specific IgA-producing cells in salivary glands (see contributions by Hamada et al. in this volume).[12] These findings clearly support the concept that the salivary glands are major IgA effector tissues in the oral region.

In comparison to the IgA inductive sites, the immunological and cellular characteristics of the IgA effector tissues are less well known. In this regard, single cell analysis of immunocompetent cells which reside in the salivary glands and intestine have not received significant attention. This is unfortunate since the most readily obtained external secretions are saliva and gut washes, and numerous studies in humans and experimental systems have relied on these secretions as a source for analysis of mucosal S-IgA antibody responses. To this end, we have recently begun to analyze lymphocytes including T and B cells isolated from salivary glands and the intestine in order to understand their potential roles for formation of secretory IgA immune responses in these IgA effector sites.

ISOLATION AND CHARACTERIZATION OF LYMPHOCYTES FROM SALIVARY GLANDS

In order to understand the characteristics of T and B lymphocytes involved in the regulation and secretion of IgA in the salivary glands, lymphocytes were enzymatically isolated from salivary gland associated tissues (SGAT), which consist of the submandibular glands (SMG), periglandular lymph nodes (PGLN) and cervical lymph nodes (CLN). The murine SGAT are located just under the mandibular bone and individual tissues were carefully dissected. Lymphocytes from SMG were obtained by a modification of the enzymatic dissociation procedure which was originally described by others.[13-15] SMG were dissected into small fragments (approximately 1.0 x 2.0 mm) and then dissociated into single cells by use of collagenase Type II (Figure 1). This enzymatic dissociation process was performed at least 4-5 times. To obtain purified lymphocyte populations, the reverse Percoll discontinuous gradient was employed (Figure 1). Cells in 75% Percoll were transferred into the bottom of a polystyrene round bottom tube. Subsequently, 55% and 40% Percoll were carefully and sequentially layered on top of the 75% layer containing cells. After centrifugation, the interface between the 75% and 55% layers was carefully removed and the cells washed. This procedure provide >97% viable lymphocytes with a cell yield for SMG of 0.5-1x10^6 cells per mouse.[16]

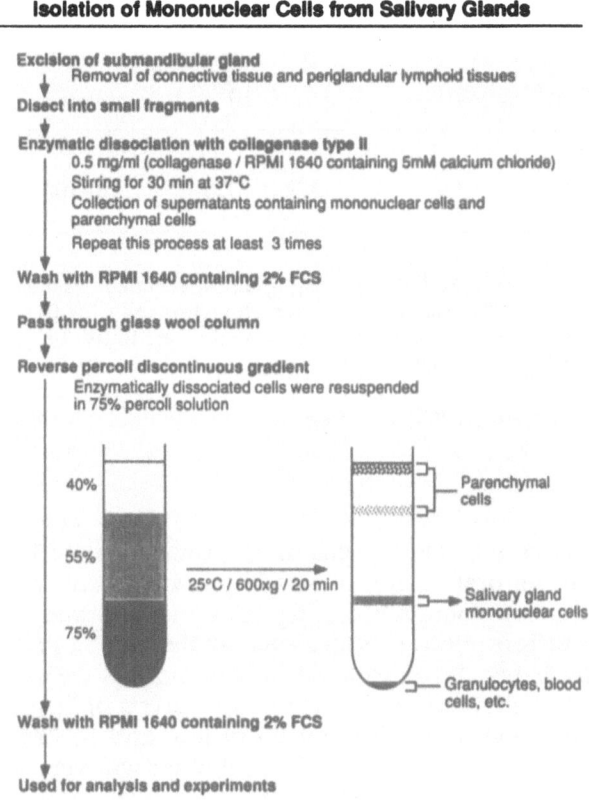

Figure 1. A protocol for the isolation of mononuclear cells from murine salivary glands by using the enzymatic dissociation and the reverse Percoll discontinuous gradient procedures.

When the DNA replication ability of these isolated lymphocytes was examined by a cell proliferation assay using T and B cell mitogens, significant proliferative responses were induced in SMG lymphocytes by both T cell-(Con A and PHA) and T cell-dependent B cell-(PWM) mitogens.[16] Lymphocytes from PGLN and CLN gave higher DNA replication responses to Con A and PHA when compared with PWM. A similar pattern of DNA replication was also seen in lymphocytes isolated from other lymphoid tissues including peripheral lymph nodes (PLN) and mesenteric lymph nodes (MLN). These findings showed that among SGAT, the lymphocytes isolated from the PGLN and CLN responded well to T cell mitogens and resemble other organized LN, while lymphocytes isolated from SMG gave similar responses to both T and B cell mitogens.[16]

Table 1. The frequency of IgM, IgG and IgA secreting cells in the murine salivary glands.[a]

| | Tissues | Isotype and Numbers of SFC/10^6 cells[b] | | |
		IgM	IgG	IgA
SGAT	SMG	0-100	0-50	9,000-11,000
	PGLN	1,500-2,000	100-150	1,000-2,500
	CLN	0-10	0-10	0-10
Small Intestine	LP	3,500-4000	50-100	35,000-45,000

[a]Cells were isolated from SGAT (salivary gland-associated tissues) including SMG (submandibular glands), PGLN (periglandular lymph nodes) and CLN (cervical lymph nodes), and LP (lamina propria) of the small intestine. Single cells were tested for the frequency of IgM-, IgG- and IgA-producing cells by the ELISPOT assay.

[b]Numbers represent the range of SFC (spot forming cells) from three different experiments.

In as much as the salivary glands are considered to be the major IgA effector tissue in the oral cavity, it was important to study the isotype and frequency of immunoglobulin-secreting cells in SMG and to compare this pattern with other IgA effector tissues such as the lamina propria (LP) region of the small intestine.[16] When isolated cells from SMG were examined by the isotype-specific ELISPOT assay, the dominant isotype of Ig-producing cells in SMG was IgA followed by small numbers of IgM and IgG spot-forming cells (SFC) (Table 1). This pattern of isotype distribution was very similar to that of the LP regions of the small intestine. However, the total number of IgA-producing cells (per 10^6 cells) in SMG was lower than that in LP (Table 1). Two distinct patterns of Ig-producing cells were noted in the two LN (e.g., PGLN and CLN) of SGAT. High numbers of IgA and IgM but not IgG SFC were seen in cells isolated from PGLN. On the other hand, essentially no Ig secreting cells were detected in CLN. In this regard, other organized lymphoid tissue such as PLN also do not contain B cells spontaneously producing antibody. Thus, this was the first demonstration that organized PGLN associated with the SGAT, in addition to the SMG, contain IgA- and IgM-producing cells. In summary, among SGAT, the SMG should be considered as a classical IgA effector tissue, while CLN possess characteristics of an organized systematic LN.[16] Interestingly, PGLN could be a unique, organized LN in the oral region, since this tissue has immunological features of both IgA effector tissue and organized systemic LN. It might be comparable to the MLN in this regard.

CD3+, CD4+ Th cells and their derived cytokines have been shown to play central roles for the terminal differentiation of sIgA+ B cells into IgA-producing plasma cells.[6] These CD3+, CD4+ Th cells are subgrouped into Th1 and Th2 cell subsets according to the profile of cytokines that are produced by the respective T cells.[17,18] CD3+, CD4+ Th cells which produce IFN-γ, IL-2 and TNF-β upon stimulation via the T cell receptor (TCR)-CD3 complex are caterogorized as Th1 cells.[17,18] On the other hand, Th2 cells secrete IL-4, IL-5, IL-6 and IL-10, an array of cytokines essential for B cell responses. In this regard, cytokines secreted by Th2-type cells, notably IL-5 and IL-6, directly act on sIgA+ B cells from GALT without any co-stimulation *in vitro* and induce them to differentiate into IgA-producing cells (see more details in Xu-Amano et al in this volume).[19,20]

When the frequency of Th1- and Th2-type cells was examined in the IgA inductive and effector sites by using IFN-γ (Th1)- and IL-5 (Th2)-specific ELISPOT assay, freshly isolated PP CD3+, CD4+ T cells contained neither IFN-γ nor IL-5-producing cells unless these cells were stimulated with T cell mitogens *in vitro*.[21] Upon mitogen stimulation, an approximately equal frequency of Th1 (IFN-γ)- and Th2 (IL-5)-type cells were seen in IgA inductive tissues. In contrast, freshly isolated CD3+, CD4+ Th cells from murine intestinal LP contained higher numbers of IL-5-producing cells when compared with IFN-γ secreting cells, suggesting that Th2-type cells are predominant in effector regions for mucosal IgA responses.[21]

Since a distinct isotype distribution of Ig-producing cells was seen among tissues associated with SGAT (Table 1), it was interesting to analyze cytokine production by T cells which reside in the respective SGAT in order to understand the contribution of Th1- and Th2- type cells for IgA B cell responses.[16] When lymphocytes were isolated from SGAT (e.g., the SMG and PGLN), all contained cells which spontaneously produce IFN-γ and IL-5 (Table 2). Among these cytokine-secreting cells, higher numbers of IL-5-producing Th2-type cells were always noted when compared with IFN-γ-secreting Th1-type cells.[16] This finding was in complete agreement with our previous finding that increased numbers of IL-5-secreting Th2-type cells are consistently found in IgA effector tissues such as LP of intestine.[21] Stimulation of these T cells from SGMG and PGLN with a T cell mitogen (e.g., Con A) resulted in enhanced numbers of cytokine-producing cells.[16] When CD4+ T cells isolated from CLN were examined for their IFN-γ and IL-5 production, freshly isolated cells did not contain cytokine-producing cells (Table 2). However, upon stimulation with Con A, an equal frequency of IFN-γ (Th1)- and IL-5 (Th2)-producing Th cells was noted.[16] This pattern of cytokine production was similar to CD4+ Th cells isolated from other organized lymphoid tissues (e.g., spleen and PLN).[21] Thus, these results demonstrated that high numbers of IL-5-producing Th2-type cells were always associated with the occurrence of increased numbers of IgA-producing cells in the salivary gland.

These findings was also confirmed at the mRNA level by dot-blot hybridization using IFN-γ- and IL-5-specific cDNA probes.[16] Significant levels

Table 2. The frequency of Th1 (IFN-γ)- and Th2 (IL-5)-type CD4+ T cells in salivary glands.

T Cells Isolated From	Numbers of Cytokine-Specific SFC (per 10^4 Cells)[a]		Levels of Cytokine-Specific mRNA (Relative Gray Scale 0-200)[b]	
	IFN-γ	IL-5	IL-5	IFN-γ
SMG	5-10	20-30	25-50	100-150
PGLN	5-10	5-10	20-50	20-50
CLN	0-5	0-2	0-15	0-5

[a]Freshly isolated lymphocytes from SMG, PGLN and CLN were tested by IFN-γ and IL-5 specific ELISPOT assays. Numbers in range represent three different experiments.

[b]Extracted mRNA from CD4+ T cells in SMG, PGLN and CLN were hybridized with IFN-γ or IL-5-specific cDNA probes. Levels of hybridization were examined by densitometry scanning and were expressed as the relative gray scale.

of IL-5-specific message were seen in mRNA isolated from SMG and PGLN CD4+ T cells (Table 2). Although the level of message was lower than that for IL-5, IFN-γ-specific mRNA was also noted in these CD4+ T cell fractions. On the other hand, neither IFN-γ– nor IL-5-specific messages were noted in mRNA prepared from CLN CD4+ T cells (Table 2). These findings further support the idea that elevated numbers of IgA-producing cells at IgA effector tissues are always associated with a higher frequency of Th2 (IL-5)-type cells over Th1 (IFN-γ)-ype CD4+ Th cells.

T CELL RECEPTOR (TCR) EXPRESSION BY SALIVARY GLAND T CELLS

Mammalian T cells recognize nominal antigen via two distinct T cell receptors (TCR), each consisting of heterodimer chains of α/β or γ/δ. In general, mature T cells that reside in the organized secondary lymphoid tissues express the α/β TCR which recognizes processed antigen together with MHC class I or II on antigen-presenting cells (APC).[22,23] The α/β TCR+ T cells can be separated into two subsets according to the co-expression of either CD4 or CD8. The α/β TCR+ and CD4+ T cells are stimulated by foreign peptide and MHC class II expressed on APC including B cells and they provide helper function for B cell responses, whereas the α/β TCR+ and CD8+ T cells behave as CTLs and recognize target cells under MHC class I restriction.[23] In contrast to the α/β TCR+ T cells, the γ/δ TCR+ T cells arise earlier in ontogeny, and small numbers

(1-10%) of T cells express, their receptors in peripheral lymphoid tissues.[24] Generally, γ/δ TCR-bearing T cells are categorized as double negative (DN) cells since they do not express either CD4 or CD8. The immunological function of γ/δ TCR+ T cells still remains unknown. However, our recent studies have provided evidence that at least two subsets of these γ/δ TCR-bearing T cells in mucosa-associated tissues are capable of producing cytokines and provide regulatory functions for IgA B cell responses.[25-27]

When different subsets of T cells were assessed according to their expression of CD4 and CD8 molecules in SGAT, three distinct subsets of T cells including those of CD4+, CD8- T cells, CD4-, CD8+ T cells and DN T cells were seen in SMG, PGLN and CLN (Table 3).[16] No CD4+, CD8+ (so called double positive [DP]) T cells were found in any of the SGAT. It was interesting to note that an approximately equal frequency of CD4+, CD8- and CD4-, CD8+ T cells was seen in the SMG (Table 3). This finding was consistent with studies of rat salivary glands where the mononuclear cells were shown to contain approximately 60% W3/13+ T cells with a CD4/CD8 ratio of 1.3.[14,15] This equal distribution of CD4+, CD8- and CD4-, CD8+ T cells in SMG was unique since other tissues including those of IgA inductive (e.g., PP) and effector (e.g., LP) tissues, systemic tissue (e.g., SP) as well as organized LN in SGAT (e.g., CLN and PGLN) all contained higher frequencies of CD4+, CD8- T cells (Table 3). Further, a relatively high number of DN T cells was found in SMG when compared with other tissues (Table 3). In this regard, up to 15% of CD3+ T cells were DN T cells in some cases while other tissues contained lower numbers of CD3+, DN T cells (Table 3).

The usage of α/β or γ/δ TCR by three different subsets of T cells in SGAT was also examined.[16] All CD4+, CD8- T cells expressed the α/β TCR (Table 3). On the other hand, approximately 25% of CD4-, CD8+ T cells isolated from SMG expressed γ/δ TCR (Table 3). Prior to this finding, γ/δ TCR-bearing CD8+ T cells were only found in intraepithelial lymphocytes (IELs).[28,29] The CD4-, CD8+ T cells which reside in PGLN and CLN of SGAT express the α/β TCR (Table 3). The γ/δ TCR-bearing T cells were also seen in the DN fraction of SMG (Table 3). In addition, although PGLN and CLN possessed small numbers of DN T cells (Table 3), all of these cells used the γ/δ form of TCR (Table 3). These findings were consistent with previous studies where DN T cells always expressed γ/δ TCR.[25-27,30] Taken together, this finding provided the first evidence that CD4-, CD8+ T cells that reside in glandular tissue contain both γ/δ TCR+ and α/β TCR+ cells.

Although the precise biological function of these CD8+, γ/δ TCR+ T cells is not yet known, it has been shown that these T lymphocytes in IELs possess cytolytic function.[28,31,32] Further, we have recently demonstrated that CD8+, γ/δ TCR+ T cells from IELs of mice orally immunized with T cell-dependent (TD) antigen possess the ability to convert oral tolerance to antigen-specific immune responses upon adoptive transfer to orally tolerized mice (see below).[25,27] Thus, CD8+, γ/δ TCR+ T cells could be an important regulatory T cell type which protects or enhances CD4+ Th cells for maximum IgA responses at IgA effector tissues in an active state of oral tolerance. The presence of CD8+, γ/δ TCR+ T cells in salivary glands might be an essential feature for the dominant IgA antibody production seen in this tissue.

Table 3. Analysis of T cell subsets and TCR expression by salivary gland T cells.

T Cell Subsets[a] and TCR Expression[b]	Tissues Examined [Frequency of Positive Cells (%)]		
	SMG	PGLN	CLN
CD4+, CD8-	12-23[a]	60-67	61-65
α/β TCR	100[b]	100	100
γ/δ TCR	0	0	0
CD4-, CD8+	18-25	9-15	23-28
α/β TCR	75	100	100
γ/δ TCR	25	0	0
CD4-, CD8- (DN)	6-16	2-4	6-8
α/β TCR	0	0	0
γ/δ TCR	100	100	100
CD4+, CD8+ (DP)	0	0	0

[a]Cells were isolated from SMG, PGLN and CLN, stained for CD3, CD4 and CD8 and then analyzed by multicolor flow cytometry. Numbers represent range of percent positive cells among total cells.

[b]Using multicolor flow cytometry, the relative frequency of α/β TCR and γ/δ TCR-bearing cells was examined in CD4+, CD8- T cells, CD4-, CD8+ T cells and DN T cells. Numbers represent the relative frequency of γ/δ TCR+ and α/β TCR+ cells in the individual T cell subsets.

IMMUNOREGULATORY FUNCTION OF γ/δ TCR+ AND α/β TCR+ INTRAEPITHELIAL T CELLS FOR IgA RESPONSES

In addition to the salivary glands, another important IgA effector site is the gastrointestinal tract. The LP regions of the intestine contain large numbers of CD4+ T cells and IgA-producing plasma cells.[33] Less well understood effector sites include the lymphocyte population which resides in the epithelial layer and the IELs which first encounter the myriad of antigens that are present in the gastrointestinal tract. It is important to better understand the function of lymphocytes in these effector regions, since these IELs, like salivary gland T cells, possess a number of unique features which are distinct from lymphocytes residing in other lymphoid tissues, including their phenotype of surface markers, use of heterodimeric chains of TCR and biologic functions. For example, 80-90% of IELs are CD3+ T cells which can be divided into four distinct subsets based on the expression of CD4 and CD8.[25,26,34] Approximately 75% of these are CD4-, CD8+ T cells followed by CD4+, CD8- (7.5%), DN (7.5%) and DP (10%) T cell subsets. Interestingly, CD4-, CD8+ T cells in IELs, like the salivary glands, contained both γ/δ TCR- (45-65%) and α/β TCR- (35-55%) bearing T cells.[26] In addition, γ/δ TCR+ T cells were also seen in the DN T cell fraction. On the other hand, CD4+, CD8- T cells and DP T cells exclusively use the α/β heterodimer chains of TCR.[26] Thus, it was of importance to examine the precise biological function of these different subsets of T cells in IELs. When one considers the physiological and anatomical features of the IELs, these different subsets of T cells are continuously exposed to environmental antigens from the gut lumen but also are present above the LP regions, where large numbers of IgA plasma cells occur.[33] Thus, IELs are in sentinel locations to potentially serve as regulatory cells for mucosal immune responses.

Since our previous study showed that CD3+ T cells in IELs are capable of producing Th1 (IFN-γ)- and Th2 (IL-5)-type cytokines,[21,26] it was important to examine the immunoregulatory function of γ/δ TCR+ and α/β TCR+ T cells in IELs for IgA responses. A model which determines the ability of different T cell subsets to abrogate a state of oral tolerance (OT) which is induced by the repeated oral administration of TD antigens, was used to elucidate the biological function of α/β TCR+ and γ/δ TCR+ T cells in IELs.[25,27] When CD3+, γ/δ TCR+ T cells from IELs of mice orally immunized with TD antigen [e.g., sheep erythrocytes (SRBC)] were adoptively transferred to mice which were unresponsive to this antigen, the conversion of OT to SRBC-specific IgM, IgG1, IgG2b and IgA responses was seen.[25,27] Further, the addition of γ/δ TCR+ T cells from the IELs of orally immunized mice to spleen cell cultures from orally tolerized mice resulted in the abrogation of OT to IgA responses (Table 4).[27] These findings were the first direct evidence that CD3+, CD4-, CD8+ and CD3+, DN T cells, which use the γ/δ TCR, possess immunoregulatory functions.[27] Of further significance was the finding that γ/δ TCR+ T cells from IELs can rescue antigen-specific antibody responses from the influence of systemic unresponsiveness, an effect that is especially pronounced for the IgA isotype.

Table 4. Unique Characteristics Of γ/δ TCR⁺ and α/β TCR⁺ T Cells In IELs for the regulation oF IgA responses.

Biological Function Tested	Addition of T Cell Subsets (IgA-Plaque Forming Cells/Culture[a])		
	None	γ/δ TCR	α/β TCR
Abrogation of Oral Tolerance[b]	25-40	500-650	30-50
Helper Function For B Cells[c]	15-30	20-35	850-1,000

[a]Numbers represent range of SRBC-specific IgA PFC per culture containing 5×10^6 cells.

[b]Splenic cells from SRBC orally tolerized mice were cultured either in the presence or absence of γ/δ TCR⁺ and α/β TCR⁺ IELs from mice orally immunized with SRBC. These cultures were then immunized with SRBC and incubated for 5 days. After incubation, nonadherent cells were harvested and tested for IgA-producing cells by the PFC assay.

[c]Splenic B cells from normal mice were cultured either in the presence or absence of the respective IEL T cell subsets from mice orally immunized with SRBC. These cultures were then similarly treated as described above.

In contrast to γ/δ TCR⁺ T cells, α/β TCR⁺ T cells from IELs of orally immunized mice could not convert OT to secondary-type responses (Table 4). However, the α/β TCR⁺ T cells possessed a strong helper function for B cell responses, since the α/β TCR⁺ IELs supported IgA responses in B cell cultures which exceeded those seen with splenic CD4⁺, α/β TCR⁺ Th cells (Table 4).[27] On the contrary, γ/δ TCR⁺ T cells which were capable of abrogating OT did not provide any helper activity in B cell cultures. These findings provided important new evidence that α/β TCR⁺ IELs are capable of helper function, while γ/δ TCR⁺ IELs contain subsets that are responsible for abrogating OT.[27] The anatomical proximity of γ/δ TCR⁺, CD8⁺ and DN T cells as well as α/β TCR⁺, CD4⁺ T cells at these IgA effector sites would make physiological sense, since this would provide maintenance of efficient IgA responses for the protection of mucosal surfaces. Further, it is also possible that γ/δ TCR⁺ T cells in the IELs could also enhance (or protect) cytotoxic T cells for the influence of suppression.

SUMMARY

The IgA effector sites such as the salivary glands and the intestinal tract contain several distinct T cell subsets which possess unique biologic characteristics. Freshly isolated CD3⁺ T cells from the salivary glands, the LP region of the small intestine and IELs all harbor T cells which spontaneously produce Th1 (IFN-γ)- and Th2 (IL-5 and IL-6)-type cytokines. Interestingly, a high frequency of IL-5-producing Th2-type cells is always associated with the occurrence of increased numbers of IgA plasma cells (e.g., the salivary glands and the LP region of the small intestine). Further, the salivary gland CD3+ T cells can be divided into three distinct subsets including those of CD4+, CD8- (12-23%), CD4-, CD8+ (18-25%) and DN (6-16%) T cells. In terms of TCR expression, CD4+, CD8- and DN T cells exclusively expressed α/β TCR and γ/δ TCR, respectively. One of the unique features of the salivary gland T cells is that like IELs, relatively high numbers of γ/δ TCR-bearing cells are seen in the CD4-, CD8+ T cell fraction. Since our study has provided important new evidence that these γ/δ TCR-bearing T cells from IELs of mice orally immunized with TD antigen possess the capability of abrogating oral tolerance to antigen-specific immune responses including the IgA isotype, one can visualize that γ/δ TCR+ T cells can be essential regulatory T cells which protect (or enhance) α/β TCR+, CD4+ Th cells for maximum IgA responses at IgA effector tissues including the salivary glands, and the gastrointestinal tract in the presence of an active state of systemic unresponsiveness or oral tolerance.

ACKNOWLEDGMENTS

This work was supported by U.S. Public Health Service Grants DE 09837, DE 08228, DE 04217, DE 08182, AI 30366, AI 19674, AI 28147, AI 18958, DK 44240 and Contract AI 15128. Hiroshi Kiyono is the recipient of NIH Research Career Development Award, DE 00237. We thank Ms. Debra H. Clisby for the preparation of this paper.

REFERENCES

1. J.F. Heremans, and H. Bazin, Antibodies induced by local antigenic stimulation of mucosal surfaces, *Ann. N.Y. Acad. Sci.* 190:268 (1971).
2. S.W. Craig, and J.J. Cebra, Peyer's patches: an enriched source of precursors for IgA-producing immunocytes in the rabbit, *J. Exp. Med.* 134:188 (1971).
3. J. Mestecky, and J.R. McGhee, Immunoglobulin A (IgA): molecular and cellular interactions involved in IgA biosynthesis and immune responses, *Adv. Immunol.* 40:153 (1987).
4. J. Mestecky, The common mucosal immune system and current strategies for induction of immune responses in external secretions, *J. Clin. Immunol.* 7:265 (1991).
5. J.R. McGhee, Mucosal immunology, Encyclopedia of Human Biology, 5:137 (1991).
6. J.R. McGhee, J. Mestecky, C.O. Elson, and H. Kiyono, Regulation of IgA synthesis and immune response by T cells and interleukins, *J. Clin. Immunol.* 9:175 (1989).
7. M.A. Taubman, and D.J. Smith, Oral immunization for the prevention of dental diseases, *Curr. Top. Microbiol. Immunol.* 146:187 (1989).
8. S.M. Michalek, and N.K. Childers. Development and outlook for a caries vaccine, *Crit. Rev. Oral Biol. Med.* 1:37 (1990).
9. C. Czerkinsky, M.W. Russell, N. Lycke, M. Lindblad, and J. Holmgren, Oral administration of a streptococcal antigen coupled to cholera toxin B subunit evokes strong antibody responses in salivary glands and extramucosal tissues, *Infect. Immun.* 57:1072 (1989).
10. M.W. Russell, and H.-Y. Wu, Distribution, persistence and recall of serum and salivary antibody responses to peroral immunization with protein antigen I/II of *streptococcus mutans* coupled to the cholera toxin B subunit, *Infect. Immun.* 59:4061 (1991).
11. T. Ogawa, H. Shimauchi, and S. Hamada, Mucosal and systemic immune responses in BALB/c mice to *Bacteroides gingivalis* fimbriae administered orally, *Infect. Immun.* 57:3466 (1989).

12. T. Ogawa, Y. Kusumoto, H. Shimauchi, H. Kiyono, J.R. McGhee, and S. Hamada, Occurrence of antigen-specific B cells following oral and parental immunization with *Porphyromonas gingivalis* fimbriae, *Intern. Immunol.* (in press, 1992).

13. M. Oudghiri, J. Seguin, and N. Deslauriers, The cellular basis of salivary immune responses in the mouse: incidence and distribution of B cells, T cells and macrophages in single-cell suspensions of the major salivary glands, *Eur. J. Immunol.* 16:281 (1986).

14. J.L. Ebersole, M.J. Steffen, and J. Pappo, Secretory immune responses in ageing rats. II. Phenotype distribution of lymphocytes in secretory and lymphoid tissues, *Immunology* 64:289 (1988).

15. J. Pappo, J.L. Ebersole, and M.A. Taubman, Phenotype of mononuclear leukocytes resident in rat major salivary and lacrymal glands, *Immunology* 64:295 (1988).

16. J. Mega, J.R. McGhee, and H. Kiyono, Cytokine and Ig producing cells in mucosal effector tissues: analysis of IL-5 and IFN-γ producing T cells, TCR expression and IgA plasma cells from mouse salivary gland associated tissues, *J. Immunol.* (in press, 1992).

17. T.R. Mosmann, and R.L. Coffman, Th1 and Th2 cells: Different patterns of lymphokine secretion lead to different functional properties, *Annu. Rev. Immunol.* 7:145 (1989).

18. N.E. Street, and T.R. Mosmann, Functional diversity of T lymphocytes due to secretion of different cytokine patterns, *FASEB J.* 5:171 (1991).

19. K.W. Beagley, J.H. Eldridge, H. Kiyono, M.P. Everson, W.J. Koopman, T. Honjo, and J.R. McGhee, Recombinant murine IL-5 induces high rate IgA synthesis in cycling IgA-positive Peyer's patch B cells, *J. Immunol.* 141:2035 (1988).

20. K.W. Beagley, J.H. Eldridge, F. Lee, H. Kiyono, M.F. Everson, W.J. Koopman, T. Hirano, T. Kishimoto, and J.R. McGhee, Interleukins and IgA synthesis. Human and murine interleukin 6 induce high rate IgA secretion in IgA-committed B cells, *J. Exp. Med.* 169:2133 (1989).

21. T. Taguchi, J.R. McGhee, R.L. Coffman, K.W. Beagley, J.H. Eldridge, K. Takatsu, and H. Kiyono, Analysis of Th1 and Th2 cells in murine gut-associated tissues. Frequencies of CD4+ and CD8+ T cells that secrete IFN-γ and IL-5, *J. Immunol.* 145:68 (1990).

22. P. Marrack, and J. Kappler, The antigen-specific, major histocompatibility complex-restricted receptor on T cells, *Adv. Immunol.* 38:1 (1986).

23. B.E. Bierer, B.P. Sleckman, S.E. Ratnofsky, and S.J. Burakoff, The biologic roles of CD2, CD4 and CD8 in T-cell activation, *Annu. Rev. Immunol.* 7:579 (1989).

24. D.H. Raulet, The structure, function and molecular genetics of the γδ T cell receptor, *Annu. Rev. Immunol.* 7:175 (1989).

25. K. Fujihashi, T. Taguchi, J.R. McGhee, J.H. Eldridge, M.G. Bruce, D.R. Green, B. Singh, and H. Kiyono, Regulatory function for murine intraepithelial lymphocytes. Two subsets of CD3+, T cell receptor-1+ intraepithelial lymphocyte T cells abrogate oral tolerance, *J. Immunol.* 145:2010 (1990).

26. T. Taguchi, W.K. Aicher, K. Fujihashi, M. Yamamoto, J.R. McGhee, J.A. Bluestone, and H. Kiyono, Novel function for intestinal intraepithelial lymphocytes: Murine CD3+, γδ TCR+ T cells produce interferon gamma and interleukin 5, *J. Immunol.* 147:3736 (1991).

27. K. Fujihashi, T. Taguchi, W.K. Aicher, J.R. McGhee, J.A. Bluestone, J.H. Eldridge, and H. Kiyono, Immunoregulatory function for murine intraepithelial lymphocytes: γδ TCR+ T cells abrogate oral tolerance while αβ TCR+ T cells provide B cell help, *J. Exp. Med.* (in press, 1992).

28. T. Goodman, and L. Lefrancois, Expression of the γ–δ T cell receptor on intestinal CD8+ intraepithelial lymphocytes, *Nature* 333:855 (1988).

29. M. Bonneville, C.A. Janeway, Jr., K. Ito, W. Haser, I. Ishida, N. Nakanishi, and S. Tonegawa. Intestinal intraepithelial lymphocytes are a distinct set of γδ T cells, *Nature* 336:479 (1988).

30. A.M. Lew, D.M. Pardoll, W.L. Maloy, B.J. Fowlkes, A. Kruisbeek, S.-F. Cheng, R.N. Germain, J.A. Bluestone, R.H. Schwartz, and J.E. Coligan, Characterization of T cell receptor gamma chain expression in a subset of murine thymocytes, *Science* 234:1401 (1987).

31. J.R. Klein, Ontogeny of the Thy-1−, Lyt-2+ murine intestinal intraepithelial lymphocyte, Characterization of a unique population of thymus-independent cytotoxic effector cells in the intestinal mucosa, *J. Exp. Med.* 164:309 (1986).

130

32. P.B. Ernst, D.A. Clark, K.L. Rosenthal, A.D. Befus, and J. Bienenstock, Detection and characterization of cytotoxic T lymphocyte precursors in the murine intestinal intraepithelial leukocyte population, *J. Immunol.* 136:2121 (1986).

33. P. Brandtzaeg, Overview of the mucosal immune system, *Curr. Top. Micro. Immunol.* 146:13 (1989).

34. R.L. Mosley, D. Styre, and J.R. Klein, CD4+, CD8+ murine intestinal intraepithelial lymphocytes, *Intern. Immunol.* 2:361 (1990).

IMMUNOLOGICAL ADJUVANTS

Anthony C. Allison and Noelene E. Byars

Institute of Immunology and Biological Sciences
Syntex Research
Palo Alto, CA 94304

INTRODUCTION

New vaccines are urgently needed. They could prevent the spread of human immunodeficiency virus; protect susceptible groups of humans against infections, including young children and the elderly; and eliminate preventable cancers, including primary hepatocellular carcinoma and nasopharyngeal carcinoma. New methods for producing antigens have been developed. These include recombinant DNA technology, exemplified by hepatitis B virus surface antigen, and glycoconjugates, exemplified by Haemophilus influenzae vaccine. Moreover, our understanding of the immune system has been greatly advanced in the last few decades. The challenge before us is how to use all of this information to produce a new generation of safe and efficacious vaccines.

While live viral and bacterial vaccines have been very useful, they can produce severe infections in immunodeficient recipients. Recently, evidence has accumulated showing that subunit antigens can elicit protective immunity as effectively as live vaccines, and so they are preferred for new-generation vaccines.

It is generally agreed that subunit antigens efficiently elicit cell-mediated and humoral immune responses only when they are administered with adjuvant formulations. Adjuvants are required to augment the formation of protective antibodies, elicit cell-mediated immunity (CMI) and generate populations of memory T- and B-lymphocytes with specificity for the vaccine antigen.

Two adjuvant formulations have a long history. One is based on mineral oil emulsions, with or without mycobacteria, and the second on adsorption of antigens to aluminum salts. In recent years three adjuvant formulations have been developed: liposomes, immune-stimulating complexes (ISCOMs) and squalene or squalane emulsions. Although several types of surface-active agents have adjuvant activity, most attention has been focused on saponin-like Quil A molecules in ISCOMs and Pluronic® block co-polymers which are used to make stable squalane emulsions. Analogs of muramyl dipeptide (MDP) and lipopolysaccharide (LPS) have been produced with the objective of preserving their adjuvant activity while minimizing their side effects. In experimental animals, promising results have been obtained with the new adjuvant

Genetically Engineered Vaccines, Edited by
J.E. Ciardi *et al.*, Plenum Press, New York, 1992

formulations, and trials of their efficacy in humans have been initiated. Because of space limitations only our own adjuvant formulation (SAF) is discussed in this chapter; for information on other adjuvants, readers are referred to Allison and Byars (1992).

AFFINITIES AND ISOTYPES OF ANTIBODIES

Adjuvants should augment the production of antibodies of sufficient quantity and affinity for antigen to elicit protection against infections and toxins. Studies of passive protection with isotype-switch variants of monoclonal antibodies show that those which activate complement well and act synergistically with antibody-dependent cytotoxic cells are most effective (IgG_{2a} antibodies in the mouse, Kaminski et al., 1986). Studies with reshaped human antibodies in vitro show the IgG1 isotype to be most effective (Reichmann et al., 1988).

Adjuvants influence the isotype of antibodies formed in response to the same antigen. In response to Freund's incomplete adjuvant (FIA), guinea pigs produce γ_1 FCA antibodies, whereas in response to the complete adjuvant (FCA), γ_2 antibodies predominate (White, 1976). Antigens given to mice in LPS elicit mainly IgG1 and IgE antibodies, whereas in Syntex Adjuvant Formulation (SAF, defined below), IgG_{2a} antibodies predominate (Kenney et al., 1989). LPS induces the production by cells in lymph nodes draining sites of injection of IL-4, which selects for IgG1 and IgE, whereas SAF elicits strong cell-mediated immunity with release of IFN-γ, which favors the formation of IgG2a antibodies (Finkelman et al., 1990). An interesting question, discussed briefly below, is whether systemic immunization, with the formation of IgG antibodies, can protect sero-mucous surfaces against infections, or whether secretory IgA antibodies are required.

CELL-MEDIATED IMMUNITY (CMI)

Helper T-lymphocytes are required for the formation of antibodies against most antigens. In addition, cytotoxic T-lymphocytes can lyse infected cells or produce mediators, such as interferon-γ, following interaction with antigen in a genetically restricted situation (Morris et al., 1982). Cytotoxic T-lymphocytes able to lyse autologous cells expressing several antigens of human immunodeficiency virus (HIV) are demonstrable in infected persons (Walker et al., 1988), although it is not yet known whether they have a protective role.

It is therefore likely that for optimal protection against some infectious agents (e.g., herpesviruses and possibly HIV), the elicitation of CMI is desirable. Tests for CMI should include not only delayed hypersensitivity but also proliferative responses to the antigen and the release of IL-2 (Byars and Allison, 1987) and IFN-γ. Cytotoxicity for autologous or syngeneic infected target cells should also be studied. If mice or rats are used, syngeneic target cells are readily available. With outbred species, such as humans and subhuman primates, B-cells transformed by Epstein-Barr virus, transfected with a vaccinia virus vector expressing the antigen under consideration (e.g., HIV antigens, Walker et al., 1988), can provide autologous target cells for studies of genetically restricted cell-mediated cytotoxicity. According to traditional wisdom, replicating viruses are required to elicit cytotoxic T-lymphocyte responses. That is not the case: recombinant envelope glycoprotein of HIV in ISCOMs elicits CD8[+] MHC class I-restricted cytotoxicity in mice (Takahashi et al., 1990). In our hands, recombinant gpD of HSV-2 in SAF was found to elicit CD4[+] class II-restricted T-lymphocytes in guinea pigs. Hence, subunit vaccines in an efficacious adjuvant can elicit cytotoxic T-cell responses.

Aluminum salts are the only adjuvants approved by regulatory authorities for human use. They have been valuable (e.g., for eliciting antibodies against bacterial toxoids). However, antigens in aluminum salts inconsistently elicit CMI and alum has not been useful with several antigens [e.g., influenza hemagglutinin (Nicholson et al., 1979) and recombinant HSV subunits (Berman et al., 1985)].

While FCA is effective in experimental animals it is too toxic to use in human and veterinary vaccines. The mycobacteria in FCA have several disadvantages. They contain tuberculin and other proteins, so that after a single injection animals become sensitized. A second injection of FCA produces a massive delayed-type hypersensitivity response. It was a useful advance when Ellouz et al. (1974) showed that the minimal adjuvant-active component of the mycobacterial cell wall is N-acetylmuramyl-L-alanine-D-isoglutamine, also known as muramyl dipeptide (MDP). When protein antigens are administered to guinea pigs in FIA with MDP, delayed-type hyerpsensitivity and antibodies of the γ2 isotype are elicited.

MDP has undesirable effects, including pyrogenicity and the capacity to induce anterior uveitis and arthritis. Over 130 MDP analogs were synthesized at Syntex Research in an attempt to separate adjuvant activity from side effects. The threonyl analog of MDP (N-acetylmuramyl-L-threonyl-D-isoglutamine, Fig. 1) shows the greatest separation of adjuvant activity from side effects so far obtained (Waters et al., 1986). This analog was therefore selected as an acceptable counterpart of mycobacteria in an adjuvant formulation (Allison and Byars, 1986). Our next challenge was to develop an alternative to the mineral oil emulsion of Freund's adjuvant that would be suitable for human use.

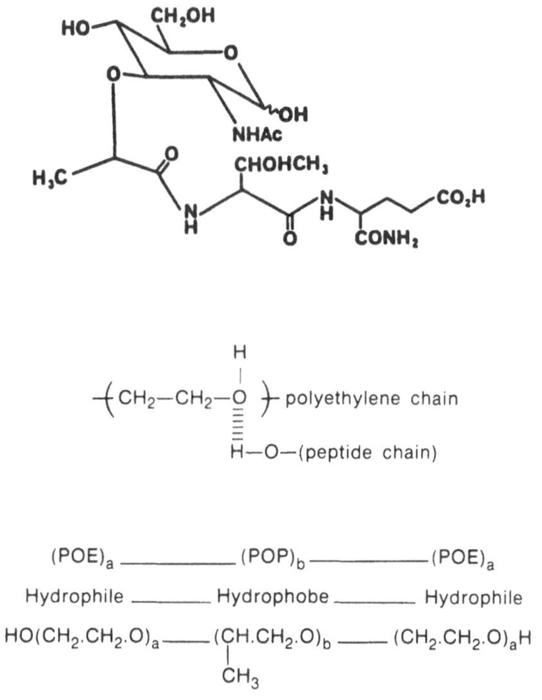

Figure 1. Structure of N-acetylmuramyl-L-threonyl-D-isoglutamine and L-121.

After much experimentation with liposomes, several oil preparations and various surface-active agents, we found that squalene or squalane emulsions, prepared with the Pluronic® block-copolymer L-121 and stabilized with a small amount of Tween 80, provided a versatile vehicle for antigens. Hunter et al. (1981) had used L-121 and related molecules with mineral oil as adjuvants. In L-121 a central block of polyoxypropylene is hydrophobic, while two flanking blocks of polyoxyethylene are hydrophilic because of hydrogen bonding with water. Since it is surface-active, L-121 associates with membranes, but it does not penetrate into membranes and disrupt their structure, unlike saponins which bind cholesterol and are cytolytic. Squalane is saturated and stable in formulation, unlike squalene, which is unsaturated and becomes oxidized. Our microfluidized squalane-L-121 emulsion is remarkably stable, even when frozen.

Our interpretation of the effectiveness of the squalane L-121 emulsion as a vehicle for antigens is as follows (Fig. 2): electron micrographs show that labeled protein antigens are concentrated on the surface of the squalane microspheres. Antigens are retained there partly because they are amphipathic and partly by hydrogen bonding to L-121. The squalane L-121 emulsion system is therefore more versatile than squalane emulsions lacking the block copolymer. It is also more versatile than liposomes, the structure of which has to be optimized for each antigen. The squalane L-121 microsphere particles also activate complement and migrate from injection sites to lymph nodes of the drainage chain. The C3b on the surface of the microspheres could target them to follicular dendritic cells, major antigen-presenting cells. A depot of antigen on follicular dendritic cells is more important for immunogenicity, as well as better for the patient, than a depot at the injection site.

Thus, the function of the squalane-L-121 emulsion is to target antigens to antigen-presenting cells. The function of the MDP analog is to induce the formation of cytokines and increase the expression of major histocompatibility genes, which are required to trigger T-cell-mediated immune responses.

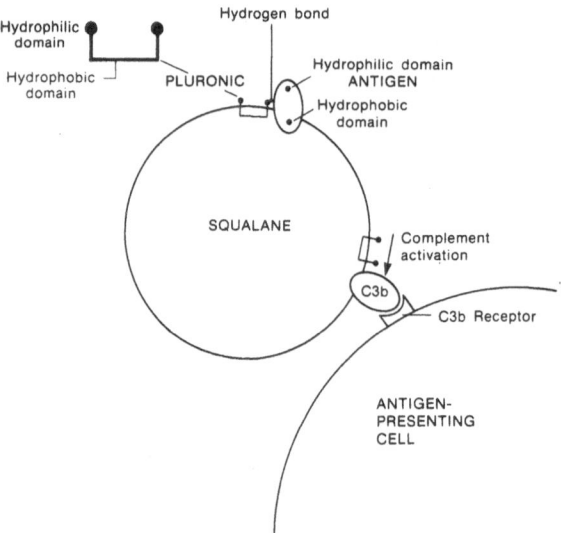

Figure 2.   Diagram of the putative structure of a microsphere in the adjuvant formulation, showing the antigen held at the interface partly because of its amphipathic character and partly because acceptor groups in the antigen form hydrogen bonds with the pluronic polymer.

USE OF SAF IN VACCINES

Examples of the use of SAF in vaccines include inactivated virus vaccines for feline leukemia virus (Braemer et al., 1984), a simian AIDS virus (Marx et al., 1986) and simian immunodeficiency virus (Murphey-Corb et al., 1989) and feline immunodeficiency virus (Yamamoto et al., 1991). HIV-1 antigens in SAF have elicited protection against virus challenge in chimpanzees (Girard et al., 1991). These are promising leads toward the development of a vaccine against HIV. In the absence of an efficacious adjuvant, little or no protection is observed.

Two examples will suffice to illustrate that adjuvants can, to some extent, overcome the effects of age and genetic restriction on immune responses. In general, infants and persons over the age of 65 show lower responses to vaccines than do older children and young adults. Indeed, it is estimated that less than one-third of old recipients of influenza hemagglutinin (HA) show antibody responses (Arden et al., 1986). Alum is not an effective adjuvant for HA (Nicholson et al., 1979). In mice, SAF augments responses to HA, especially in very young and old animals (Byars et al., 1990). If this holds true in humans, the efficacy of influenza vaccines will be improved in populations where they are most needed.

Where hepatitis B virus is prevalent (e.g., Southeast Asia and Africa), immunization of infants can prevent chronic infection, which is associated with hepatitis and hepatocellular carcinoma. Mice of some genetic constitutions are low responders to HBsAg (Milich and Chisari, 1982): we have found that the low response can be overcome by the use of SAF. A good adjuvant can thus improve immune responses when they are deficient because of age or genetic constitution.

SAF is also efficient at eliciting anti-idiotypic antibodies and protecting against B-lymphomas in mice (Campbell et al., 1989). This procedure is currently being used to elicit anti-idiotypic immune responses in humans.

PROTECTION OF SERO-MUCOUS SURFACES BY SYSTEMIC IMMUNITY

Mucosal immunity is interesting academically, and it is likely that locally produced secretory antibodies combining with food proteins play a role in preventing allergies. Research should certainly continue on ways to improve secretory immune responses. Nevertheless, the question arises whether systemic immunity can protect hosts against infections of sero-mucous surfaces. This question is especially relevant in the context of vaccination, because technology now exists for eliciting good circulating IgG responses, CMI and memory in both B- and T-cell compartments, whereas this cannot yet be consistently achieved for mucosal immunity. IgG antibodies of the appropriate subclass are highly protective for reasons described above. While IgM is confined to the vascular compartment, IgG is present in tissue fluid, including the submucosa, and is found in normal intestinal lymph (Barrowman, 1978). When infection is initiated in a sero-mucous surface, local inflammation increases vascular permeability and recruits leukocytes which, together with IgG antibodies, can limit the spread of the infection. Perhaps the first aim of vaccination should be to limit such infections rather than prevent them altogether, which is difficult to achieve at present.

We shall review a few examples from our own experience supporting the assertion that systemic vaccination can protect sero-mucous surfaces from the nose to the vagina. Bordetella bronchiseptica, a bacterium that adheres to the upper respiratory epithelium of young piglets, multiplies and produces local inflammation and nasal cartilage destruction. Consequent impaired growth of the animals is of commercial importance. Our colleague Simon Lee found that systemic immunization of sows can protect their young offspring from nasal colonization with *B. bronchiseptica*. This must be due to maternally derived IgG antibodies and not secretory IgA antibodies in the offspring.

Another colleague, Jean-Louis Virelizier, administered to mice enough cyclophosphamide to abolish their capacity to produce antibodies, systemic or secretory. In these immunosuppressed mice, passive immunization with IgG antibodies directed against the variant-specific determinants of influenza virus hemagglutinin efficiently protected the mice against infection with the virus (Virelizier et al., 1979). The efficacy of SAF in improving antibody responses to influenza HA, especially in very young and old recipients, has been documented (Byars et al., 1990).

Herpes simplex virus (HSV), an example of a virus affecting the oral and genital cavities, has been discussed several times at this meeting. Our aim has been to produce an effective vaccine against HSV using as antigen a recombinant surface glycoprotein with determinants shared between HSV-1 and HSV-2. The model is the guinea pig challenged vaginally with a virulent strain of HSV-2. The virus multiplies prolifically in the vaginal mucosa, spreads in most animals to the dorsal root ganglia (a source of virus in recurrent infections) and produces lethal infections of the central nervous system in about 30% of recipients. Following subcutaneous immunization of guinea pigs with 100 ng gD2t in SAF, vaginal lesions are markedly decreased (though not abolished altogether, Fig. 3). Lethal infections and infections of the dorsal root ganglia are also prevented on vaccination with recombinant gD2 (Table 1). If these findings can be extended to humans, an effective vaccine against genital HSV will have been developed. Vaccination of infants against HSV-1 may likewise decrease stomatitis following primary infection and prevent nerve ganglionic infections and recurrence.

Table 1. Efficacy of vaccination of guinea pigs with recombinant glycoprotein D2 of herpes simplex virus (HSV-2) against genital infection with HSV-2 (unpublished observations of N. Byars, E. Fraser-Smith and A.C. Allison).

Group	No. of Animals	% Dead	% With Genital Lesions	Mean Lesion Score	% With Ganglion Infection
Untreated	40	30	93	2.5	30[3]
Vaccinated[1]	10	0	40	0.08[2]	0

[1]30 µg glycoprotein D was given per animal once at 5 weeks and once at 7 weeks of age.

[2]Lesion scores are measures of the severity of the infection calculated from the overall mean from the daily mean lesion scores over a 14-day period. The difference of the vaccinated from the control group was statistically significant (p<0.05, Mann Whitney U test).

[3]This is an underestimate because it does not include animals that died, presumably from central nervous system infections. The probable infection rate of nerve ganglia was 60%.

We have also been able to immunize susceptible subhuman primates, cottontop tamarins (Sanguinus oedipus), against Epstein-Barr virus by using a major surface glycoprotein (gp340) in SAF (Morgan et al., 1989). The objective of this research in North America, Europe and Japan is to prevent infectious mononucleosis resulting from

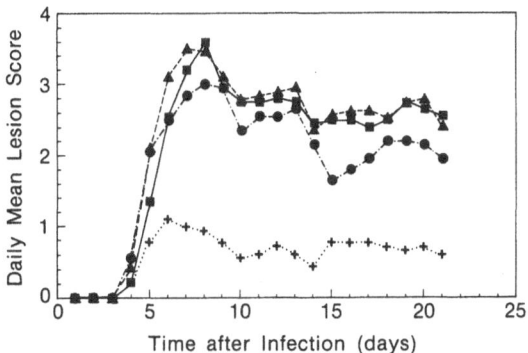

Figure 3. Local lesions following vaccination and vaginal challenge of guinea pigs with HSV-2. Squares - untreated controls. Triangles - SAF only. Filled circles - 100 ng gD2t in phosphate-buffered saline. Crosses - 100 ng gD2t in SAF (N. Byars et al., 1992, submitted for publication).

orally acquired EBV infections. The objective in developing countries such as the People's Republic of China and several African countries is to immunize young children, thereby preventing early EBV infections which are associated with nasopharyngeal carcinoma and Burkitt lymphoma (see Morgan et al., 1989).

If vaccines against periodontal bacteria are to be developed, perhaps systemic immunization, raising in humans IgG1 or IgG3 antibodies of high affinity, may be all that is required. Interacting with complement and leukocytes, such antibodies may be able to reduce bacterial colonization of gingival pockets.

PROSPECTS

Adjuvants with the efficacy of FCA, but without unacceptable side effects limiting human use, have been developed. When used with subunit antigens, including those produced by recombinant technology, the adjuvants elicit protective immunity, including efficacy against some infections of sero-mucous surfaces. Possible applications to infections of the oral cavity, including HSV and EBV, deserve early consideration. The same kind of approach, using LPS or other antigens of gingival bacteria, may be useful in periodontal disease.

REFERENCES

Allison, A.C., and Byars, N.E., 1986, An adjuvant formulation that selectively elicits the formation of antibodies of protective isotypes and cell-mediated immunity, J. Immunol. Methods 95:157.

Allison, A.C., and Byars, N.E., 1992, Adjuvants for new generation vaccines, *In* Vaccines: New Approaches to Immunological Problems, R.W. Ellis, ed. Butterworths, Stoneham, MA (in press).

Arden, N.H., Patriarca, P.A., and Kendal, A.P., 1986, Experiences in the use and efficacy of influenza vaccine in nursing homes, *In* Options for Control of Influenza, A.P. Kendal and P.A. Patriarca, eds., Alan R. Liss, New York, p. 155.

Barrowman, J.A., 1978, "Physiology of the Gastrointestinal Lymphatic System," University Press, Cambridge, p. 145.

Berman, P.W., Gregory, T., Crase, P., and Lasky, L.A., 1985, Protection from genital herpes simplex type 2 infection by vaccination with cloned glycoprotein D, Science 227:1490.

Braemer, A., Peterson, M., Renneke, G., Bass, E., Allison, A.C., Byars, N.E., and Fraser, D., 1984, Effect of inactivated FeLV vaccines on the development of persistent viremia, Proc. 65th Conf. Res. Workers in Animal Diseases, p. 75.

Byars, N.E., and Allison, A.C., 1987, Adjuvant formulation for use in vaccines to elicit both cell-mediated and humoral immunity, Vaccine 5:223.

Byars, N.E., Allison, A.C., Harmon, M.W., and Kendal, A.P., 1990, Enhancement of antibody responses to influenza B virus hemagglutinin by use of a new adjuvant formulation, Vaccine 8:49.

Campbell, M.J., Esserman, L., Byars, N.E., Allison, A.C., and Levy, R., 1989, Development of a new therapeutic approach to B-cell malignancy. The induction of immunity by the host against cell surface receptor on the tumor, Intern. Rev. Immunol. 4:251.

Ellouz, F., Adam, A., Ciorbaru, R., and Lederer, E., 1974, Minimal structural requirements for adjuvant activity of bacterial peptidoglycans, Biochem. Biophys. Res. Comm. 59:1317.

Girard, M., Kieny, M.-P., Renter, A., Barre-Sinoussi, F., Nara, P., Kobe, H., Kusumi, K., Chaput, A., Reinhart, T., Muchmore, E., Ronco, J., Kaczonek, M., Garrard, E., Gluckman, J.-C., and Fultz, P.N., 1991, Immunization of chimpanzees confers protection against challenge with human immunodeficiency virus, Proc. Natl. Acad. Sci. (USA) 88:542.

Finkelman, F.D., Holmes, J., Katona, I.M., Urban, J.F., Jr., Beckmann, P., Park, L.S., Schooley, K.A., Coffman, R.L., Mossman, T.R., and Paul, W.E., 1990, Lymphokine control of in vivo immunoglobulin isotype selection, Annu. Rev. Immunol. 8:303.

Hunter, R.L., Strickland, F., and Kezdy, F., 1981, The adjuvant activity of nonionic block polymer surfactants. I. The role of hydrophile-lipophile balance, J. Immunol. 127:1244.

Kaminski, M.S., Kitamura, K., Maloney, D.G., Campbell, M.J., and Levy, R., 1986, Importance of antibody isotype in monoclonal anti-idiotype therapy of murine B cell lymphoma. A study of hybridoma class-switch variants, J. Immunol. 136:1123.

Kenney, J.S., Hughes, B.W., Masada, M.P., and Allison, A.C., 1989, Influence of adjuvants on the quantity, affinity, isotype and epitope specificity of murine antibodies, J. Immunol. Methods 121:157.

Marx, P.A., Pedersen, N.C., Lerche, N.W., Osborn, K.G., Lowestine, L.J., Lackner, A.A., Maul, D.H., Kwang, H.-S., Kluge, J.D., Zaiss, C.P., Sharpe, V., Spinner, A.P., Allison, A.C., and Gardner, M.B., 1986, Prevention of simian acquired immunodeficiency syndrome with a formalin-inactivated Type D retrovirus vaccine, J. Virol. 60:431.

Morgan, A.J., Allison, A.C., Finerty, S., Scullion, F.T., Byars, N.E., and Epstein, M.A., 1989, Validation of a first generation Epstein-Barr virus vaccine preparation suitable for human use, J. Med. Virol. 29:74.

Morris, A.G., Lin, Y.-L., and Askonas, B.A., 1982, Immune interferon release when a cloned cytotoxic T-cell line meets its correct influenza-infected target, Nature 295:150.

Milich, D.R., and Chisari, F.V., 1982, Genetic regulation of the immune response to hepatitis B surface antigen (HBsAg), I. H-2 restriction of the murine humoral response to the a and d determinants of HBsAg, J. Immunol. 129:320.

Murphey-Corb, M., Martin, L.N., Davison-Fairburn, B., Montelaro, R.C., Miller, M., West, M., Okawa, S., Baskin, G.B., Zhang, J.-Y., Putney, S.D., Allison, A.C., and Eppstein, D.A., 1989, A formalin inactivated whole simian immunodeficiency virus vaccine confers protection in macaques, Science 246:1293.

Nicholson, K.G., Tyrrell, D.A.J., Harrison, P., Potter, C.W., Jennings, R., Clark, A., Schild, G.C., Wood, J.M., Yells, R., Seagrott, V., Huggens, A., and Anderson, S.G., 1979, Clinical studies of monovalent inactivated whole virus and subunit A/USSR/77 (H_1N_1) vaccine: serological and clinical reactions, J. Biol. Stand. 7:123.

Reichmann, L., Clark, M., Waldmann, H., and Winter, G., 1988, Reshaping human antibodies for therapy, Nature 332:323.

Takahashi, H., Takeshita, T., Morein, B., Putney, S., Germain, R.N., and Berzofsky, J.A., 1990, Induction of $CD8^+$ cytotoxic T-cells by immunization with purified HIV-1 envelope protein in ISCOMS, Nature 344-873.

Virelizier, J.-L., Allison, A.C., and Schild, G.C., 1979, Immune responses to influenza virus in the mouse, Brit. Med. Bull. 35:65.

Waters, R.V., Terrell, T.G., and Jones, G.H., 1986, Uveitis induction in the rabbit by muramyl dipeptides, Infect. Immun. 51:816.

Walker, B.D., Flexner, C., Paradis, T.J., Fuller, T.C., Hirsch, M.S., Schooley, R.T. and Moss, B., 1988, HIV-1 reverse transcriptase is a target for cytotoxic T-lymphocytes in infected individuals, Science 240:64.

White, R.G., 1976, The adjuvant effect of microbial products on the immune response, Annu. Rev. Microbiol. 30:579.

Yamamoto, J.K., Okuda, T., Ackley, C.D., Louie, H., Penibroke, E., Zochlinski, H., Munn, R.J., and Gardner, M.B., 1991, Experimental protection against feline immunodeficiency virus, AIDS Res. and Hum. Retroviruses 7:911.

M CELL-MEDIATED ANTIGEN TRANSPORT AND MONOCLONAL IGA ANTIBODIES FOR MUCOSAL IMMUNE PROTECTION

Marian R. Neutra[1] and Jean-Pierre Kraehenbuhl[2]

[1]Harvard Medical School
 GI Cell Biology, Children's Hospital
 300 Longwood Avenue, Boston, MA 02115
[2]Swiss Institute for Experimental Cancer Research
 CH-1066, Epalinges, Switzerland

INTRODUCTION

The intestinal mucosa is lined by a simple columnar epithelium that separates both inductive and effector cells of the mucosal immune system from the external environment. Although this tight cellular monolayer forms an effective barrier to most lumenal microbes and macromolecules, its specialized transport activities play key roles in immune protection of the mucosal surface. M cells in the lymphoid follicle-associated epithelium deliver samples of lumenal antigens into the organized mucosal lymphoid tissue, while most other epithelial and glandular cells selectively export dimeric or polymeric IgA into secretions and confer protease resistance to these effector molecules by addition of secretory component (SC).

Research in our laboratories is aimed at defining the characteristics of M cells that influence the efficiency of antigen transport, and analyzing the protective roles of specific sIgA antibodies. A first step in our approach is to apply macromolecular tracers and microbial pathogens to the Peyer's patch mucosa, and evaluate M cell binding and transport directly using morphological methods[1,2,3]. The IgA response is then indirectly analyzed by immunodetection of specific IgA in secretions and by generation of specific IgA-producing hybridomas from Peyer's patch cells[2,4,5]. Mucosally derived hybridomas provide monoclonal IgA antibodies of defined antigen specificities that can then be tested for their protective capacities *in vivo* and *in vitro*, as described below.

M CELLS AND ANTIGEN TRANSPORT

The epithelium over mucosal lymphoid follicles differs dramatically from the epithelium lining the rest of the mucosa in that it lacks basolateral receptors for polymeric immunoglobulins and thus does not conduct export of sIgA[6]. The follicle-associated epithelium contains M cells[7,8] that lack highly organized apical brush borders and the thick layer of membrane hydrolases of intestinal absorptive enterocytes[1,9]. They conduct continuous endocytosis of material from the lumen from abundant coated[1] and uncoated[8] pits in their broad apical membranes. The architecture of these cells is unique: the basolateral plasma membrane is deeply invaginated to form a large, intraepithelial pocket containing T cells, B cells and macrophages[10]. The basolateral "pocket" domain of the plasma membrane thus lies within a few microns of the apical membrane and endocytic vesicles rapidly traverse this short distance to deliver samples of lumenal material into the intraepithelial pocket. The route of transcytotic vesicle traffic in M cells is unlike that in other types of epithelial cells in that none is directed to dense lysosomes deep in the cell, or to the lateral or basal cell surfaces.

Genetically Engineered Vaccines, Edited by
J.E. Ciardi *et al.*, Plenum Press, New York, 1992

Adherence

Since transcytosis of antigens by M cells is apparently the first step in mounting a secretory immune response, it is significant that macromolecules that adhere to M cell apical membranes are transported into the basolateral pocket at least 50 times more efficiently than non-adherent ones[1]. This is consistent with the observation that adherent microorganisms, toxins and lectins are most effective in evoking specific sIgA antibodies[11]. M cell membranes display a variety of glycoconjugates that can serve as receptors for lectins[1,9]. A single component unique to M cells has yet to be identified, but the fact that certain pathogenic bacteria and viruses adhere selectively to M cells implies that unique membrane components are indeed exposed on these cells. For example, reovirus, a mouse pathogen, adheres exclusively to M cells *in vivo* [12], and the human pathogens poliovirus[13] and HIV-1[3] adhere to M cell surfaces in mucosal explants. Several types of pathogenic gram-negative bacteria adhere preferentially (but not exclusively) to M cell surfaces; these include some (but not all) pathogenic strains of *Escherichia coli*[14], *Shigella* [15],*Vibrio cholerae* [2,16], and *Salmonella*[17]. It is possible that a primitive recognition mechanism such as lectin-carbohydrate interactions allow the M cell to deliver a wide range of bacteria for sampling in the mucosal immune system.

Such adherence may reflect the M cell's lack of the thick glycocalyx of complex glycoprotein enzymes rather than unique membrane components. To test this possibility, we conducted studies involving the application of cholera toxin to Peyer's patch mucosa, either as single toxin molecules coupled to rhodamine or as polyvalent complexes on 15 nm colloidal gold particles (R. Weltzin, H. M. Amerongen, W.I. Lencer, and M. R. Neutra, unpublished). The toxin-rhodamine conjugate bound to ganglioside receptors on apical membranes of both M cells and absorptive epithelial cells, whereas the colloidal gold conjugate was able to bind only to M cells. This suggests that M cell membrane glycoconjugates may be particularly accessible to polyvalent, particulate ligands such as viruses and bacteria; if so, it could explain in part the special ability of M cells to bind microorganisms.

Endocytosis and transport

Adherent macromolecules and small particles are endocytosed by M cells via clathrin-coated pits and vesicles[1], while soluble proteins enter in the fluid content of coated and uncoated vesicles[1,8]. Endocytosed tracers of all kinds enter an apical endosomal system and within 10-15 minutes these vesicles begin to fuse with the pocket membrane, delivering their content into the intraepithelial pocket. It is not known whether M cell apical endosomes contain endosomal hydrolases or mannose-6-phosphate receptors as evidence of enzyme delivery, but the membranes of some apical vesicles contain the glycoprotein lgp120 that is a marker for late endosomes and lysosomes in many cells[18]. This is of interest because proteolytic processing of immunogens in the transepithelial pathway could have important consequences for the mucosal immune response.

Uptake of adherent *V. cholerae* involves a process resembling phagocytosis, in which the M cell apical membrane is separated from the bacterial outer membrane by a uniform 10 to 20 nm gap (J. Mack and M.R. Neutra, unpublished). This interaction involves recruitment of unknown M cell membrane components that interact directly or indirectly with actin, since actin filaments are recruited to the site of interaction. Bacteria (and viruses) are transported and released into the intraepithelial pocket morphologically unaltered and presumably alive since some, such as reovirus and *Salmonella,* go on to invade mucosal cells[19,20]. In the case of the non-invasive pathogen *V. cholerae*, however, M cell transport is advantageous to the host, as it results in processing and presentation of bacterial antigens, and initiation of a protective mucosal immune response. Since delivery of bacteria and virus into the pocket is followed by release of the organisms from the M cell membrane, it is likely that initial binding is accomplished through multiple, low-affinity interactions that are readily reversed by a change in pH or ion content in the vesicles or the pocket.

Release of microorganisms and macromolecules in the pocket delivers them directly to the cells of the immune system. MHC class II complexes were recently detected in M cell apical vesicles[18], but it is not yet known whether these cells are capable of presenting foreign antigens. Other cells that express MHC class II (B cells, T cells, macrophages) are present in the pocket and thus immune responses could be initiated in this sequestered site. In addition, a network of MHC class II-positive dendritic cells and macrophages, T helper cells, and IgM-positive cells are present immediately below the follicle-associated epithelium to complete the processing and presentation of antigens[21]. These mucosal inductive sites may be functionally isolated from serum antibodies, since the capillaries in Peyer's patch tissues are non-fenestrated and impermeable to circulating tracer proteins[22], and we have observed that intravenous HRP does not enter the M cell pocket. The functional isolation of Peyer's patch mucosa would in part explain the relative autonomy of the mucosal immune response, and the fact that induction of secretory immune responses generally requires application of immunogens to mucosal surfaces. This also helps to explain why circulating IgG antibodies are often unable to prevent invasion of the Peyer's patches by pathogens[23].

M CELLS AS INVASION SITES

While M cells were apparently designed to deliver microorganisms and antigens to the antigen processing and presenting cells of the mucosal immune system, their ability to rapidly transport adherent materials creates functional breaches in the epithelial barrier. The pathogenic bacteria and viruses that exploit this pathway to invade the mucosa and the body differ widely in their subsequent infection strategies. For example, *Salmonella typhi* in humans and *S. typhimurium* in mice invade macrophages and other cells in the mucosa, spread to the liver and spleen, and eventually cause a lethal systemic disease[20,24,25]. In contrast, *Shigella flexneri*, after delivery across the epithelium by M cells[15], can invade enterocytes via their basolateral surfaces and spread laterally within the epithelium[26]. As noted above, reovirus exploits the M cell pathway[12,19] and then infects epithelial cells from the basolateral side[27]. This allows the virus to use both the M cell and surrounding enterocytes as viral factories[28]. Reovirus also proliferates in other mucosal cells and targets itself to mucosal nerves through which it eventually spreads to the central nervous system[19,29]. We have recently shown that retroviruses can also invade via M cells. Members of our group have observed that mouse mammary tumor virus (MTV) in infected milk is taken up across M cells (H. Amerongen and P. Michetti, unpublished); later, MTV induces clonal deletion of T cells expressing a specific V-beta element of the T cell receptor[30]. It is apparently the infected lymphocytes that are responsible for transmission of virus to the mammary gland, the ultimate target tissue[31]. Colleagues in our laboratories have also identified the M cell as a potential site of entry of HIV-1[3]. The virus selectively binds to M cells in Peyer's patches from mice and rabbits, and is transported into the intraepithelial pocket that contains T cells. Thus, for many microorganisms, M cells represent an efficient invasion site. Protection of this vulnerable epithelial site may be accomplished by secretory IgA antibodies, however, as detailed below.

PRODUCTION AND EVALUATION OF MONOCLONAL IgAs

There is considerable evidence that sIgA in intestinal secretion is associated with protection against enterotoxigenic and invasive enteric pathogens, but whether sIgA alone can provide protection against both surface-colonizing and invasive microorganisms had not been established previously. This is difficult to prove in immunized animals because "sampling" of bacteria and viruses by cells in organized mucosal lymphoid tissue generally results in production of serum IgG and IgA and, in some cases, cell-mediated immunity as well as secretory IgA. The presence of all of these types of immune protection, as well as non-immune protection mechanisms, complicates the analysis of IgA action. In addition, the relative protective capacities of individual sIgAs of well-defined antigen specificities have not been systemically tested or compared in the intestine *in vivo*. Dissection of the exact roles of sIgA in protection against the various bacterial, viral, protozoal and parasitic pathogens that colonize the intestinal lumen, infect intestinal mucosa, or use the intestinal epithelium as an invasion route requires monospecific IgA antibodies.

We have found that fusion of Peyer's patch cells provides a very efficient means of generating IgA hybridomas[4]. Our attempts at direct injection into Peyer's patch tissue were less effective, presumably because the immune response is most efficient when M cells deliver the immunogens directly to organized assemblies of antigen-processing and presenting cells. After two or three oral boosts, cells are obtained by excision of Peyer's patch mucosa and digestion with collagenase. Released cells are fused with mouse myeloma cells, seeded in 96-well plates and screened by ELISA, Western blot, and/or immunoprecipitation. At least 50 percent of the pathogen- or antigen-specific hybridomas generated using this method were of the IgA type. Indeed, after mucosal immunization with *S. typhimurium,* all of the stable anti-*Salmonella* hybridomas obtained produced IgA antibodies[5].

To function in our protection assays, it is important that the monoclonal IgAs are structurally and functionally analogous to the IgAs produced by normal subepithelial plasma cells. By gradient SDS-PAGE and immunoblot analysis, almost all of IgA hybridomas derived from Peyer's patch cells produce IgA primarily in dimeric form, with some monomers and high polymers[2,4,5]. Metabolically radiolabeled IgA from hybridoma cultures were injected IV into rats and were shown to be efficiently transported into bile[4]. Transported IgA was recovered from bile as dimers and polymers whose molecular weight was increased by about 80 kD, consistent with the addition of SC. Thus, the monoclonal IgAs are recognized by polymeric Ig receptors and are transported into secretions by epithelial cells *in vivo* via the normal, receptor-mediated transepithelial transport system.

EVALUATION OF PROTECTION BY MONOCLONAL IgA

Monoclonal IgA antibodies against respiratory pathogens have been shown to provide protection after passive application to the respiratory mucosa[32] or active secretion following systemic injection[33]. Monoclonal IgA lacks secretory component and would be readily degraded in the GI tract[34,35]. Dimeric IgA is more stable than monomeric IgG but the survival of dimeric monoclonal IgA without SC in the stomach and intestine is not known. In the gut, there is also the difficult problem of achieving effective, predictable antibody distribution over the mucosa of the entire intestine after an oral dose and correctly timing the oral IgA to coincide with local microbial challenge. Furthermore, normally secreted IgA antibodies are delivered into intestinal crypts and directly onto epithelial surfaces, but orally administered monoclonal IgA might never reach these sites.

Members of our laboratories have developed a convenient method to achieve continuous, widespread delivery of monoclonal IgA into secretions. It was known that hybridoma cells can form immunoglobulin-producing tumors *in vivo* [36], and that circulating dimeric IgA binds to epithelial poly-Ig receptors and is transported into secretions[37]. IgA-producing hybridoma cells are injected subcutaneously on the upper backs of BALB/c mice and palpable tumors with blood and lymphatic drainage form within 1 to 2 weeks. As the tumors grow, increasing levels of specific monoclonal IgA in serum are accompanied by increasing levels in secretions, whereas antibodies from IgG-producing hybridoma tumors enter secretions in low amounts despite very high blood levels[2]. The IgA hybridoma "backpack" tumor method allows monoclonal IgA to be continually delivered into secretions via the normal physiologic route, and with addition of secretory component. In "backpack" tumor-bearing mice, however, blood and tissue levels of monoclonal IgA antibodies are very high[2]. This is in contrast to a normal mucosal immune response where dimeric IgA is produced locally by subepithelial plasma cells and blood levels of IgA are relatively low[38]. Total secreted intestinal IgA levels are not elevated in backpack mice, however; this suggests that the amount of IgA secreted is limited by the capacity of the receptor-mediated epithelial transport system[39]. In tumor-bearing mice, the secreted monoclonal IgA may represent a relatively large proportion of total lumenal IgA. Thus, the backpack tumor method allows us to assess the protective capacities of specific sIgA antibodies against pathogens that colonize or invade the mouse intestine in the absence of other immune protection mechanisms.

Vibrio cholerae

V. cholerae colonizes the surface of absorptive cells in the intestinal mucosa and causes diarrheal disease by secreting cholera toxin that binds with high affinity to the ganglioside, GM1[40,41]. *V. cholerae* infection, or immunization with bacterial cells and toxins, evokes sIgA directed against both toxin and bacterial surface components including lipopolysaccharide (LPS), and this response is associated with protection against subsequent challenge[41-43]. Both cholera toxin and *V. cholerae* bind to M cells, are efficiently transported into Peyer's patch mucosa, and evoke specific IgA lymphoblasts. We generated hybridomas that produce monoclonal IgAs directed against a LPS carbohydrate epitope specific to the Ogawa strain, and showed that the epitope is exposed on the bacterial surface[2]. Monoclonal IgA antibodies were also generated against the binding (B) subunit of cholera toxin (CTB)[39]. We have tested the protective capacities of these antibodies against virulent organisms and against cholera holotoxin both *in vivo* and *in vitro*.

Using the hybridoma "backpack" tumor system in suckling mice, we demonstrated that secretion of monoclonal sIgA directed against the Ogawa strain-specific LPS epitope protected mice against oral challenge with 100 times the LD_{50} of virulent organisms[2]. Secretion of irrelevant IgAs from tumors was not protective, and mice with anti-LPS tumors were not protected against a lethal dose of the Inaba strain[2]. Unexpectedly, secretion of anti-CTB IgA from backpack tumors failed to protect against the same challenge[39]. Anti-LPS IgA presumably aggregated vibrios in the lumen and prevented colonization, whereas anti-toxin IgA permitted colonization to occur and was then unable to intercept toxin secreted from colonies directly onto epithelial cells. These results are consistent with previous observations in immunized animals and trials in human volunteers[41-46].

Anti-LPS and CTB monoclonal IgAs were also tested for their ability to provide suckling mice with passive oral protection against live organisms or purified toxin. Oral administration of a single dose of 5 μg of anti-LPS IgA prevented diarrheal disease after oral challenge with *V. cholerae*, but protection was observed only when the bacterial challenge was given within 3 hours after the IgA[39]. In contrast, oral delivery of large doses (up to 100 μg) of anti-toxin failed to protect against oral challenge with live vibrios. Large doses of anti-CTB IgA did provide some protection against oral challenge with toxin alone, however, although this probably involved mixing of toxin and IgA in the stomach. These studies directly demonstrated that anti-LPS IgA is much more efficient than anti-toxin IgA in providing protection against *V. cholerae*. Indeed, in oral cholera vaccine trials, antibodies against the bacterial cells were found to be an important component in protection[42,45,46].

Salmonella typhimurium

We used a *Salmonella* model to test whether secretory IgA alone can prevent infection of the mucosa by an invasive organism and hence prevent systemic disease[5]. *S.typhimurium* in mice, like *S.typhi* in humans, causes lethal systemic disease by invading the intestinal mucosa via M cells and enterocytes[17,20, 47]. BALB/c mice were orally immunized with attenuated mutant strains of *S. typhimurium*, and 48 stable IgA-producing hybridoma clones were obtained. For our studies, a hybridoma was selected that produced monoclonal dimeric and polymeric IgA directed against a surface carbohydrate epitope present on wild-type organisms, and that efficiently agglutinated live organisms *in vitro*. As control, an *S. typhimurium* mutant was isolated that lacked the epitope and "escaped" IgA-mediated aggregation *in vitro*, but was nevertheless fully virulent *in vivo*.

The monoclonal IgA was tested for protection in adult mice using the hybridoma backpack tumor method[5]. Backpack tumor-bearing mice challenged orally with up to 10^7 wild-type organisms were protected against disease, whereas those challenged with the "escape" mutant were consistently infected. We also tested whether the abnormally high levels of specific monoclonal IgA present in the blood and tissues of tumor-bearing mice might have provided protection by neutralizing the bacteria after invasion of the mucosa. Mice with IgA hybridoma tumors, however, were not protected with intraperitoneal challenge with *S. typhimurium*. This established that the secreted monoclonal IgA was responsible for providing the protection observed[5].

Reovirus

To determine whether IgA can prevent entry of pathogens that invade exclusively via the M cell pathway, monoclonal anti-reovirus IgA antibodies were also tested *in vivo*. In these studies, we considered the possibility that invasion could be enhanced by IgA rather than prevented. This is because immunoglobulin binding sites on M cells[4] might facilitate uptake of IgA-coated virus if contact with the follicle-associated epithelium occurred. Reovirus type 1 is a mouse pathogen that adheres specifically to M cells and exploits M cell transport to invade the Peyer's patch mucosa, infect mucosal cells and spread systemically[12,19,29]. Monoclonal IgA antibodies directed against two major outer capsid proteins of reovirus (sigma 3 and mu1c) were produced[4]. Adult mice were given IgA hybridoma backpack tumors of three different cell clones, and they secreted monoclonal sIgA antibodies against one of the reovirus proteins (sigma 3 or mu1c) or an irrelevant antigen. When challenged orally with an infectious dose of reovirus, levels of virus in Peyer's patch mucosa (24 hours after challenge) were dramatically reduced in mice secreting monoclonal anti-mu1c IgA as compared to control mice secreting irrelevant monoclonal IgA. In most mice secreting anti-sigma 3 IgA, however, levels of virus were not lowered[48]. Sigma 3 protein is cleaved from the viral surface by pancreatic proteases during passage of virus through the intestinal lumen whereas most of the mu1c protein remains exposed on the virus[29]. This explains the relative inability of anti-sigma 3 IgA to provide protection. These studies underline the importance of the microbial alterations that normally occur in the intestinal lumen prior to epithelial adherence and invasion, in rational design of effective oral vaccines.

CONCLUSIONS

Our studies have demonstrated that IgA hybridomas are readily derived by fusion of Peyer's patch cells after mucosal immunization with bacteria and viruses that are efficiently transported by M cells. The monoclonal IgA antibodies produced by these hybridomas are analogous to IgA produced by normal mucosal plasma cells in that they are primarily dimeric, and can be transported and secreted via the normal receptor-mediated epithelial transport system. Using a hybridoma "backpack" tumor method in BALB/c mice, we have demonstrated that secretion of sIgA directed against a single, surface-exposed epitope can be sufficient to prevent mucosal surface colonization and diarrheal disease by *V. cholerae*, whereas anti-cholera toxin sIgA alone is not protective. Secretion of sIgA alone can also prevent mucosal infection and systemic disease by *S. typhimurium*, apparently by preventing mucosal contact and invasion. Invasion of the M cells of Peyer's patch mucosa by reovirus can be prevented only by secretion of sIgA directed against a viral surface antigen that survives intestinal passage. This experimental approach provides direct evidence for the importance of secretory IgA in the protection of mucosal surfaces against viral and bacterial pathogens that infect or invade the intestinal mucosa.

ACKNOWLEDGMENTS

We gratefully acknowledge the important contributions of members of our laboratories to the work summarized here: Richard Weltzin, Louis Winner III, Julie Mack, Helen Amerongen, Pierre Michetti, Felice Apter and Wayne Lencer. The collaboration of Bernard Fields, Lynda Morrison, John Mekalanos, Michael Mahan and Jim Slauch was also essential to this work. We thank Betty Ann McIsaac for preparation of this manuscript.

REFERENCES

1. M.R. Neutra, T.L. Phillips, E.L. Mayer, and D.J. Fishkind, Transport of membrane-bound macromolecules by M cells in follicle-associated epithelium of rabbit Peyer's patch, *Cell Tissue Res.* 247:537 (1987).
2. L.S. Winner III, J. Mack, R.A. Weltzin, J.J. Mekalanos, J.P. Kraehenbuhl, and M.R. Neutra, New model for analysis of mucosal immunity: intestinal secretion of specific monoclonal immunoglobulin A from hybridoma tumors protects against *Vibrio cholerae* infection, *Infect. Immun.* 59:977 (1991).
3. H.M. Amerongen, R.A. Weltzin, C.M. Farnet, P. Michetti, W.A. Haseltine, and M.R. Neutra, Transepithelial transport of HIV-1 by intestinal M cells: a mechanism for transmission of AIDS, *J. Acquir. Immune Defic. Syndr.* 4:760 (1991).
4. R.A. Weltzin, P. Lucia-Jandris, P. Michetti, B.N. Fields, J.P. Kraehenbuhl, and M.R. Neutra, Binding and transepithelial transport of immunoglobulins by intestinal M cells:demonstration using monoclonal IgA antibodies against enteric viral proteins, *J. Cell Biol.* 108:1673 (1989).
5. P. Michetti, M.J. Mahan, J.M. Slauch, J.J. Mekalanos, and M.R. Neutra, Monoclonal secretory IgA protects mice against oral challenge with the invasive pathogen *Salmonella typhimurium, Infect. Immun.* 60:1786 (1992).
6. J. Pappo and R.L. Owen, Absence of secretory component expression by epithelial cells overlying rabbit gut-associated lymphoid tissue, *Gastroenterology* 95:1173 (1988).
7. D.E. Bockman and M.D. Cooper, Pinocytosis by epithelium asociated with lymphoid follicles in the bursa of Fabricius, appendix, and Peyer's patches, An electron microscopic study, *Am. J. Anat.* 136:455 (1973).
8. R.L. Owen, Sequential uptake of horseradish peroxidase by lymphoid follicle epithelium of Peyer's patch in the normal unobstructed mouse intestine: an ultrastructural study. *Gastroenterology* 72:440 (1977).
9. R.L. Owen and D.K. Bhalla, Cytochemical analysis of alkaline phosphatase and esterase activities, and of lectin-binding and anionic sites in rat and mouse Peyer's patch M cells, *Am. J. Anat.* 168:199 (1983).
10. T.H. Ermak, H.J. Steger, and J. Pappo, Phenotypically distinct subpopulations of T cells in domes and M-cell pockets of rabbit gut-associated lymphoid tissues, *Immunol.* 71:530 (1990).
11. H.J. DeAizpurua and G.J. Russell-Jones, Oral vaccination: Identification of classes of proteins that provoke an immune response upon oral feeding, *J. Exp. Med.* 167:440 (1988).
12. J.L. Wolf, D.H. Rubin, R. Finberg, R.S. Kauffman, A.H. Sharpe, J.S. Trier, and B.N. Fields, Intestinal M cells: a pathway for entry of reovirus into the host, *Science* 211:471 (1981).
13. P. Sicinski, J. Rowinski, J.B. Warchol, Z. Jarzcabek, W. Gut, B. Szczygiel, K. Bielecki, and G. Koch, Poliovirus type 1 enters the human host through intestinal M cells, *Gastroenterology* 98:56 (1990).
14. L.R. Inman and J.R. Cantey, Specific adherence of *Escherichia coli* (strain RDEC-1) to membranous (M) cells of the Peyer's patch in *Escherichia coli* diarrhea in the rabbit, *J. Clin. Invest.* 71:1 (1983).
15. J.S. Wassef, D.F. Keren, and J.L. Mailloux, Role of M cells in initial antigen uptake and in ulcer formation in the rabbit intestinal loop model of shigellosis, *Infect. Immun.* 7:58 (1989).
16. R.L. Owen, N.F. Pierce, R.T. Apple, and W.C. Cray, Jr., M cell transport of *Vibrio cholerae* from the intestinal lumen into Peyer's patches: a mechanism for antigen sampling and for microbial transepithelial migration, *J. Infect. Dis.* 153:1108 (1986).
17. S. Kohbata, H. Yokobata, and E. Yabuuchi. Cytopathogenic effect of *Salmonella typhi* GIFU 10007 on M cells of murine ileal Peyer's patches in ligated ileal loops: An ultrastructural study, *Microbiol. Immunol.* 30:1225 (1986).
18. C.H. Allan, D.L. Mendrick, and J.S. Trier, M cells contain acidic compartments and express MHC class II determinants, *Gastroenterology* 102:A589 (1992).
19. A.H. Sharpe and B.N. Fields, Pathogenesis of viral infections. Basic concepts derived from the reovirus model, *N. Engl. J. Med.* 312:486 (1985).
20. A.W. Hohmann, G. Schmidt, and D. Rowley, Intestinal colonization and virulence of *Salmonella* in mice, *Infect. Immun.* 22:763 (1978).
21. T.H. Ermak and R.L. Owen, Differential distribution of lymphocytes and accessory cells in mouse Peyer's patches, *Anat. Rec.* 215:244 (1986).
22. C.H. Allan and J.S. Trier, Structure and permeability differ in subepithelial villus and Peyer's patch follicle capillaries, *Gastroenterology* 100:1172 (1991).
23. K.L. Tyler, H.W. Virgin, R. Bassel-Duby, and B.N. Fields, Antibody inhibits defined stages in the pathogenesis of reovirus serotype 3 infection of the nervous central system, *J. Exp. Med.* 170:887 (1989).
24. S.L. Gorbach, Infectious diarrhea, in : "Gastrointestinal Disease: Pathophysiology, Diagnosis Management," M.H. Sleisenger and J.S. Fordtran, ed., W.B. Saunders Company, Philadelphia (1989).
25. P.B. Carter and F.M. Collins, The route of enteric infection in normal mice, *J. Exp. Med.* 139:1189 (1974).
26. T. Vasselon, J. Mounier, M.C. Prevost, R. Hellio, and P.J. Sansonetti, Stress fiber-based movement of *Shigella flexneri* within cells, *Infect. Immun.* 59:1723 (1991).
27. D. Rubin, Reovirus serotype 1 binds to the basolateral membrane of intestinal epithelial cells, *Microb. Pathogen.* 3:215 (1987).

28. D.M. Bass, J.S. Trier, R. Dambrauskas, and J.L. Wolf, Reovirus type I infection of small intestinal epithelium in suckling mice and its effect on M cells, *Lab. Invest.* 58:226 (1988).
29. M.L. Nibert, D.B. Furlong, and B.N. Fields, Mechanisms of viral pathogenesis. Distinct forms of reoviruses and their roles during replication in cells and host, *J. Clin. Invest.* 88:727 (1991).
30. H. Acha-Orbea, H. Shakhov, L. Scarpellino, E. Kolb, V. Muller, A. Vessaz-Shaw, R. Fuchs, K. Blochlinger, P. Rollini, J. Billotte, M. Sarafidou, H.R. MacDonald, and H. Diggelmann, Clonal deletion of V beta 14 positive T cells in mammary tumor virus transgenic mice, *Nature* 350:207 (1991).
31. A. Tsubara, M. Inaba, S. Imai, A. Murakami, N. Oyaizu, R. Yasumizu, Y. Ohnishi, H. Tanaka, S. Morii and S. Ikehara, Intervention of T cells in transportation of mouse mammary tumor virus (milk factor) to mammary gland cells *in vivo*, *Cancer Res.* 48:6555 (1988).
32. M.B. Mazanec, J.G. Nedrud, and M.E. Lamm, Immunoglobulin A monoclonal antibodies protect against Sendai virus, *J. Virol.* 61:2624 (1987).
33. K.B. Renegar and P.A.J. Small. Passive transfer of local immunity to influenza virus infection by IgA antibody. *J. Immunol.* 146:1972 (1991).
34. B.J. Underdown and K.J. Dorrington, Studies on the structural and conformational basis for the relative resistance of serum and secretory immunoglobulin A to proteolysis, *J. Immunol.* 112:949 (1974).
35. E. Lindh. Increased resistance of immunoglobulin A dimers to proteolytic degradation after binding of secretory component, *J. Immunol.* 114:284 (1975).
36. M.C. Harris, S.D. Douglas, G.B. Kolski, and R.A. Polin, Functional properties of anti-group B streptococcal monoclonal antibodies, *Clin. Immunol. Immunopathol.* 24:342 (1982).
37. G.D.F. Jackson, I. Lemaitre-Coelho, and J.P. Vaerman, Transfer of MOPC-315 IgA to secretions in MOPC-215 tumour-bearing and normal BALB/c mice, in "Protides of the biological fluids," H. Peters, ed., Pergamon Press, New York (1977).
38. J. Mestecky, Immunobiology of IgA, *Am. J. Kidney Dis.* 12:378 (1988).
39. F.M. Apter, W.I. Lencer, J.J. Mekalanos, and M.R. Neutra, Analysis of epithelial protection by monoclonal IgA antibodies directed against cholera toxin B subunits *in vivo* and *in vitro*, *J. Cell Biol.* 115:399a (1991).
40. J.F. Miller, J.J. Mekalanos, and S. Falkow, Coordinate regulation and sensory transduction in the control of bacterial virulence, *Science* 243:916 (1989).
41. M.M. Levine, J.B. Kaper, R.E. Black, and M.L. Clements, New knowledge on pathogenesis of bacterial enteric infections as applied to vaccine development, *Microbiol Rev.* 47:510 (1983).
42. M.M. Levine, J.B. Kaper, D. Herrington, G. Losonsky, J.G. Morris, M.L. Clements, R.E. Black, B. Tall, and R. Hall, Deletion mutants of *Vibrio cholerae* 01 prepared by recombinant techniques: insights on pathogenesis and immunity. *Infect. Immun*, 56:161 (1988).
43. A.M. Svennerholm, M. Jertborn, L. Gothefors, A.M.M.M. Karim, D.A. Sack, and J. Holmgren, Mucosal antitoxic and antibacterial immunity after cholera disease and after immunization with a combined B subunit-whole cell vaccine, *J. Infect. Dis.* 149:884 (1984).
44. C.O. Tacket, B. Forrest, R. Morona, S.R. Attridge, J. LaBrooy, B.D. Tall, M. Reymann, D. Rowely, and M.M. Levine, Safety, immunogenicity, and efficacy against cholera challenge in humans of a typhoid-cholera hybrid vaccine derived from *Salmonella typhi* Ty21a, *Infect. Immun.* 58:1620 (1990).
45. J. Holmgren, J. Clemens, D.A. Sack, and A.M. Svennerholm, New cholera vaccines, *Vaccine*, 7:94-96 (1989).
46. J.B. Kaper,*Vibrio cholerae* vaccines, Reviews of Infectious Diseases, 11 (Suppl. 3), 5568 (1989).
47. A. Takeuchi. Electron microscope studies of experimental *Salmonella infection*, I. Penetration into the intestinal epithelium by *Salmonella typhimurium*, *Am J. Pathol.* 50:109 (1967).
48. R. Weltzin, L. Morrison, L.S. Winner III, B.N. Fields, J.P. Kraehenbuhl, and M.R. Neutra, *In vivo* secretion of specific monoclonal IgA protects mice against an epithelially-transported enteric virus, *J. Cell Biol.* 109:295a (1989).

A MECHANISM OF PASSIVE IMMUNIZATION WITH MONOCLONAL

ANTIBODIES TO A 185,000 M_r STREPTOCOCCAL ANTIGEN

T. Lehner, J. K-C. Ma and C. G. Kelly

Department of Immunology
United Medical and Dental Schools of
Guy's and St. Thomas' Hospitals
London, United Kingdom

ABSTRACT

The cell surface streptococcal antigen (SA) I/II of 185,000 M_r is an immunodominant molecule that expresses one or more adhesion determinants. A series of 14 monoclonal antibodies (MAb) to defined parts of SA I/II were generated and some of these were used in passive immunization of macaques. Topical administration of selected MAb to the teeth of macaques prevented colonization of endogenous or implanted exogenous *Streptococcus mutans* for a period of 1 year. Significant reduction of both smooth surface and fissure caries was found in macaques who had MAb (Guy's 1) applied to their teeth, as compared with saline-treated animals.

A series of *in vivo* passive immunization experiments was then carried out in 57 human subjects. Topical application of MAb to SA I/II prevented colonization of both artificially implanted exogenous strains of *S. mutans,* as well as natural recolonization by indigenous *S. mutans.* The properties of the protective MAb were then investigated and the epitope specificity within the SA I/II molecule was found to be essential but not the isotype specificity of the immunoglobulin (Ig). The requirement for complement activating and the phagocyte binding sites of the Fc fragment of MAb was not essential, as the F(ab') 2 fragment of the MAb was as protective as the intact IgG, but the Fab fragment failed to prevent recolonization of *S. mutans.* Prevention of recolonization was specifically restricted to *S. mutans,* as the proportion of other organisms, such as *S. sanguis,* failed to show a significant change. The surprising feature of these experiments was that protection of re-colonization of *S. mutans* lasted up to 2 years, although MAb was applied for only 3 weeks and functional MAb was detected on the teeth only 3 days following application of the MAb. The long-term protection could therefore not be accounted for by a persistence of MAb on the teeth, but may be due to a shift in the microbial balance in which other bacteria occupy the ecological niche vacated by *S. mutans,* resulting in colonization resistance to *S. mutans.*

Gene cloning and sequencing the SA from *S. mutans, S. sobrinus* and *S. sanguis* identified a conserved region (residues 955-1213) which on Southern hybridization and partial DNA sequence analysis was also found in 19 α-haemolytic oral streptococci. The results suggest that the SA molecule may constitute a family of adhesins in oral α haemolytic streptococci. Preliminary mapping of adhesion epitopes with 14 MAb to SA I/II, some of which have been tested in humans for prevention of adherence of *S. mutans,* suggests that the conserved region may express some of the immunodominant adhesion epitopes.

Genetically Engineered Vaccines, Edited by
J.E. Ciardi *et al.,* Plenum Press, New York, 1992

INTRODUCTION

Colonization of *Streptococcus mutans* and the development of dental caries have been prevented by active immunization with whole cells of *S. mutans*.[1-4] A number of cell surface glycoproteins have been identified[5] of which the streptococcal antigen (SA) with an M_r of 185,000 has been extensively studied and variably termed SA I/II[6], B[7], P1[8], spa P[9] or PAc.[10] The antigen is expressed on the cell surface of *S. mutans* [11,12]; it may function as an adhesin[13,14] and as a virulence factor.[15] The antigen is immunogenic in the non-human primate, eliciting serum IgG, IgA and IgM antibodies and CD4+ T cell proliferative responses.[16,17] Serum antibodies and T cell proliferative responses are commonly found in human subjects naturally immunized by *S. mutans*.[16-18] Furthermore, a synthetic peptide derived from a 3800 M_r SA[19] which cross-reacts and co-purifies with the 185k M_r SA is immunogenic in macaques[20] and prevents colonization of *S. mutans*, after local gingivo-mucosal immunization.[21] The 185K M_r antigen has been recently cloned[9,10] and sequenced.[10,22] The immunodominant T cell and B cell epitopes and the adhesin determinant(s) can now be determined.

The rationale for local passive immunization was the finding that systemic passive immunization with IgG antibodies to *S. mutans* protects rhesus monkeys from dental caries[23]. We have formally demonstrated that ^{125}I-labeled serum IgG, administered intravenously, will pass through the gingival tissue and is detectable in gingival fluid within 30 minutes of injection[24]. Furthermore, specific IgG antibodies to *S. mutans* were demonstrated in the gingival fluid of animals that developed specific serum IgG antibodies after systemic immunization with *S. mutans*[25]. IgG monoclonal antibodies (MAb) to the SA I/II were then prepared[26] and applied to the teeth of rhesus monkeys. This resulted in decreased colonization of the teeth by *S. mutans* and prevented the development of dental caries[27]. We also found a protective effect of MAb by local passive immunization of cynomolgus monkeys and challenging of the teeth with single large doses of exogenous *S. mutans* [28]. These experiments were followed by passive immunization studies in human subjects[29]. Topical application of MAb to SA I/II in humans prevented colonization of both artificially implanted exogenous strains of *S. mutans*, as well as natural recolonization by indigenous *S. mutans* .[30,31]

The objectives of this paper are to review the role of MAb in preventing colonization of *S. mutans* and the development of dental caries in non-human primates. The mechanism and specificity of MAb in preventing colonization of exogenous and indigenous *S. mutans* in humans will then be discussed. An attempt will be made to relate the structure and deduced sequence of SA I/II to the putative adhesion epitope(s), as determined by *in vivo* prevention of colonization of *S. mutans* by MAb. Finally, the long-term colonization resistance to *S. mutans* will be discussed.

SUBJECTS AND METHODS

Experimental Animals and Method of Immunization

Young rhesus monkeys were offered a human-type diet, insofar as they develop indigenous *S. mutans* (serotype c) in the dental plaque, which this leads to the development of dental caries[2]. The teeth were examined at about 2-month intervals with a probe and mirror as well as X rays. The mean (± standard error) number of carious lesions per animal is given as the caries score. Four monkeys were immunized 12 times (indicated by the arrows in Fig. 1) with the MAb to SA I/II (Guy's 1). The teeth were dried and about 100 µl (containing 1 mg of IgG) of MAb was applied to the smooth and occlusal surfaces of all teeth (about 5 µl of MAb per tooth). Five control monkeys had only saline applied to their teeth.

Human Subjects

A group of 42 healthy volunteers from staff and students at Guy's Hospital was studied. The selected subjects were not taking any medication and were free of active caries. Throughout the investigation they were allowed to continue with their routine oral hygiene procedures and diet. For each experiment, the subjects were matched for age, sex, caries index (DMFS) and occupation. Gingival indices[32] and plaque indices[33] were measured before and after immunization. At the same time, venous blood, whole saliva and gingival fluid washings were collected, as described previously.[29]

Monoclonal Antibodies

The preparation of 14 MAb against SA I/II was described previously.[26,34] An IgG fraction of each of four MAb was then purified.[29] In addition, an unrelated MAb (raised against *Campylobacter jejuni*), kindly donated by Dr. D. Newell, was prepared as a control. The properties, IgG subclass and antibody binding capacity of the four MAb are given in Table 1. It is important to appreciate that Guy's 1 bound only *S. mutans* and Guy's 9 only *S. sobrinus,* whereas Guy's 11 and 13 bound both *S. mutans* and *S. sobrinus* cells. The MAb to *C. jejuni* failed to recognize either streptococcus.

Table 1. Properties Of Monoclonal Antibodies Used In Passive Immunization.

Mab Guy's	Isotype	%[125]I binding* to cells of *S. mutans*	%[125]I binding* to cells of *S. sobrinus*	Serotype Specificity	*S. mutans* Colonization
1	IgG2a	16.4(3.2)	0.3(0.2)	c,e,f	Prevented
13	IgG1	11.2(1.2)	12.4(3.5)	a,c,d,e,f,g	Prevented
11	IgG2b	12.1(1.3)	11.3(1.5)	a,c,d,e,f,g	Nil
9	IgG2a	0.4(0.1)	10.3(1.9)	d,g	Nil

*Mean (±SD)

Prevention of Implantation of Exogenous *S. mutans* by MAb

Prevention of implantation of a streptomycin-resistant *S. mutans* strain (K2) was investigated in 22 subjects.[29] These were divided into two matched groups; 14 subjects were given MAb Guy's 11 (n=7), with matched subjects having saline (n=7). The other matched group was given MAb Guy's 1 (n=4) and the controls received the MAb to *C. jejuni* (n=4). MAb were applied after scaling the teeth on days 0, 1 and 7; 5 μl of MAb at a concentration of 10 mg per ml was applied to the buccal surface and fissures of each tooth.[29] *S. mutans* was implanted on day 8, using 10^9 c.f.u. of the organism.

Prevention of Recolonization by Indigenous *S. mutans*

Recolonization by indigenous *S. mutans* was studied in selected subjects who had *S. mutans* cultured from their dental plaque. Detectable *S. mutans* was first eliminated with chlorhexidine.[30] MAb (5 μl) was applied to the buccal and lingual surfaces and fissures of each tooth after the chlorhexidine treatment was completed, on days 0, 3, 7, 10, 14 and 21. The group consisted of 11 subjects, of whom 4 were given MAb Guy's 1 or Guy's 13; the controls had saline (n=4). The specificity of the MAb was examined in three subjects given MAb to *S. sobrinus* (Guy' 9).

Effect of F(ab')$_2$ and Fab Fragments as Compared with Intact IgG on Recolonization by *S. mutans*

The F(ab')$_2$ fragment of MAb Guy's 13 was prepared by pepsin digestion, and the Fab fragment by papain digestion.[31] The activities of intact IgG and the F(ab')$_2$ and Fab digests of MAb Guy's 13 were tested by the solid-phase radioimmunoassay, in which purified protein antigen SA I/II was bound to the microtiter plates. IgG F(ab')$_2$, and Fab preparations of MAb Guy's 13 bound equally well to SA I/II from *S. mutans* (11.5 - 11.7%). Similarly, in the fluid-phase assay, the three MAb preparations bound equally well to the whole streptococcal cells (11.6 - 13.1%). This confirms that the digestion and purification procedures had no effect on binding of the MAb preparations to both purified and native cell surface antigens. The absence of intact MAb IgG in the F(ab')$_2$ and Fab preparations was confirmed by radioimmunoassay in which ^{125}I-labeled goat anti-mouse IgG Fc antibody failed to detect bound MAb in these preparations. Recolonization by indigenous *S. mutans* was then studied as described above. A matched group of 15 subjects was immunized as follows: 3 subjects had MAb Guy's 13 F(ab')$_2$ fragment, 3 subjects had Guy's 13 Fab fragment, 2 subjects had Guy's 13 IgG, 3 subjects had the control Guy's 11 IgG applied, and 4 subjects had saline applied to their teeth.

Bacteriology for *S. mutans* and *S. sanguis*

Dental plaque was collected from smooth and approximal surfaces of the first incisors and molars and from the fissures of the first and second molars. Unstimulated saliva was also collected for a period of 4 minutes. These samples were taken at decreasing intervals over a period of up to 100 or 360 days (see Figs. 1 and 2). They were plated onto TYC agar for *S. sanguis* or TYC agar supplemented with sucrose (20%) and bacitracin (0.1 units/ml) for *S. mutans*[35] and horse blood agar for the total anaerobic count. The TYC-based plates were incubated anaerobically at 37°C for 3 days and the horse blood plates were incubated anaerobically at 37°C for 5 days. The levels of streptococci were expressed as a percentage of the total anaerobic count on horse blood agar.

Radioassay for Antibodies to MAb and SA I/II

The possibility of a local or systemic immune response to the murine MAb was tested for by antibodies in gingival fluid, saliva and serum. IgG, IgM, and IgA antibodies were assayed by incubation with ^{125}I-labeled, affinity-purified goat anti-human IgG, IgM or IgA antibodies (Nordic Pharmaceutical Ltd Maidenhead, England) on microtitre plates coated with MAb or with SA I/II (1 μg/ml). All the results are given as a mean ± SD; serum dilutions were 1:1000, salivary dilutions were 1:50 and gingival fluid dilutions were 1:2.

RESULTS

Prevention of Colonization of *S. mutans* and Development of Dental Caries in Rhesus Monkeys

Colonization of both smooth surfaces and fissures of teeth by *S. mutans* was significantly lower or absent in animals whose teeth were treated with MAb than in saline-

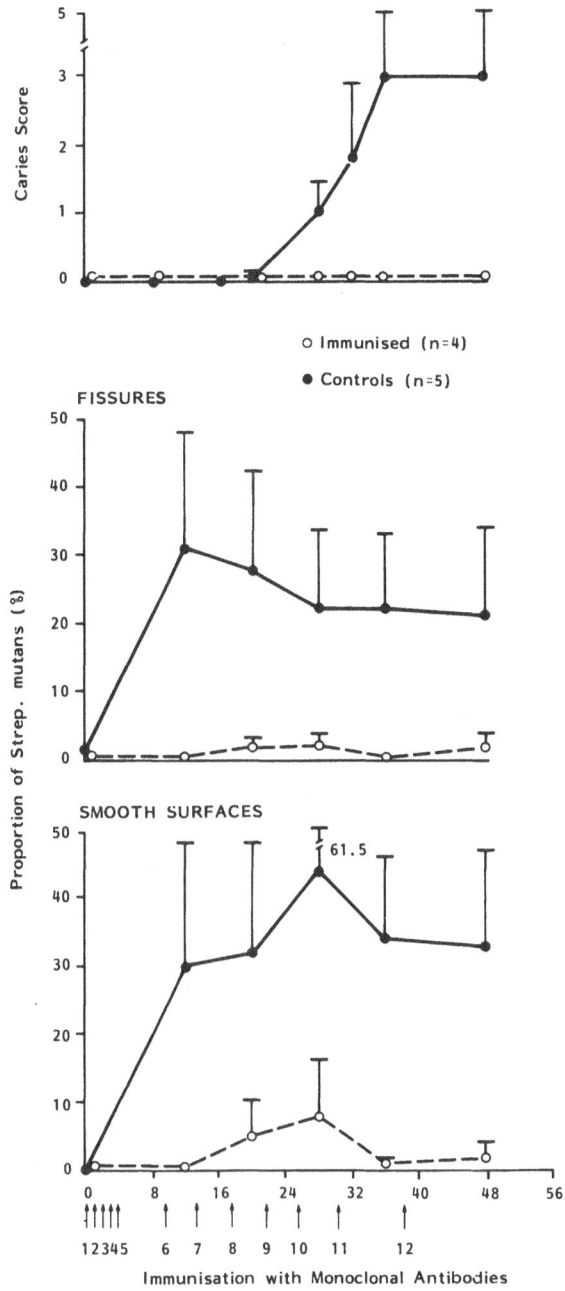

Figure 1. Effect of local passive immunization with MAb (Guy's 1) on the caries score and the proportion (as a percentage) of *S. mutans* in fissures and smooth surfaces of teeth in five control animals (●) and four immunized animals (o).

treated animals (Fig. 1). The proportion of *S. mutans* in the fissures and smooth surfaces of teeth of control monkeys was between 21% and 44%, as compared with 0% to 8% in immunized monkeys. Application of MAb (Guy's 1) to the deciduous teeth of rhesus monkeys prevented the development of caries over a period of 1 year, as compared with a mean (± standard error) of 3.0 (± 2.0) carious lesions per control animal (Fig. 1). There were eight smooth-surface and seven fissure carious lesions in the control animals, so that both smooth-surface and fissure caries were prevented in the immunized animals.

155

Prevention of Implantation by Exogenous *S. mutans* in Humans

Colonization by *S. mutans* was detected at all sites examined in the control group receiving MAb to *C. jejuni*. In smooth-surface and fissure plaque and saliva, the implantation levels of the exogenous *S. mutans* were significantly raised (see reference 29, Fig. 2). A significantly lower level of implantation was seen in the subjects receiving MAb Guy's 1. At least 5-fold lower levels of implanted *S. mutans* were detected in smooth-surface plaque and saliva. No implanted *S. mutans* was found in the fissures. MAb Guy's 13 induced similar protection to that obtained with Guy's 1 when compared with the control subjects (having saline) who showed significantly higher levels of implantation in all three sample sites (data not shown; see reference 30). The finding that two different MAbs (Guy's 1 and Guy's 13) can prevent implantation of *S. mutans* suggests that more than one adhesion epitope can be found in the SA molecule.

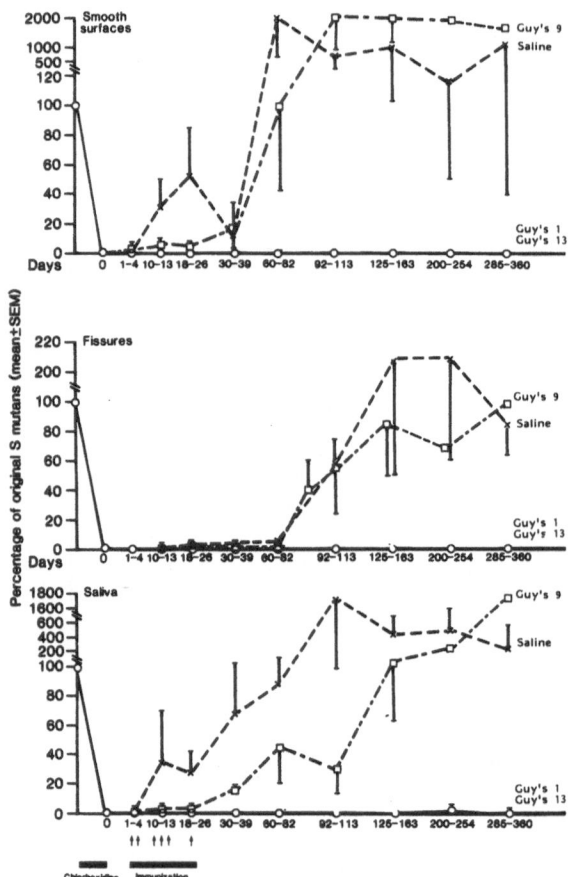

Figure 2. Proportion of recolonizing indigenous *S. mutans* in smooth-surface and fissure plaque and saliva. The proportion of *S. mutans* is expressed as the mean ± SEM percentage of the pre-experimental levels of *S. mutans*. Data for MAb Guy's 1 and MAb Guy's 13 group (o) are combined (n=4); saline group (x) (n=4) and MAb Guy's 9 group (□) (n=3).

Prevention of Recolonization by Indigenous *S. mutans*

Chlorhexidine treatment of subjects preselected for the presence of *S. mutans* eliminated all detectable *S. mutans* (Fig. 2). Subsequent treatment of these subjects either with saline or MAb to *S. sobrinus* (Guy's 9) resulted in rapid recolonization by *S. mutans* in saliva, smooth-surface plaque and then fissures (Fig. 2).[30] However, in subjects given the specific MAb to *S. mutans* (Guy's 1 or 13), recolonization was not detected in the saliva or plaque 360 days after the MAb was applied. Indeed, these subjects remained free of *S. mutans* for a period of 2 years.

Comparative Effect of Intact IgG, F(ab')₂ and Fab Fragment on Recolonization of *S. mutans*

Chlorhexidine treatment reduced the *S. mutans* levels in plaque and saliva to undetectable levels in all subjects (Fig. 3). In subjects sham-immunized with saline or in those who had MAb Guy's 11 applied, *S. mutans* reappeared in both plaque and saliva shortly after chlorhexidine treatment ceased. The level of *S. mutans* continued to rise and reached 100% or more by the end of 100 days of the experimental period. In contrast, subjects who had been immunized with either intact IgG (Guy's 13) MAb or the F(ab')₂ fragment did not recolonize with *S. mutans* in plaque or saliva (Fig. 3). However, in the group that had been immunized with the Guy's 13 Fab fragment, no protection was seen and recolonization by *S. mutans* occurred at a similar rate to that of the control subjects. It appears that while intact IgG and the F(ab')₂ fragment can prevent recolonization of *S. mutans*, the Fab fragment lost this function.[31]

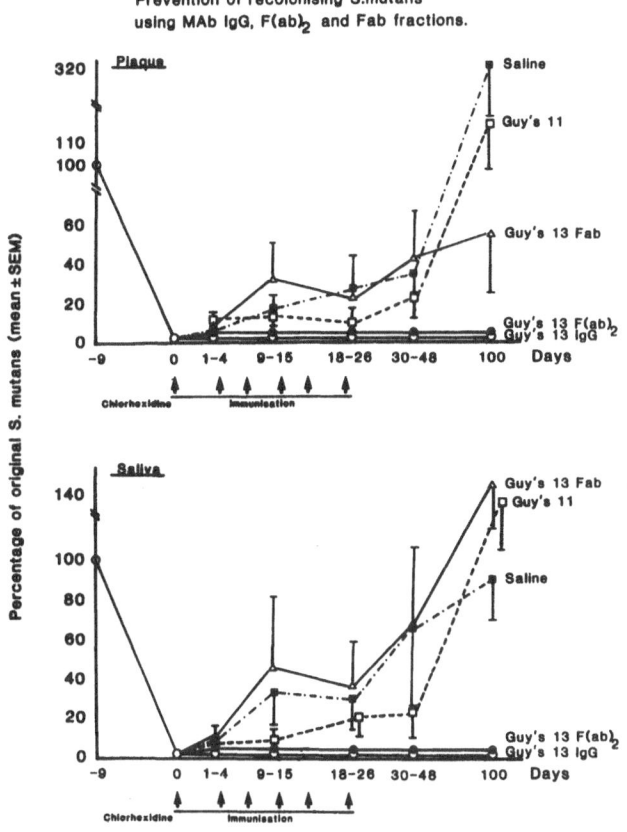

Figure 3. The effect of application of intact MAb IgG (Guy's 13 and Guy's 11) and F(ab')₂ or Fab fragments on the proportion of recolonizing indigenous *S. mutans* in plaque and saliva. The *S. mutans* proportions are expressed as the mean ± SEM percentage of the pre-experimental levels of *S. mutans*. The pre-experimental levels at day 9 are the mean proportions of four samples taken over 14 days.

Effect of MAb on Other Oral Microorganisms

MAb Guy's 13 had no significant effect on colonization of indigenous *S. sanguis* (Fig. 4) or veillonella species (data not presented), as compared with the control saline or MAb Guy's 11.[31] In the same experiment, *S. mutans* failed to recolonize the teeth of subjects who received MAb Guy's 13, unlike those receiving saline or MAb Guy's 11. The results are consistent with the activity of MAb (Guy's 13) being specific against *S. mutans* and having little or no effect on *S. sanguis* or veillonella species.

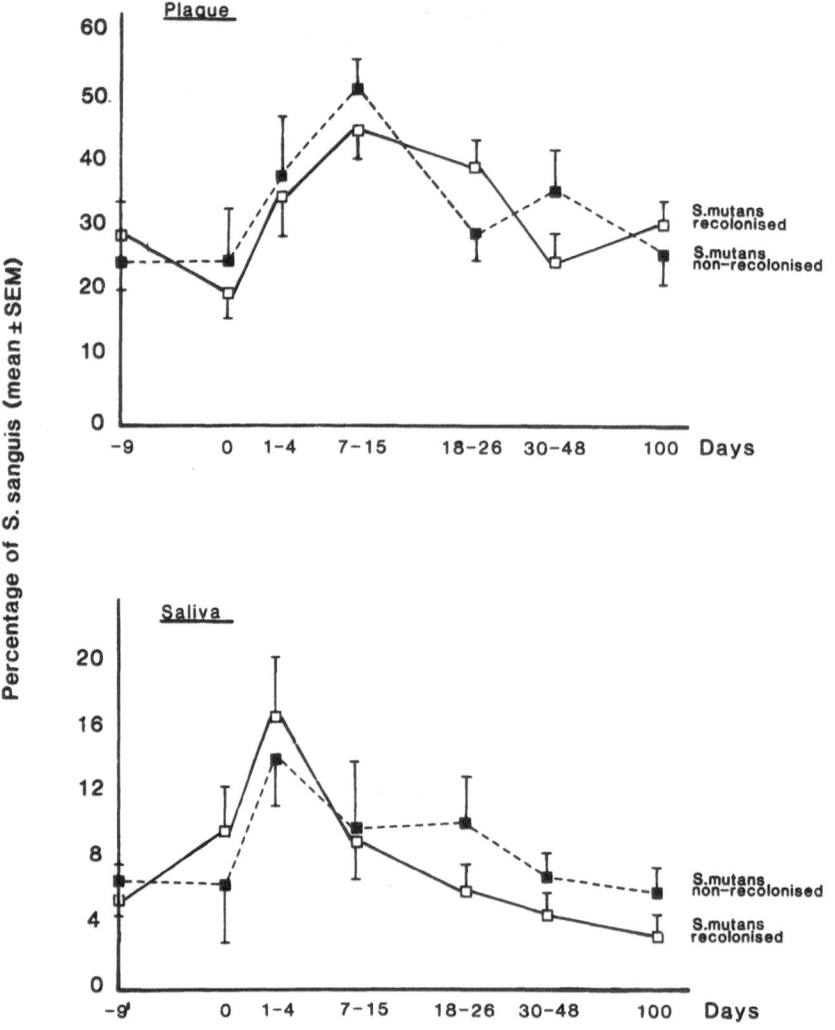

Proportion of S.sanguis
in plaque and saliva.

Figure 4. The proportions of *S. sanguis* are presented for the (i) *S. mutans*-recolonized group which received saline, MAb Guy's 11, or Guy's 13 Fab fragment (□) and the (ii) *S. mutans* non-recolonized group which received MAb Guy's 13 IgG of F(ab')$_2$ fragment (■). The results are expressed as the percentage of *S. sanguis* in the total anaerobic count. Pre-experimental levels (day 9) are the mean proportions of four samples taken over 14 days.

Unwanted Effect of MAb

The subjects did not report any ill effect from the MAb, nor were there any clinical signs detected as a result of passive immunization. The mean gingival index (± SD) for the MAb-immunized group, including all subjects receiving Guy's 1, 9, 11 or 13 (n=24), was 0.6 ± 0.1 before immunization and 0.6 ± 0.1 after MAb treatment was completed.[36] The control subjects who received saline only (n=17) had gingival indices of 0.7 ± 0.2 before and 0.6 ± 0.1 after sham immunization. Similarly, the plaque indices remained unchanged; in the MAb-immunized group, the mean plaque index was 0.9 ± 0.3 before immunization and 0.8 ± 0.2 after immunization, whereas the control group had plaque indices of 0.8 ± 0.2 and 0.6 ± 0.2, respectively. The possibility that antibodies might be elicited to the MAb applied to the teeth was tested in serum, gingival fluid and saliva. Serum anti-MAb antibodies of the IgG, IgM and IgA classes, as well as salivary IgA and gingival fluid IgG antibodies, were unchanged in subjects who received either three or six applications of MAb[36]. There was no significant difference in serum IgG, IgM and IgA, salivary IgA or gingival fluid IgG antibodies to SA I/II between the experimental groups and controls.

DISCUSSION

Monoclonal antibodies have been used in bone marrow transplantation and immunotherapy of leukemia.[37] However, passive immunization with MAb has not been used in the prevention of microbial colonization in humans. Topical application of an IgG$_{2a}$ class of MAb to the 185K M_r streptococcal cell surface antigen (Guy's 1) prevented colonization of indigenous *S. mutans* and the development of dental caries in rhesus monkeys.[27] Further experiments in cynomolgus monkeys were carried out to examine the effect of MAb on the implantation of an exogenous *S. mutans*. Indeed, colonization of *S. mutans* and the development of dental caries were prevented by MAb Guy's 1.[28] In these pre-clinical experiments, the MAb were applied initially 12 times and then reduced in the second experiment to 4 applications over a period of 20 weeks. Significant reduction of *S. mutans* was maintained throughout the experimental period of about 1 year, as compared with sham-immunized macaques. These results were consistent with those using hyperimmunized cow's milk, which prevented oral colonization by *S. mutans* in rodents.[38]

A series of *in vivo* experiments was then pursued in 42 human subjects. Implantation of an exogenous *S. mutans* was prevented or significantly reduced by three applications of MAb (Guy's 1 or 13) over a period of only 7 days[29,30]. This mode of transmission of *S. mutans* may be relevant clinically, as streptococci are acquired by children from their parents through saliva, kissing and sharing feeding implements.[39] A similar mechanism of transmission of *S. mutans* may apply between adults.[40] Recolonization by indigenous *S. mutans* has also been prevented in pre-selected subjects with *S. mutans* present in their plaque and saliva[30].

The microbial species specificity of the *in vivo* prevention of re-colonization was specific for *S. mutans*, as MAb Guy's 13 had no significant effect on colonization of indigenous *S. sanguis* or veillonella species. Serotype specificity was also evident, as MAb Guy's 9 against *S. sobrinus* (serotypes d and g) were ineffective in preventing colonization by *S. mutans* (serotypes c, e and f). Adhesion epitope specificity is particularly significant in view of the findings that the Guy's 11 and Guy's 13 each recognizes a common serotype determinant among serotypes a, c, d, e, f and g[34], yet Guy's 13 prevents colonization of *S. mutans* but Guy's 11 fails. However, isotype specificity of the MAb was not critical, as IgG$_1$ (Guy's 13) and IgG$_{2a}$ (Guy's 1) were equally protective.

The requirements of the Fc portion of IgG were then examined by preparing F(ab')$_2$ and Fab fragments from MAb Guy's 13 and comparing their functions *in vivo* with those of the intact IgG. The F(ab')$_2$ fragment of MAb Guy's 13 was as effective as the intact IgG in preventing recolonization of *S. mutans*. F(ab')$_2$ differs from IgG in that the Fc portion, which carries the CH2 and CH3 domains, has been removed. The function of these domains is in binding to phagocytes as well as activating complement. This suggests that the mode of action of the MAb in preventing streptococcal colonization might be independent of these two biological properties. The Fab fragment of IgG, however, failed to prevent colonization of *S. mutans* although binding to SA I/II or whole cells *in vitro* was unaffected. This may be due to Fab being a monovalent molecular form unlike the bivalent F(ab')$_2$ and IgG molecules. Alternatively, the smaller antibody fragment may be more susceptible to proteolysis than the larger F(ab')$_2$ of the IgG molecule. It is also noteworthy that *in vitro* studies revealed that intact IgG and F(ab')$_2$, but not the Fab fragment, caused long chaining and aggregation of streptococcal cells.[31]

Eliminating *S. mutans* by an antimicrobial agent (e.g., chlorhexidine) to undetectable levels led to recolonization of the indigenous *S. mutans* after antimicrobial treatment was discontinued. However, two MAb to SA I/II (Guy's 1 and 13) prevented recolonization, and surprisingly, the subjects remained free of *S. mutans* 2 years after the MAb were applied. Because functional MAb were detected *in vivo* on the teeth for only up to 3 days after application of the MAb[31], the long-term protection could not be accounted for by a persistence of MAb on the teeth.

We have proposed an hypothesis for the mechanism of MAb[27-30], which can be divided into four phases (Fig. 5): (1) The MAb adheres to the salivary glycoprotein pellicle on the tooth surface. (2) Any *S. mutans* encountered is bound specifically to the MAb and may inhibit direct adherence of the organisms to the tooth. (3) *S. mutans* cells may undergo chaining and aggregation and may be shed from the tooth surface. Alternatively, *S. mutans* may fail to proliferate normally and may be phagocytosed and killed by the local neutrophils. Because the Fc portion of IgG is not essential in preventing recolonization of *S. mutans*, Fc-dependent opsonization, allowing Fc-receptor dependent binding to neutrophils, does not appear to be essential in this mode of prevention. (4) The long-term prevention of recolonization of *S. mutans* suggests a shift in the microbial balance, in which other bacteria occupy the ecological niche vacated by *S. mutans,* resulting in colonization resistance to *S. mutans*. However, this shift in microbial ecology might be difficult to demonstrate quantitatively. Indeed, we have not detected a significant change with *S. sanguis* or veillonella species, but these are only two of many candidate organisms that might be involved.

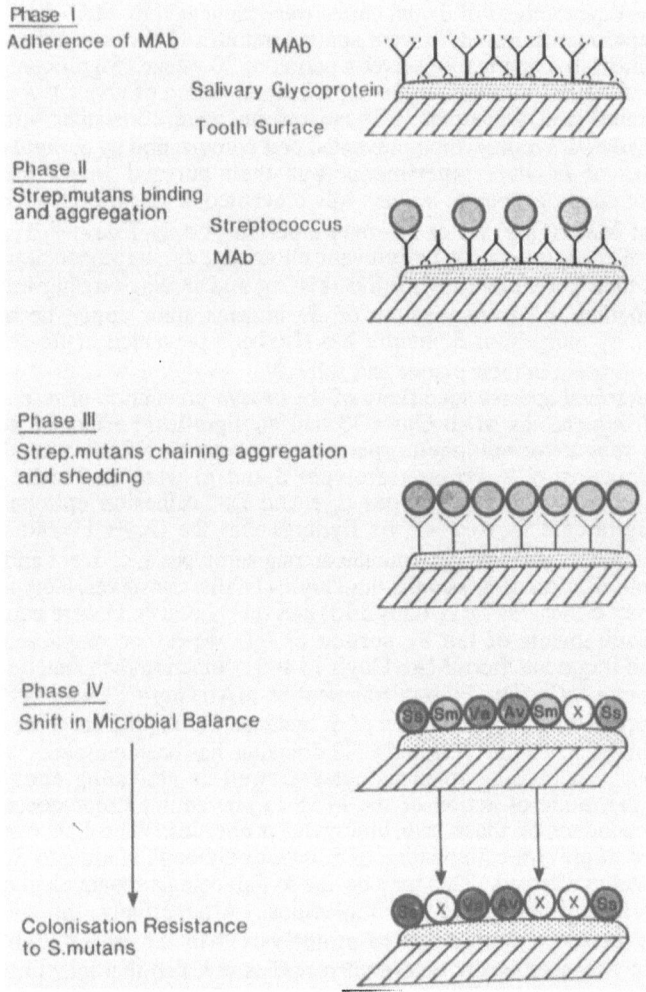

Figure 5. Four-phase hypothesis of the mechanism of prevention of colonization of *Streptococcus mutans* by monoclonal antibodies to cell surface adhesion determinants.

160

Gene cloning and sequencing of SA I/II from *S. mutans* (serotype c)[22,41], serotype f,[42] *S. sobrinus* [43] and *S. sanguis (S. gordonii)*[44] identified five regions from the N to the C terminus of the SA I/II molecule: (1) alanine-rich, (2) variable, (3) proline-rich, (4) conserved and (5) a wall-spanning region (Fig. 6). The conserved region [955-1213] shows 75-80% homology between *S. mutans, S. sobrinus* and *S. gordonii*. However, Southern hybridization with five overlapping DNA probes and partial DNA sequence analysis revealed that the conserved region is also found in 19 α-haemolytic oral streptococci including *S. sanguis, S. sobrinus, S. intermedius, S. oralis* and *S. anginosus*.[45] These results suggest that the conserved region of the 185K M_r antigen (residues 955-1213) may constitute an adhesion family of oral streptococcal cell surface antigens. Indeed, MAb Guy's 13, which prevented colonization by *S. mutans*, has a broad cross-reactivity among the mutans group of streptococci and binds weakly to *S. sanguis*[34] and *S. intermedius* as well as other α-haemolytic streptococci (unpublished data). Furthermore, preliminary results of mapping adhesion epitopes with 14 MAb to SA I/II, some of which have been tested *in vivo* for prevention of adherence of *S. mutans*, suggest that the conserved region of this molecule may represent some of the immunodominant adhesion epitopes.

GENE CLONING AND SEQUENCING OF S. MUTANS, S. SOBRINUS AND S. SANGUIS OF THE STREPTOCOCCAL ANTIGEN I/II REVEALED 5 REGIONS
SOUTHERN HYBRIDISATION WITH 5 DNA PROBES AND PARTIAL DNA SEQUENCE ANALYSIS REVEALED A CONSERVED REGION (RESIDUES 963-1213) AMONG α HAEMOLYTIC STREPTOCOCCI

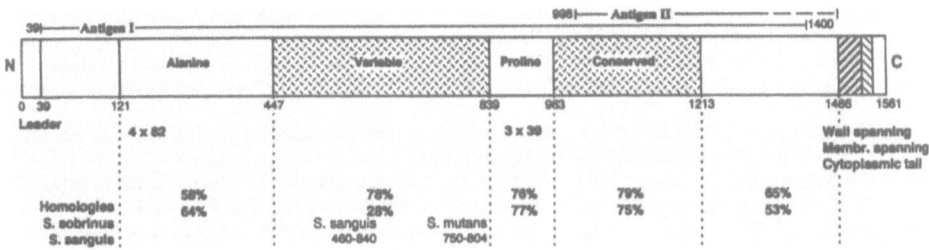

Figure 6. The five mapped regions of streptococcal antigen I/II, derived from the comparative analysis of the amino acid sequences of *S. mutans, S. sobrinus* and *S. sanguis*.

ACKNOWLEDGMENTS

This project was funded by a grant from the Medical Research Council and the Newland Pedley Fund.

REFERENCES

1. M.A. Taubman and D.J. Smith, Effects of local immunization with *Streptococcus mutans* on induction of salivary immunoglobulin A antibody and experimental dental caries in rats, *Infect. Immun.* 9:1029 (1974).
2. T. Lehner, S.J. Challacombe, and J. Caldwell, Immunological and bacteriological basis for vaccination against dental caries in rhesus monkeys, *Nature (Lond.)* 254:517 (1975).
3. W.H. Bowen, B. Cohen, M.F. Cole, and G. Colman, Immunization against dental caries, *Br. Dent. J.* 139:45 (1975).

4. S.M. Michalek, J.R. McGhee, J. Mestecky, R.R. Arnold, and L. Bozzo, Ingestion of *Streptococcus mutans* induces secretory immunoglobulin A and caries immunity, *Science* 192:1238 (1976).
5. M.W. Russell and T. Lehner, Characterization of antigens extracted from cells and culture fluids of *Streptococcus mutans* serotype c, *Arch. Oral. Biol.* 23: 7 (1978).
6. M.W. Russell, L.A. Bergmeier, E.D. Zanders, and T. Lehner, Protein antigens of *Streptococcus mutans*: purification and properties of a double antigen and its protease-resistant component, *Infect. Immun.* 28:486 (1980).
7. R.R.B. Russell, Wall associated antigens of *Streptococcus mutans*, *J. Gen. Microbiol.* 114:109 (1979).
8. H. Forrester, N. Hunter, and K.W. Knox, Characteristics of a high molecular weight extracellular protein of *Streptococcus mutans*, *J. Gen. Microbiol.* 129:2779 (1983).
9. S.F. Lee, A. Progulske-Fox, and A.S. Bleiweis, Molecular cloning and expression of a *Streptococcus mutans* major cell surface protein antigen, P1 (I/II), in *Escherichia coli*, *Infect. Immun.* 56:2114 (1988).
10. N. Okahashi, C. Sasakawa, M. Yoshikawa, S. Hamada, and T. Koga, Molecular characterization of a surface protein antigen gene from serotype c *Streptococcus mutans*, implicated in dental caries, *Molec. Microbiol.* 3:673 (1989).
11. E.D. Zanders and T. Lehner, Separation and characterization of a protein antigen from cells of *Streptococcus mutans*, *J. Gen. Microbiol.* 122:217 (1981).
12. I. Moro and M.W. Russell, Ultrastructural localization of protein antigens I/II and III in *Streptococcus mutans*, *Infect. Immun.* 41:410 (1983).
13. B.C. McBride, M. Song, B. Krasse, and J. Olsson, Biochemical and immunological differences between hydrophobic and hydrophilic strains of *Streptococcus mutans*, *Infect. Immun.* 44:68 (1984).
14. S.F. Lee, A. Progulsek-Fox, G.W. Erdos, D.A. Piacentini, G.Y. Ayakawa, P.J. Crowley, and A.S. Bleiweis, Construction and characterization of isogenic mutants of *Streptococcus mutans* deficient in major surface protein antigen P1 (I/II), *Infect. Immun.* 57:3306 (1989).
15. R. Curtiss III, Genetic analysis of *Streptococcus mutans* virulence, *Curr. Top. Microbiol. Immunol.* 118:253 (1985).
16. T. Lehner, The relationship between human helper and suppressor factors to a streptococcal protein antigen, *J. Immunol.* 129:1936 (1982).
17. T. Lehner, J. Caldwell, and J. Avery, Sequential development of helper and suppressor functions, antibody titres and functional avidities to a streptococcal antigen in rhesus monkeys, *Eur. J. Immun.* 14:814 (1984).
18. A. Childerstone, J. Haron, and T. Lehner, T cell interactions generated by synthetic peptides covalently linked to a carrier, *Eur. J. Immunol.* 19:169 (1989).
19. A.S.N. Giasuddin, T. Lehner, and R.W. Evans, Identification, purification and characterisation of a streptococcal protein with a molecular weight of 3800, *Immunology* 50:651 (1983).
20. T. Lehner, J. Caldwell, and A.S.M. Giassudin, Comparative immunogenicity and protective effect against dental caries of a low (3800) and a high (185,000) molecular weight protein in rhesus monkeys (*Macaca mulatta*), *Arch. Oral. Biol.* 30:207 (1985).
21. T. Lehner, A. Mehlert, and J. Caldwell, Local active gingival immunization by a 3,800-molecular streptococcal antigen in protection against dental caries, *Infect. Immun.* 52:682 (1986).
22. C. Kelly, P. Evans, L. Bergmeier, S.F. Lee, A. Progulske-Fox, A.C. Harris, A. Aitken, A.S. Bleiweis, and T. Lehner, Sequence analysis of the cloned streptococcal cell surface antigen I/II, *FEBS Letters* 258:127 (1989).
23. T. Lehner, M.W. Russell, S.J. Challacombe, C.M. Scully, and J. Hawkes, Passive immunization with serum immunoglobulins against dental caries in rhesus monkeys, *Lancet* i:693 (1978).
24. S.J. Challacombe, M.W. Russell, J. Hawkes, L. Bergmeier, and T. Lehner, Passage of immunoglobulins from the plasma to the oral cavity in rhesus monkeys, *Immunology* 35:923 (1978).
25. R. Smith and T. Lehner, A radioimmunoassay for serum and gingival crevicular fluid antibodies to a purified protein of *Streptococcus mutans*, *Clin. Exp. Immunol.* 43:417 (1981).
26. R. Smith, T. Lehner, and P.C.L. Beverley, Characterization of monoclonal antibodies to *Streptococcus mutans* antigenic determinants I/II, and III and their serotype specificities, *Infect. Immun.* 46:168 (1984).
27. T. Lehner, J. Caldwell, and R. Smith, Local passive immunization by monoclonal antibodies against streptococcal antigen I/II in the prevention of dental caries, *Infect. Immun.* 50:796 (1985).
28. T. Lehner, J. Ma, K. Grant, and R. Smith, Passive dental immunization by monoclonal antibodies against *Streptococcus mutans* antigen in the prevention of dental caries, *in*: T Lehner and G. Cimasoni eds., "Borderland Between Caries and Periodontal Disease", Editions Medicine et Hygiene, Geneva, p.347 (1986).
29. JK-C. Ma, R. Smith, and T. Lehner, Use of monoclonal antibodies in local passive immunization to prevent colonization of human teeth by *Streptococcus mutans*, *Infect. Immun.* 55:1274 (1987).
30. JK-C. Ma, M. Hunjan, R. Smith, and T. Lehner, Specificity of monoclonal antibodies in local passive immunization against *Streptococcus mutans*, *Clin. Exp. Immunol.* 77:331 (1989).

31. JK-C. Ma, M. Hunjan, R. Smith, C. Kelly, and T. Lehner, An investigation into the mechanism of protection by local passive immunization with monoclonal antibodies against *Streptococcus mutans*, *Infect. Immun.* 58:3407 (1990).

32. H. Löe and J. Silness, Periodontal disease in pregnancy, I. Prevalence and severity, *Acta. Odontal. Scand.* 21:532 (1963).

33. J. Silness and H. Löe, Periodontal disease in pregnancy, II. Correlation between oral hygiene and periodontal condition, *Acta. Odontal. Scand.* 22:121 (1964).

34. R. Smith and T. Lehner, Characterization of monoclonal antibodies binding to common protein epitopes on the cell surface of *Streptococcus mutans* and *Streptococcus sobrinus*, *Oral Microbiol. Immunol.* 4:153 (1989).

35. W.H. Van Palenstein Helderman, M. Ijsseldijk, and J.H.J. Huis in't Veld, A selective medium for the two major subgroups of the bacterium *S. mutans* isolated from human dental plaque and saliva, *Archs. Oral Biol.* 28:599 (1983).

36. JK-C. Ma and T. Lehner, Prevention of colonization of *Streptococcus mutans* by topical application of monoclonal antibodies in human subjects, *Archs. Oral Biol.* 35:115S (1990).

37. R. Kurrle, K.H. Enssie, and F.R. Seiler, Monoclonal antibodies directed to leucocyte differentiation antigens for therapeutic use, *Behring Inst. Mitt.* 82:154 (1988).

38. S.M. Michalek, R.L. Gregory, C.C. Harmon, J. Katz, G.J. Richardson, T. Hilton, S.T. Filler, and J.R. McGhee, Protection of gnotobiotic rats against dental caries by passive immunization with bovine milk antibodies to *Streptococcus mutans*, *Infect. Immun.* 55:2341 (1987).

39. B. Kohler and D. Bratthall, Intrafamilial levels of *Streptococcus mutans* and some aspects of the bacterial transmissions, *Scand. J. Dent. Res.* 86:35 (1978).

40. M. Svanberg and B. Krasse, Oral implantation of saliva treated *Streptococcus mutans* in man, *Archs. Oral Biol.* 26:197 (1981).

41. T. Koga, N. Okahashi, I. Takahashi, T. Kanamoto, H. Asakawa, and M. Iwaki, Surface hydrophobicity, adherence and aggregation of cell surface protein antigen mutants of *Streptococcus mutans* serotype c, *Infect. Immun.* 58:289 (1990).

42. J.A. Ogier, M. Scholler, Y. Lepoivre, A. Pina, P. Sommer, and J.P. Klein, Complete nucleotide sequence of the *sr* gene from *Streptococcus mutans* OMZ175, *FEMS Microbiol. Lett.* 68:223 (1990).

43. R.J. LaPolla, J.A. Haron, C.G. Kelly, W.R. Taylor, C. Bohart, M. Hendricks, J. Pyati, R.T. Gratt, JK-C. Ma, and T. Lehner, Sequence and structural analysis of surface protein antigen I/II (Spa A) of *Streptococcus sobrinus*, *Infect. Immun.* 59:2677 (1991).

44. D.R. Demuth, E.E. Golub, and D. Malamud, Streptococcal-host interactions. Structural and functional analysis of a *Streptococcus sanguis* receptor for a human salivary glycoprotein, *J. Biol. Chem.* 265:7120 (1990).

45. JK-C. Ma, C.G. Kelly, G. Munro, R.A. Whiley, and T. Lehner, Conservation of the gene encoding streptococcal antigen I/II in oral streptococci, *Infect. Immun.*, 59:2686 (1991).

DELIVERY OF ANTIGENS BY RECOMBINANT AVIRULENT

SALMONELLA STRAINS

Teresa A. Doggett and Roy Curtiss III

Washington University
Department of Biology
St. Louis, MO 63130

INTRODUCTION

Most infectious agents penetrate and/or colonize the mucosal surface of the digestive, respiratory or genitourinary tracts. Inhalation or ingestion of antigens brings them into contact with the lymphoreticular tissues at these sites, namely the bronchial-associated lymphoid tissue (BALT) and the gut-associated lymphoid tissue (GALT). In this review we shall concentrate on the immune responses initiated in the GALT and the delivery of antigens to the lymphoid tissue of the gastrointestinal (GI) tract using avirulent recombinant *Salmonella*.

ROLE OF MUCOSA-ASSOCIATED LYMPHOID TISSUE AND SECRETORY IgA

One of the most important protective humoral immune responses occurs locally at the site of invasion in the form of secretory IgA (SIgA). Although both IgM and IgG isotypes have been detected in intestinal secretions following oral administration of antigen[1,2], they constitute only a small proportion of the total immunoglobulin found at or on mucosal surfaces; the predominant immunoglobulin isotype is IgA. The induction of SIgA responses in the gut involves the uptake of the antigen by specialized membranous or microfold (M) cells[3] located in the epithelium covering the dome area of Peyer's patches (PP), which are localized lymphoid follicles situated along the length of the GI tract. The M cells are actively pinocytotic and deliver the phagocytosed antigen, without processing, to the underlying lymphoid tissue which consists of distinct follicles composed of B cell zones containing germinal centres and a parafollicular T cell-dependent area. This T cell zone contains all the major T cell subsets, as well as macrophages and dendritic cells. The B cell zones have a high proportion of B cells that express IgA at their surface and are the precursors of mature IgA-producing plasma cells. Antigen-sensitized IgA+ B cells leave the PP via the efferent lymphatics and enter the systemic circulation and circulate through the body, returning to the mucosal tissues and secretory glands where they differentiate into plasma cells under the influence of antigen, T cells and cytokines, and secrete IgA antibodies. These sensitized B cells not only return to their site of induction but also seed other distant mucosae[4,5], a phenomenon termed the common mucosal immune system[6,7]. The appearance of antigen-senstized B cells in the mucosae is accompanied by the production of specific SIgA antibodies in secretions at sites that were not in contact with the antigen[8].

Unlike its serum counterpart, which is predominantly monomeric and primarily derived from the bone marrow, SIgA is a polymeric immunoglobulin and is composed of two or more IgA monomeric units, the J chain and the secretory component (SC). The presence of

Genetically Engineered Vaccines, Edited by
J.E. Ciardi *et al*., Plenum Press, New York, 1992

specific IgA antibodies in mucosal secretions has been shown to diminish absorption of protein antigens[9,10] and biologically active antigens, such as bacterial toxins and enzymes[9], that may be present at mucosal surfaces. The association of the polymeric IgA with SC has also been shown to provide some degree of protection against the proteolytic environment into which SIgA is translocated[9,11] and thus provides functional advantages to SIgA compared with the other immunoglobulin isotypes that are found on mucosal surfaces. SIgA is therefore the primary defense mechanism against infectious agents that enter the host via mucosal surfaces, and antibodies directed against specific virulence determinants of pathogenic organisms play an important role in mucosal immunity. We can utilize this system in the development of oral vaccines to potentially protect the host against a variety of infectious agents. While parenteral immunization with inactivated whole-cell preparations is effective in inducing a protective serum IgG response and delayed-type hypersensitivity reactions (DTH), it is ineffective in inducing significant mucosal SIgA responses unless preceded by a mucosal priming[12,13]. The precise mechanism by which the mucosal response is generated following parenteral boosting in not known. Live oral vaccines based on bacterial strains attenuated by the introduction of genetically defined mutations, particularly mutants of *Salmonella*, have shown great promise, not only in protecting against salmonellosis[14-18] but also in using *Salmonella* as carriers for plasmids that code for heterologous antigens, such as virulence determinants of mucosal pathogens[19-27]. The heterologous antigens expressed by these recombinant strains would be delivered directly to the lymphoid population in the GALT, with the subsequent induction of specific mucosal immune responses directed against both the carrier and the expressed antigen/s, as well as the production of serum antibodies.

ATTENUATED *SALMONELLA* MUTANTS

There are a number of *Salmonella* serotypes that cause disease in humans and animals. Some, *S. typhi* and *S. paratyphi* A, are restricted in their host range and infect only one species, while others have a broader host range causing a severe invasive disease in more than one animal species. These *Salmonella* of 'animal origin' are usually restricted to the intestinal tract when they infect man, such as *S. typhimurium* which causes typhoid-like symptoms in mice and is a major cause of salmonella food-poisoning in humans.

Numerous strategies have been developed to attenuate *Salmonella*[14-18,28-30] so that they are avirulent yet retain their tissue tropism[15,31-33]. Some of these mutants are effective as live oral vaccines, being able to protect against infection by the virulent wild-type *Salmonella* parent[15,28-33]. However, problems such as phenotype instability, the ability to overcome the avirulent phenotype by utilizing compounds from the host's diet, or overattenuation such that immunization did not induce protective immunity were observed in earlier attempts to develop oral vaccines. These problems have now been overcome with considerable success. One class of *Salmonella* mutants are the *galE* mutants[30,34], which, when grown on media lacking galactose, are deficient in their lipopolysaccharide (LPS) O side-chains and are phenotypically rough. However, they become phenotypically smooth when grown with galactose but lyse within 24 hours if maintained on galactose, due to the accumulation of toxic levels of phosphorylated galactose. *S. typhimurium galE* mutants, although quickly cleared by the murine immune system, are capable of inducing long-lasting protective immunity in mice to challenge with the virulent parent strain[30]. *S. typhi* Ty21a, also a *galE* mutant, when administered orally in gelatin capsules has been shown to protect against challenge and provide long-lasting immunity[35,36]. Nencioni and colleagues[37] were also able to demonstrate that immunization with Ty21a enhanced the natural cell-mediated immunity not only to *S. typhi* but also to *S. paratyphi* A and B. Forrest and colleagues[38] demonstrated that oral immunization of human volunteers with Ty21a resulted in significant levels of specific IgA antibodies in intestinal secretions together with high systemic levels of specific IgA and IgG antibodies.

Another approach is the use of auxotrophic mutants, such as those described by Bacon and associates[28,29]. These mutagen-induced auxotrophs were dependent on aromatic compounds and purines and were avirulent in mice. They were also capable of reverting to prototrophy and thus constituted a safety problem. Hoiseth and Stocker[14] overcame this problem by using transposon-mutagenesis, resulting in the isolation of non-reverting deletion mutants, *aroA*. These were dependent on aromatic acids, pABA and dihydroxybenzoic acid, grew poorly in vivo and were avirulent in mice. *S. typhimurium aroA* mutants have been shown to be effective as live oral vaccines in several animal species[14,16] as have *aroA* and *aroC* mutants of *S. dublin*[39]. All mutations within the chorismate pathway -- that is, *aroA*, *aroC* and *aroD* -- reduce virulence to a similar degree[16,39,40]. Purine mutants are also avirulent, particularly if they are blocked after the synthesis of inosine monophohate (IMP). *PurA* mutants can persist in mice without causing splenomegaly but fail to induce significant protective immunity and may therefore be overattenuated[16]. Other mutants, such as those that are defective in the production of aspartate semialdehyde dehydrogenase (Δ*asd*) and that have a requirement for diamino-pimelic acid, threonine and methionine, are totally avirulent by all routes of administration. These mutants survive for only a short period in the GALT and induce only moderate secretory immune responses[41].

Since the potential for the reversion of the attenuating phenotype exists, it is generally considered that bacterial strains to be used as candidates for oral vaccines should contain two or more attenuating mutations that map at separate regions of the chromosome, thus limiting the potential danger of reversion. Some of these double mutants, such as the *aroA purA* double mutants of *S. typhimurium*[16] and *S. typhi*[42], are so attenuated that they are no longer protective. However, double mutations in the chorismate pathway, such as the *S. typhimurium aroA aroC* constructed by Dougan et al.[39], are as avirulent and immunogenic as the *aroC* strain. More recently it was demonstrated that 7-day-old calves orally immunized with an *aroA aroD* mutant of a calf-virulent *S. typhimurium* strain, ST4/74, could be protected against experimental infection with the virulent parent strain[17].

Other mutants that are being investigated for their possible use as oral vaccines are those with a deletion in the *cya* and *crp* genes[15]. Thus they lack the enzyme adenylate cyclase and the cyclic AMP receptor protein, which are required for the expression of numerous genes for carbohydrate and amino acid transport or utilization. Δ*cya* Δ*crp S. typhimurium* strains are avirulent and immunogenic in BALB/c mice and provide protection against challenge with 10^4 times the wild-type lethal dose via the oral route, with immunity lasting at least 4 months[43]. These strains retain their ability to invade and persist in the GALT but have a diminished ability to reach the spleen and liver[15]. It has been shown that Δ*cya* Δ*crp* strains have longer generation times *in vitro* than their wild-type parents[15], possibly contributing to their avirulence *in vivo* by allowing the host defense mechanisms a greater opportunity to deal with the invading organisms. Δ*cya* Δ*crp* mutations have also been introduced into other *Salmonella*, for example *S. choleraesuis* (S.M. Kelly et al., unpublished data) and *S. enteriditis* , the latter being avirulent and immunogenic in chickens[44]. An *S. typhi* Ty2 Δ*cya* Δ*crp* strain has undergone testing in human volunteers and was compared to *aroA aroD* and *aroA aroC* mutants for its ability to induce humoral immune reponses. The results of these tests indicated that *aroA aroD* mutants induced significantly higher serum antibodies than either *aroA aroC* or Δ*cya* Δ*crp* mutants. The latter two strains induced a mild vaccinemia in a few individuals following oral immunization (M. Levine et al., unpublished data) and may require additional mutations to render them more avirulent and to allow an increase in dosage for the induction of protective immune responses.

Among the many other *Salmonella* mutants that have been shown to be of possible use as live oral vaccines are the *S. typhimurium phoP* mutants. These mutants were isolated during the preparation of *S. typhimurium* strains lacking phosphatase activities that could be used for Tn*phoA* mutagenesis. Doses equivalent to 10^4 times the LD$_{50}$ of the parent strain could be administered either i.p. or perorally without illeffect. These strains were able to protect immunized mice against challenge with the virulent parent strain, despite the low-level colonization of the PP and *phoP* mutant's reduced ability to reach the spleen[33].

EXPRESSION OF HETEROLOGOUS ANTIGENS BY AVIRULENT *SALMONELLA*

The ability of attenuated *Salmonella* strains to induce protective immunity following oral administration made them attractive as potential candidates for the development of multivalent vaccines by introducing genetic determinants of potentially protective antigens from other pathogens that infect man or animals. One of the first reported uses of recombinant avirulent

Salmonella was the expression of *Shigella sonnei* O antigen in *S. typhi* Ty21a[19]. The *S. sonnei* plasmid specifying the O antigen was conjugally transferred to an *S. typhi* Ty21a recipient, which was found to produce both the *S. sonnei* O antigen and *S. typhi* somatic antigens. Mice immunized with the recombinant strain were protected against challenge with both *S. sonnei* and *S. typhi*[19] with the induction of secretory and systemic antibodies to *S. sonnei* O antigen and *S. typhi*[45]. Human volunteers orally immunized with the recombinant *S. typhi* contained antibody-secreting cells (ASC) to O antigens of both *S. typhi* and *S. sonnei* in their blood[46]. More recently, Forrest and LaBrooy[47] have shown that human volunteers orally immunized with *S. typhi* Ty21a expressing the *V. cholerae* O antigen produced intestinal IgA antibodies to both *S. typhi* and *V. cholerae* LPS. These responses were dependent upon the variation in the galactose-dependent O antigen side-chain of *S. typhi*[46]. *S. typhi* grown in a medium deficient of galactose prior to administration had a greater surface expression of *V. cholerae* O antigen which induced significantly higher anti-cholera IgA intestinal antibody response. This observation may be of importance in the use of not only *S. typhi* for the expression of heterologous antigens at the surface of this vector but also other bacterial vector systems.

The narrow host range of *S. typhi* means that it has limited use in the development of oral vaccines since it is not a natural pathogen for animals other than man. The emphasis has now turned to *S. typhimurium*, which is capable of inducing typhoid-like symptoms in mice and has a broader host range. *galE* mutants of *S. typhimurium* that expressed K88, a fimbrial antigen associated with enterotoxigenic *E. coli* (ETEC) of swine, have been shown to induce both secretory and humoral responses to K88 in immunized mice[21]. The same antigen expressed by *aroA S. typhimurium* mutants was also found to induce serum antibodies in orally immunized mice, and intravenous boosting with the recombinant strain resulted in serum antibodies able to agglutinate an *E. coli* K-12 strain expressing the K88 antigen[48]. *S. typhimurium aroA* mutants have been used to express a variety of other heterologous antigens such as LT-B. Mice orally immunized with 10^{10} viable *aroA S. typhimurium* expressing LT-B were found to have high serum and intestinal antibody titers to LT-B, both of which could be boosted with an oral dose of the vaccine strain expressing LT-B[32]. Serum from mice orally immunized with this recombinant *S. typhimurium* was also capable of neutralizing the activity of the heat-labile toxin in tissue culture assays[31].

Brown et al.[23] used the *S. typhimurium aroA* vaccine strain SL3261 to express β-galactosidase (GZ) encoded on a recombinant plasmid, which is expressed as an intracellular protein. Mice intravenously immunized with this recombinant strain developed circulating antibodies to GZ indicating that the immune system recognized intracellular antigens as well as those expressed at the cell surface[23]. More interestingly they demonstrated DTH responses to GZ when purified antigen was injected into the footpads of immunized mice after booster immunizations. Other investigators have used *aroA* mutants of *S. typhimurium* or *S. dublin* to express a variety of heterologous antigens from a number of pathogens, such as *Bordetella pertussis*[49], hepatitis B[50], tetanus toxin[51], M protein of *Streptococcus pyogenes*[52], and Influenza A nucleoprotein[53]. Still others have used Δ*cya* Δ*crp S. typhimurium* strains to express surface protein antigen A (SpaA) of *Streptococcus sobrinus*[24], *Plasmodium falciparum* blood stage antigens[27], *Brucella abortus* 31kDa protein[25], M protein of *S. equi*[54], and hepatitis B viral antigens[50].

One of the potential problems associated with live vaccine strains is the stable maintainenance of recombinant clones and loss of the plasmid during colonization of the vaccinated animal. This problem has been addressed with the construction of a cloning vector encoding the *asd* gene of *Streptococcus mutans*[24] or *S. typhimurium*[55], which complements the *asd* chromosomal deletion and restores wild-type virulence in Δ*asd S. typhimurium*. Loss of this plasmid results in lysis of the *Salmonella*, due to their inability to synthesize diaminopimelic acid, an essential cell wall constituent in gram-negative bacteria. A Δ*cya* Δ*crp* Δ*asd S. typhimurium* strain has been constructed to use the Asd+ vector for cloning genes for heterologous antigens such as SpaA of *S. sobrinus*[24]. The *spaA* fragment encoding the major antigenic determinant of SpaA[56] was cloned into the *S. typhimurium* Asd+ vector and introduced into a Δ*cya* Δ*crp* Δ*asd S. typhimurium* strain either as a single or a three-tandem repeat of the *spaA* fragment. Significant serum IgG anti-SpaA antibodies, as well as intestinal IgA titers, could be detected in mice orally immunized with *S. typhimurium* expressing the SpaA protein encoded by the three-tandem repeat of the *spaA* gene, (T.A. Doggett, unpublished data).

The majority of these investigations have involved the demonstration of serum antibodies and intestinal antibody responses. Few have demonstrated the induction of cellular immunity or protection against the pathogens supplying the colonization/virulence antigens expressed by the recombinant *Salmonella* vaccine strains. Among those that have demonstrated protection against pathogens using recombinant *Salmonella* vaccines expressing these heterologous antigens are Black et al.[45], who were able to protect human volunteers against shigellosis by immunization with *S. typhi* Ty21a expressing *S. sonnei* O antigen; Poirier et al.[52], who were able to show protective immunity against *S. pyogenes* in mice immunized with *aroA S. typhimurium* expressing *S. pyogenes* M protein; and Fairweather et al.[51], who orally immunized mice with an *S. typhimurium* mutant expressing a 50kDa fragment of tetanus toxin and were able to induce protective immunity against a lethal challenge with tetanus toxin. In all of these experiments, protection was accompanied by high serum antibody titers to the heterologous antigen. However, Sadoff et al.[57] were able to demonstrate the induction of protective immunity to *Plasmodium berghei* without the induction of a humoral immune response, while significant DTH responses to sporozoite antigens were observed, indicating that T cell epitopes were capable of inducing protective cell-mediated immunity. More recently, Aggarwal and colleagues[58] were able to demonstrate that protection to *P. berghei* was due to the induction of class I-restricted CD8+ cytotoxic T lymphocytes (CTL) directed against the circumsporozoite peptide epitope expressed by *S. typhimurium*, while mice immunized with an *aroA S. typhimurium* vaccine strain expressing *Leishmania major* gp63 antigen developed, without detectable DTH reactivity, humoral antibody as well as proliferative T cell responses to gp63[26]. These activated T cells were mainly CD4+ that secrete IL-2 and IFNγ and which are class II- restricted.

FUTURE DEVELOPMENTS

One of the major issues in the development of oral vaccines is the cellular placement of expressed antigens within the vaccine strain. Forrest and LaBrooy[47] have demonstrated that antigens expressed at the surface, such as *V. cholerae* O antigen, are masked by *Salmonella* LPS side-chains, resulting in poor recognition of the expressed antigen by the immune system. This problem may be overcome by utilizing semirough strains thereby reducing the masking effect of the *Salmonella* LPS. Another approach involves the insertion of genes encoding virulence determinants into the gene coding for flagellar-antigen determinants, an approach that has resulted in the induction of humoral immune responses[59,60]. Recently, Newton et al.[61] were able to demonstrate that mice immunized with an *aroA S. dublin* strain expressing flagella carrying a streptococcal M protein epitope were protected against challenge with type 5 streptococci while also producing opsonizing antibodies to M5.

Enhancement of the mucosal immune responses to antigens expressed by recombinant *Salmonella* has led to an interest in using CT-B and LT-B, both of which have been found to be potent stimulators of the mucosal immune system. When administered orally in conjunction with a poor mucosal antigen, CT-B has been shown to elicit secretory immune responses to itself and enhanced responses to antigen[62], although the response is dependent upon the H-2 haplotype of the immunized animal[63]. CT has also been shown to abrogate the appearance of systemic hypersenstivity to ovalbumin[64] and to prime T cells in the spleen, mesenteric lymph nodes, PP and intestinal lamina propria[65]. LT-B is believed to function in a similar manner. Many groups are now working on the construction of fusion peptides of LT-B or CT-B to be introduced into *Salmonella* vaccine strains. Fusion proteins of LT-B and antigens such as preS-2 B cell epitope and T cell epitopes of hepatitis B virus, HBV[66] and heat stable toxin (ST) of *E. coli*[1] have been generated. Parenteral immunization of mice with a crude lysate of *S. typhimurium* expressing LT-B/ST elicited a serum response to ST that was greater than that observed when ST and LT were given separately. The LT-B/preS-2 HBV protein, when expressed by either *aroA* or *galE S. typhimurium* mutants, resulted in elevated titers to preS-2 protein in immunized mice, compared with control animals[50]. We have also constructed LT-B fusions with *dex* and *spaA* gene of *S. sobrinus* (E.K. Jagusztyn-Krynicka, unpublished data) and are in the process of evaluating the expression of these fusion proteins by Δ*cya* Δ*crp* Δ*asd S. typhimurium* vaccine strains and the induction of humoral and cellular immune responses in immunized mice to the expressed chimeric protein.

It has been noted that i.p. immunization of mice with the *aroA S. typhimurium* strain SL3235 resulted in immunosuppression of certain immune functions but was accompanied by protection against challenge with wild-type organisms[67]. This immunosuppression was shown to be due to two cell populations of monocytic lineage[68] which function by inhibiting the production of IL-2 by splenocytes from infected animals[69]. This suppression was, however, reversed by the addition of IL-4 to splenic cell cultures and is believed to act directly at the level of the macrophages[69], thereby reducing or down-regulating the production of inflammatory products by activated macrophages. What must now be investigated is the effect of *S. typhimurium* vaccine strains on the GALT when given orally.

The mechanisms that invoke protective immunity are complex and the utilization of attenuated *Salmonella* vaccine strains to express heterologous antigens requires evaluation of these responses, particularly in the GALT. It is evident that humoral immunity may not be of great importance in protection against some pathogens but requires activation of cell-mediated immune responses, while the induction of both cellular and humoral responses may be necessary for protection against other pathogens. *Salmonella* vaccines provide an excellent opportunity to evaluate the induction of mucosal immunity and the methods by which it can be manipulated to provide protection against a number of human and animal pathogens.

ACKNOWLEDGMENTS

Our grateful appreciation to S. Tinge and J. Maurer for their critical reviewing of this manuscript. This work was supported, in part, by USPHS Grants R37-DE06673 and R01-DE06669 from NIDR.

REFERENCES

1. J.D. Clements and L.Cardenas, Vaccines against enterotoxigenic bacterial pathogens based on hybrid Salmonella that express heterologous antigens, *Res. Microbiol.* 141:981 (1990).
2. T.J. Stabel, J.E. Mayfield, L.B. Tabatabai, and M.J. Wannemeuhler, Swine immunity to an attenuated *Salmonella typhimurium* mutant containing a recombinant plasmid which codes for production of a 31-kilodalton protein of *Brucella abortus, Infect. Immun.* 59:2941 (1991).
3. R.L. Owen and A.L. Jones, Epithelial cell specialization within human Peyer's patches: an ultrastructural study of intestinal lymphoid follicles, *Gastroenetrology* 66:189 (1974).
4. C. Czerkinsky, M.W. Russell, N. Lycke, M. Lindblad, and J. Holmgren, Oral administration of a streptococcal antigen coupled to cholera toxin B subunit evokes strong antibody responses in salivary glands and extramucosal tissues, *Infect. Immun.* 57:1072 (1989).
5. C. Czerkinsky, A.M. Svennerholm, M. Quiding, R. Jonsson, and J. Holmgren, Antibody-producing cells in peripheral blood and salivary glands after cholera vaccination of humans, *Infect. Immun.* 59:996 (1991).
6. A.J. Husband and J.L. Gowans, The origin and antigen-dependent distribution of IgA-containing cells in the intestine, *J. Exp. Med.* 148:1146 (1978).
7. M.R. McDermott and J. Bienenstock, Evidence for a common mucosal immunologic system. I. Migration of B immunoblasts into intestinal, respiratory and genital tissues, *J. Immunol.* 122:1892 (1979).
8. P.C. Montgomery, K.M. Connelly, and C.A. Skandera, Remote-site stimulation of secretory IgA antibodies following bronchial and gastric stimulation, *in*: "Advances in Experimental Medicine and Biology," J.R. McGhee, J. Mestecky and J.L. Babb, eds., Plenum Press, New York (1978).
9. M. Kilian, J. Mestecky, and M.W. Russell, Defense mechanisms involving Fc-dependent functions of immunoglobulin A and their subversion by bacterial immunoglobulin A proteases, *Microbial. Rev.* 52:296 (1988).
10. W. A. Walker, K.J. Isselbacher, and K.J. Bloch, Intestinal uptake of macromolecules: effect of oral immunization, *Science* 177:608 (1972).
11. J. Mestecky and J.R. McGhee, Immunoglobulin A (IgA): molecular and cellular interactions involved in IgA biosynthesis and immune response, *Adv. Immunol.* 40:153 (1987).
12. A.M. Svennerholm, L.A. Hanson, J. Holmgren, B.S. Lindblad, B. Nilsson, and F. Querrshi, Different secretory immunoglobulin A antibody responses to cholera vaccination in Swedish and Pakastani women, *Infect. Immun.* 30: 427 (1980).
13. A.M. Svennerholm, L.A. Hanson, J. Holmgren, B.S. Lindblad, S.R. Khan, A. Nilsson, and B. Svennerholm, Milk antibodies to live and killed polio vaccines in Pakastani and Swedish mothers, *J. Infect. Dis.* 143:707 (1981).

14. S.K. Hoiseth and B.A.D. Stocker, Aromatic-dependent *Salmonella typhimurium* are non-virulent and effective as live vaccines, *Nature* 291:238 (1981).
15. R. Curtiss III and S.M. Kelly, *Salmonella typhimurium* deletion mutants lacking adenylate cyclase and cyclic AMP receptor protein are avirulent and immunogenic, *Infect. Immun.* 55: 3035 (1987).
16. D. O'Callaghan, D. Maskell, F.Y. Liew, C.S.F. Easmon, and G. Dougan, Characterization of aromatic- and purine-dependent *Salmonella typhimurium*: attenuation, persistence and ability to induce protective immunity in BALB/c mice, *Infect. Immun.* 56:419 (1988).
17. P.W. Jones, G. Dougan, C. Hayward, P. Collins, and S.N. Chatfield, Oral vaccination of calves gainst experimental salmonellosis using a double *aro* mutant of *Salmonella typhimurium*, *Vaccine* 9:29 (1991).
18. T.K.S. Mukkur, B.A.D. Stocker, and K.H. Walker, Genetic manipulation of *Salmonella* serotype *Bovismorbificans* to aromatic-dependence and evaluation of its vaccine potential in mice, *J. Med. Microbiol.* 34:57 (1991).
19. S.B. Formal, L.S. Baron, D.J. Kopecko, C. Powell, and C.A. Life, Construction of a potent bivalent vaccine strain: introduction of *Shigella sonnei* form I antigen genes into the *galE S. typhi* Ty21a typhoid vaccine strain, *Infect. Immun.* 34:746 (1981).
20. J.D. Clements and S. El-Morshidy, Construction of a potential live oral bivalent vaccine for typhoid fever and cholera-*Escherichia coli*-related diarrheas, *Infect. Immun.* 46:564 (1984).
21. G. Stevenson and P.A. Manning, Galactose epimeraseless (*galE*) mutant G30 of *Salmonella typhimurium* is a good potential live oral vaccine carrier for fimbrial antigens, FEMS *Microbiol. Lett.* 28:317 (1985).
22. D.J. Maskell, K.J. Sweeney, D. O'Callaghan, C.E. Hormaeche, L.Y. Liew, and G. Dougan, *Salmonella aroA* mutants as carriers of the *Escherichia coli* heat-labile enterotoxin B subunit to the murine systemic and secretory immune systems, *Microbial. Pathogen.* 2:211 (1987).
23. A.Brown, C.E. Hormaeche, R. Demarco de Hormaeche, M Winther, G. Dougan, D.J. Maskell, and B.A.D. Stocker, An attenutated *aroA Salmonella typhimurium* vaccine elicits humoral and cellular immunity to cloned β-Galactosidase in mice, *J. Infect. Dis.* 155:86 (1987).
24. K.Nakayama, S.M. Kelly, and R. Curtiss III, Construction of an Asd+ expression cloning vector: stable maintenance and high level expression of cloned gene genes in a *Salmonella* vaccine strain, *Biotechnology* 6:693 (1988).
25. T.J. Stabel, J.E. Mayfield, L.A. Tabatabai, and M.J. Wannemuehler, Oral immunization of mice with attenuated *Salmonella typhimurium* containing a recombinant plasmid which codes for production of a 31-kilodalton protein of *Brucella abortus*, *Infect. Immun.* 58:2048 (1990).
26. D.M. Yang, N. Fairweather, L.L. Button, W.R. McMaster, L.P. Kahl, and F.Y. Liew, Oral *Salmonella typhimurium* (Aro-) vaccine expressing a major leishmanial surface protein (gp63) preferentially induces T helper 1 cells and protective immunity against leishmaniasis, *J. Immunol.* 145:2281 (1990).
27. J. Schorr, B. Knapp, E. Hundt, H.A. Kupper, and E. Amann, Surface expression of malarial antigens in *Salmonella typhimurium*: induction of serum antibody response upon oral vaccination of mice, *Vaccine* 9:675 (1991).
28. G.A. Bacon, T.W. Burrows, and M. Yates, The effects of biochemical mutation on the virulence of *Bacterium typhosum*: the virulence mutants, *Brit. J. Exp. Pathol.* 32:714 (1950).
29. G.A. Bacon, T.W. Burrows, and M. Yates, The effects of biochemical mutation on the virulence of *Bacterium typhosum*: the loss of virulence of certain mutants, *Brit. J. Exp. Pathol.* 32:85 (1951).
30. R. Germanier and E. Furer, Isolation and characterization of GalE mutant Ty21a of *Salmonella typhi*: a candidate strain for a live, oral typhoid vaccine, *J. Infect. Dis.* 131:553 (1975).
31. D.J. Maskell, F.Y. Liew, K.Sweeney, G. Dougan, and C.E. Hormaeche, Attenuated *Salmonella typhimurium* as live oral vaccines and carriers for delivering antigens to the secretory immune system, *in*: "Vaccines 86," Cold Spring Harbor Laboratory, Cold Spring Harbor (1986).
32. G. Dougan, C.E. Hormaeche, and D.J. Maskell, Live oral *Salmonella* vaccines: potential use of attenuated strains as carriers of heterologous antigens to the immune system, *Parasite Immunol.* 9:151 (1987).
33. J. E. Galan and R. Curtiss III, Virulence and vaccine potential of *phoP* mutants of *Salmonella typhimurium*, *Microbial Pathogen.* 6:433 (1989).
34. R. Germanier and E. Furer, Immunity in experimental salmonellosis: II. Basis of the avirulence and protective capacity of *galE* mutants of *Salmonella typhimurium*, *Infect. Immun.* 4:663 (1971).
35. M.M. Levine, J.B. Kaper, R.E. Black, and M.L. Clements, New knowledge on pathogenesis of bacterial enteric infections as applied to vaccine development, *Microbiological Rev.* 47:510 (1983).
36. M.H. Whadan, C. Serie, Y. Cersier, S. Sallam, and R. Germanier, A controlled field trial of live oral *Salmonella typhi* Ty21a oral vaccine against typhoid: 3-year results, *J. Infect. Dis.* 145:292 (1982).
37. L. Nencioni, L. Villa, M.T. De Margistris, M. Romano, D. Boraschi, and A. Tagliabue, Cellular immunity against *Salmonella typhi* after live oral vaccine, *Adv. Exp. Med. Biol.* 216B:1669 (1987).

38. B.D. Forrest, J.D. LaBrooy, P. Robinson, C.E. Dearlove, and S.J. Shearman, Specific immune response in the human respiratory tract following oral immunization with live typhoid vaccine, *Infect. Immun.* 59:1206 (1991).

39. G. Dougan, S. Chatfield, D. Pickard, J. Bester, D. O'Callaghan, and J.D. Maskell, Construction and characterization of vaccine strains of *Salmonella* harboring mutations in two different *aro* genes, *J. Infect. Dis.* 158:1329 (1988).

40. I.A. Miller, S. Chatfield, G. Dougan, J. DeSilva, H.S. Joysey, and C. Hormaeche, Bacteriophage P22 as a vehicle for introducing cosmid gene banks between smooth strains of *Salmonella typhimurium*: use in identifying a role for *aroD* in attenuating virulent *Salmonella* strains, *Mol. Gen. Genet.* 215:312 (1989).

41. R. Curtiss III, R. Goldschmidt, S. Kelly, M. Lyons, S.M. Michalek, R. Pastian, and S. Stein, Recombinant avirulent *Salmonella* for oral immunization to induce mucosal immunity to bacterial pathogens, *in*: "Vaccines- New Concepts and Developments," H. Kohler and P.T. LoVerde, eds., Longman Scientific and Technical, Harlow, England (1987).

42. M.M. Levine, D. Herrington, J.R. Murphy, J.G. Morris, G. Losonsky, B. Tall, A.A. Lindberg, S. Svenson, S. Bagar, M.F. Edwards, and B.A.D. Stocker, Safety, infectivity, immunogenicity and *in vivo* stability of two attenuated auxotrophic mutant strains of *Salmonella typhi*, 541 Ty and 543 Ty, as live oral vaccines in humans, *J. Clin. Invest.* 79:888 (1987).

43. R. Curtiss III, K. Nakayama, and S.M. Kelly, Recombinant avirulent *Salmonella* vaccine strains with stable maintenance and high level expression of cloned genes in vivo, *in*: "Immunology and Immunopathology of the Alimentary Canal," B. Albini, R.J. Genco, D.L. Ogra, and M.M. Weiser, eds., Marcel Dekker, New York (1989).

44. R. Curtiss III, S.B. Porter, M. Munson, J.O. Hassan, C.R. Gentry-Weeks, and S.M. Kelly, Nonrecombinant and recombinant avrilulent live vaccines for poultry, *in*: "Colonization Control of Human Bacterial Enteropathogens in Poultry," L.C. Blankenship, J.S. Baily, N.A. Cox, N.J. Stern, and R.J. Meinersmann, eds., Academic Press, New York (1991).

45. R.E. Black, M.M. Levine, M.L. Clements, G. Losonsky, D. Herrington, S. Berman, and S.B. Formal, Prevention of shigellosis by a *Salmonella typhimurium-Shigella sonnei* bivalent vaccine, *J. Infect. Dis.* 155:1260 (1987).

46. L. van de Verg, D.A. Herrington, J.R. Murphy, S.S. Wasserman, S.B. Formal, and M.M. Levine, Specific immunoglobulin A-secreting cells in peripheral blood of humans following oral immunization with a bivalent *Salmonella typhi-Shigella sonnei* vaccine or infection by pathogenic *S. sonnei*, *Infect. Immun.* 58:2002 (1990).

47. B. Forrest and J.T. LaBrooy, In vivo evidence of immunological masking of the *Vibrio cholerae* O antigen of a hybrid *Salmonella typhi* Ty21a-*Vibrio cholerae* oral vaccine in humans, *Vaccine* 8:29 (1991).

48. G.Dougan, R. Sellwood, D. Maskell, K. Sweeney, J. Beesley, and C. Hormaeche, *In vivo* properties of a clones K88 determinant, *Infect. Immun.* 52:344 (1986).

49. N.C. Molina and C.D. Parker, Murine antibody response to oral infection with live *aroA* recombinant *Salmonella dublin* vaccine strains expressing filamentous haemagglutanin antigen from *Bordatella pertussis*, *Infect. Immun.* 58:2523 (1990).

50. F. Schödel, D.R. Milich, and H. Will, Hepatitis B virus nucleocapsid/preS-2 fusion proteins expressed in attenuated *Salmonella* for oral vaccination, *J. Immunol.* 145:4317 (1990).

51. N.F. Fairweather, S.N. Chatfield, A.J. Makoff, R.A. Strugnell, J. Bester, D.J. Maskell, and G. Dougan, Oral vaccination of mice against tetanus by use of a live attenuated *Salmonella* carrier, *Infect. Immun.* 58:1323 (1990).

52. T.P. Poirier, M.A. Kehoe, and E.H. Beachey, Protective immunity evoked by oral administration of attenuated *aroA Salmonella typhimurium* expressing cloned streptococcal M protein, *J. Exp. Med.* 168: 25 (1988).

53. J.P. Tite, X-M. Gao, C.M. Hughes-Jenkins, M. Lipscombe, D. O'Callaghan, G. Dougan, and F.Y. Liew, Antiviral immunity induced by recombinant nucleoprotein of influenza A virus: III. Delivery of recombinant nucleoprotein to the immune sytem using attenuated *Salmonella typhimurium* as a live carrier, *Immunology* 70:540 (1990).

54. J.E. Galan, J.F. Timoney, and R. Curtiss III, Expression and localization of the *Streptococcus equi* M protein in *Escherichia coli* and *Salmonella typhimurium*, *in*: "Proceeedings of the Fifth International Congress on Equine Infectious Diseases," Kentucky University Press, Lexington (1989).

55. J.E. Galan, K. Nakayama, and R. Curtiss III, Cloning and characterization of the *asd* gene of *Salmonella typhimurium*: use in stable maintenance of recombinant plasmids in *Salmonella* vaccine strains, *Gene* 94:29 (1990).

56. R.M. Goldschmidt, M. Thoren-Gordon, and R. Curtiss III, Regions of the *Streptococcus sobrinus spaA* gene encoding major antigenic determinants of antigen I, *J. Bacteriol.* 172:3988 (1990).

57. J.C. Sadoff, W. R. Ballou, L.S. Baron, W.R. Majarian, R.N. Brey, W.T. Hockmeyer, J.F. Young, S.T. Cryz, J. Ou, G.H. Lowell, and J.D. Chulay, Oral *Salmonella typhimurium* vaccine expressing circumsporozoite protein protects against malaria, *Science* 240:336 (1988).

58. A. Aggarwal, S. Kumar, R. Jaffe, D.Hone, M. Gross, and J. Sadoff, Oral *Salmonella*: malaria circumsporozoite recombinants induce specific CD8[+] cytotoxic T cells, *J. Exp. Med.* 172:1083 (1990).

59. S.M.C. Newton, C. O. Jacob, and B.A.D. Stocker, Immune response to cholera toxin epitope inserted in *Salmonella* flagellin, *Science* 244:70 (1989).

60. J.Y. Wu, S.M.C. Newton, B.A.D. Stocker, and W.S. Robinson, Expression of immunogenic epitopes of hepatitis B surface antigen with hybrid flagellin proteins by a vaccine strain of *Salmonella, Proc. Natl. Acad. Sci.* 86:4726 (1989).

61. S.M.C. Newton, M. Kotb, T.P.Poirier, B.A.D. Stocker, and E.H.Beachey, Expression and immunogenicity of a streptococcal M protein epitope inserted in *Salmonella* flagellin, *Infect. Immun.* 59:2158 (1991).

62. C.O. Elson and W. Ealding, Genetic control of the murine immune response to cholera toxin, *J. Immunol.* 135:930 (1985).

63. C.O. Elson and W. Ealding, Generalized systemic and mucosal immunity in mice after mucosal stimulation with cholera toxin, *J. Immunol.* 132:2736 (1984).

64. R.A. Kay and A. Ferguson, The immunological consequences of feeding cholera toxin. I. Feeding cholera toxin suppresses the induction of systemic delayed-type hypersensitivity but not humoral immunity, *Immunology* 66:410 (1989).

65. N. Lycke and J. Holmgren, Adoptive transfer of gut mucosal antitoxin memory by isolated B cells one year after oral immunization with cholera toxin, *Infect. Immun.* 57: 1137 (1989).

66. F. Schödel, G. Enders, and H. Will, Expression of hepatitis B virus core T-cell epitopes and preS-2 B-cell epitope as fusion protein with LT-B in *Salmonella* for oral vaccination, *in*: "Progress in Hepatitis B Immunization," P. Coursaget and M.J. Tong, eds., Colloque INSERM, John Libbey Eurotext Ltd. 194:43 (1990).

67. T.K. Eisenstein, L.M. Killar, B.A.D. Stocker, and B.M. Sultzer, Cellular immunity induced by avirulent *Salmonella* in LPS-defective C3H/HeJ mice, *J. Immunol.* 133:958 (1984).

68. B.K. Al-Ramadi, M.A. Brown, D.M. Mosser, and T.K. Eisenstein, Immunosuppression by attenuated *Salmonella*. Evidence for mediation by macrophage precursors, *J. Immunol.* 146:2737 (1991).

69. B.K. Al-Ramadi, Y.W. Chen, J.J. Meissler Jr., and T.K. Eisenstein, Immunosuppression by attenuated *Salmonella*. Reversal by IL-4, *J. Immunol.* 147:1954 (1991).

USE OF RECOMBINANT BCG AS A VACCINE DELIVERY VEHICLE

C. Kendall Stover[1], Vidal F. de la Cruz[1], Geetha P. Bansal[1], Mark S. Hanson[1], Thomas R. Fuerst, William R. Jacobs, Jr [2] and Barry R. Bloom[2]

[1] MedImmune Inc., Gaithersburg, MD 20878
[2] Albert Einstein College of Medicine, Bronx, NY 10461

INTRODUCTION

The tuberculosis vaccine strain bacille Calmette Guerin (BCG) of *Mycobacterium bovis* is still one of the most widely used human vaccines in the world. Additional advantages also make BCG attractive as a live recombinant vaccine vehicle. These advantages include: i) BCG is the most widely used vaccine, having been given to over 2,500 million people since 1948, with a low incidence of serious complications (case fatality rate of $0.19/10^6$); ii) can be given at or any time after birth, and is unaffected by maternal antibodies; iii) BCG is given as a single inoculum and sensitizes for 5 to 50 years to tuberculoproteins; iv) is a potent adjuvant in experimental animals and humans; v) BCG can be administered as an oral vaccine; vi) is the most heat stable of the existing live vaccines; and vii) is very inexpensive to produce. Major obstacles impeding the development of BCG as a vaccine vehicle are its slow growth and the lack of useful genetic systems to genetically manipulate mycobacteria. These obstacles have now been largely overcome by the development of versatile extrachromosomal and integrative *E. coli*/mycobacteria shuttle vector systems[1-4] (Fig. 1). BCG *hsp*60 and *hsp*70 expression sequences have been employed in these vectors for the introduction and high-level expression of a variety of foreign antigens (bacterial, parasitic and viral) in recombinant BCG (rBCG). Most notable of these are the antigens of HIV-1 which have generally been relatively difficult to express at high levels in other bacterial systems (Fig. 2). The availability of both extrachromosomal (multicopy) and integrative (single-copy) expression vectors has enhanced the versatility of our BCG expression vectors as extrachromosomal expression from high copy number vectors is, in some instances, lethal to BCG (e.g., HIV-1 gp120)[1,2]. In other cases, episomal and integrative expression is comparable (e.g., HIV-1 *gag*, Fig. 2).[1,2] In general, the best expression of foreign antigens in BCG has been achieved with bacterial genes with some expressed as well as *lacZ* (data not shown). These studies have established the value of the BCG *hsp* regulatory sequences in driving expression of foreign genes in mycobacteria and suggest that homologous *hsp* sequences may be generally useful for enhancing expression of foreign genes in other recombinant bacteria.

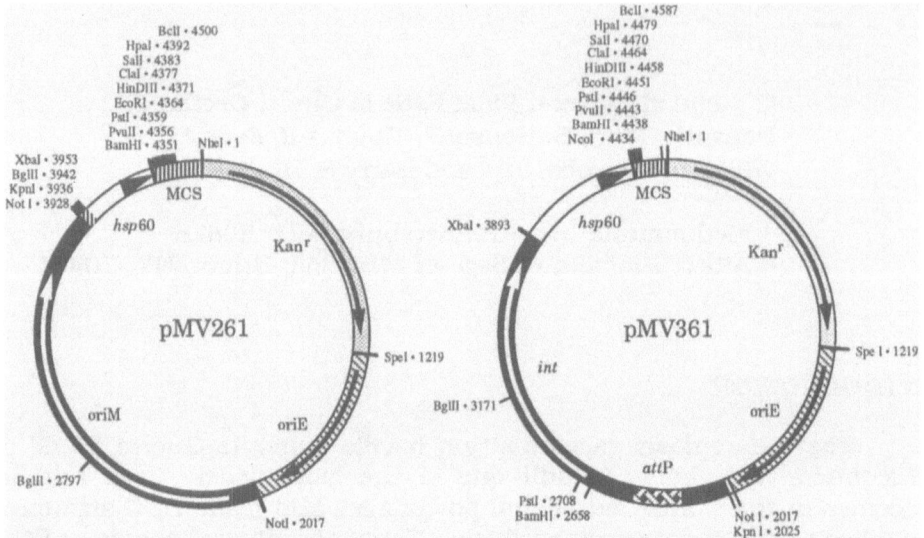

Figure 1. Cassetted extrachromosomal and integrative BCG-*E. coli* shuttle expression vectors. Vectors pMV261 and pMV361 were constructed by the directional assembly of DNA cassettes made by polymerase chain reaction (PCR) amplification of minimal essential DNA segments. pMV261, which is extrachromosomal in mycobacteria, is composed of four PCR cassettes including the TN903 derived *aph* gene encoding aminoglycoside 3'-phosphotransferase (kanamycin resistance) as a selectable marker, an *Escherichia coli* (*E. coli*) plasmid replicon derived from pUC19 (Erep.), the modified 1.8 kb mycobacterial plasmid replicon derived from pAL5000 (oriM), and the *hsp60* 5'-regulatory region. Commonly used restriction sites in the TN903 *aph* (*ClaI* and *HindIII*) and Mrep open reading frames (*EcoRI, SalI,* and *NcoI*) were eliminated by PCR mutagenesis at wobble codons so as not to change the amino acid sequence of encoded proteins. A synthetic multiple cloning site (MCS) including 18 commonly used unique restriction sites and terminators for translation and transcription was added to facilitate further manipulation of mycobacterial expression sequences and foreign antigen genes. The four DNA cassettes were directionally ligated to result in a 4 kb shuttle vector with all essential plasmid ORFs oriented unidirectionally to prevent potential interference between opposing transcripts. Plasmid pMV361, an integrative vector for mycobacteria, was constructed by substitution of the oriM cassette in pMV261 with a 1.9 kb *XbaI-PvuII* cassette encoding the L5 *att/int*.

ANTIBODY RESPONSES TO ANTIGENS EXPRESSED AT LOW LEVELS IN rBCG

Initial animal studies employing the *hsp* expression vector systems to drive high-level expression of the model antigen β-galactosidase (β-gal) had shown that these vectors are stable, that rBCG grows *in vivo*, and that long-lasting humoral and cellular (both CD8 and

CD4) immune responses could be elicited to a recombinant antigen expressed and delivered in a rBCG vaccine.[3,5] While it was not surprising to observe cellular immune responses to the recombinant β-gal antigen delivered by rBCG immunization, the striking anti-β-gal antibody response was unexpected as BCG is not noted for its ability to elicit substantial antibody responses to native BCG proteins. Indeed, the antibody response we observed to BCG proteins in mice immunized with the rBCG::*lacZ* recombinant peaked early, then rapidly waned, and was minimal compared to the antibody to rBCG expressed β-gal.[1,2] These data suggested that greater antibody responses may be achieved to recombinant proteins expressed by rBCG than to antigens of BCG itself. While studies with the experimental rBCG::*lacZ* vaccine were encouraging, the striking expression level (greater than 15% total BCG protein) achieved for the *lacZ* gene in BCG may be unachievable for most disease-relevant antigens. Therefore it was questionable whether one could achieve similar antibody responses to a recombinant antigen expressed at lower levels.

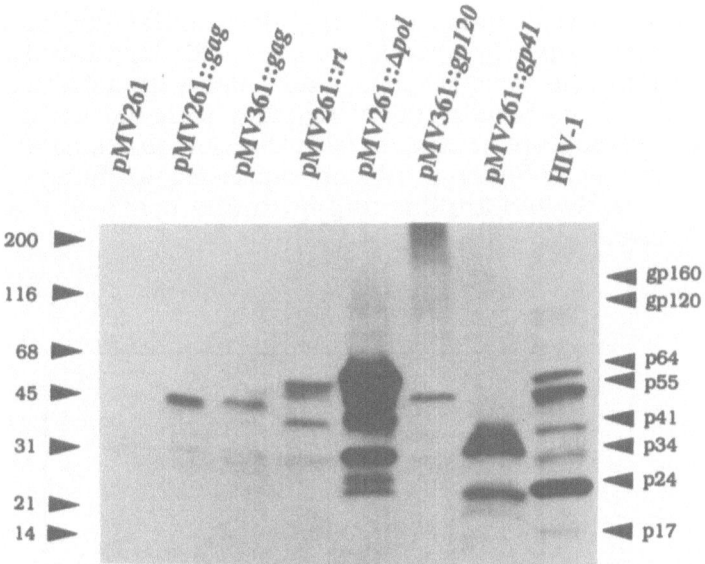

Figure 2. Expression of HIV-1 antigens (IIIB isolate) in recombinant BCG. Genes encoding HIV-1 Gag, Pol/ prot (polymerase minus protease), gp120, and gp41 HIV-1 proteins were amplified by PCR and cloned into extrachromosomal *hsp*60 and *hsp*70 (pMV261, pMV273), or integrative *hsp*60 (pMV361) expression vectors and transformed into BCG. Lysates of recombinant BCG were analyzed for expression by Western blot analysis of 4-20% gradient SDS PAGE gels using human sera from an HIV-1-infected patient and ^{125}I labelled *S. aureus* protein A. Arrows in left margin indicate molecular weight standards. Arrows in right margin indicate apparent molecular weights of HIV-1 proteins (p) and glycosylated proteins (gp). BCG recombinant gene products are not glycosylated and therefore run at lower apparent molecular weights. Reprinted by permission.[1]

To determine whether a humoral response can develop to a recombinant antigen expressed at much lower levels than *lacZ* in rBCG, BALB/c mice were immunized by a single intradermal injection with

10^6 cfu rBCG expressing the gp41 portion of the HIV-1 envelope (rBCG::HIV-1gp41) or rBCG::*lacZ* (Fig. 3). Only mice immunized with rBCG::HIV-1gp41 developed an antibody response to HIV-1 envelope. While this antibody response was slow to develop, it was substantial and still appeared to be increasing at 16 weeks. To determine whether a protective antibody response to a conformational epitope could be elicited by rBCG, a DNA segment encoding the protective fragment C of the tetanus toxin (toxC) was cloned into BCG expression vectors. The 50 kDa fragment C is known to mediate binding of tetanus toxin to the cellular ganglioside receptor and is so far the smallest tetanus toxin fragment shown to protect mice against tetanus toxin challenge.[6] Expression of the ToxC polypeptide was found to be approximately 250-fold less than that of β-gal used in previous rBCG immunization studies.[1]   NIH swiss mice were immunized by intraperitoneal (IP) or intradermal (ID) injection with a single dose of live or heat killed (HK) rBCG (10^6 cfu) carrying plasmid pMV261::*toxC* (Fig. 4).[1]   In these initial studies, similar antibody responses were observed with live rBCG immunizations by both the IP and ID routes. Furthermore, live IP and ID immunized mice were either completely or partially protected when challenged at 12 weeks with 100 lethal doses of tetanus toxin (data not shown).[7] Naive mice or mice immunized with heat-killed rBCG::*toxC* did not exhibit anti-toxin responses and died within 16 hours of toxin challenge.[7] These results, taken together with the rBCG::HIV-1gp41 immunization data are consistent with the view that sustained immune responses to recombinant antigens expressed as minor proteins in rBCG are possible and are due to growth and persistence of the rBCG vaccine *in vivo*.

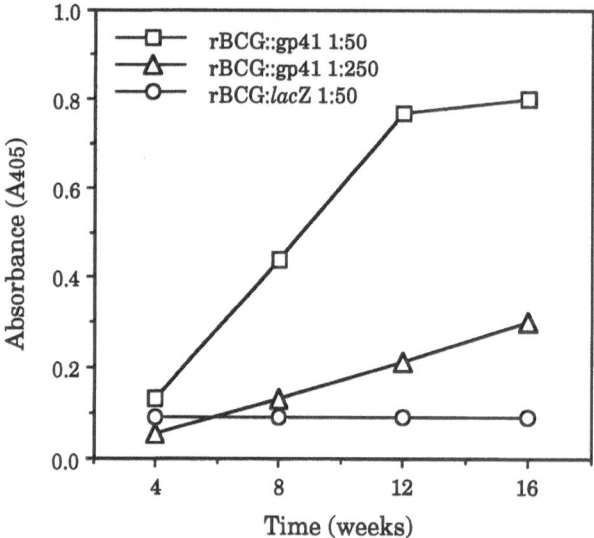

Figure 3. Antibody production to HIV-1 envelope in mice immunized with rBCG::HIV-1 gp41. BALB/c mice were immunized with 10^6 cfu of rBCG::HIV-1gp41 or rBCG::*lacZ* by intradermal injection.   Sera were obtained monthly, pooled, and analyzed by ELISA with HIV-1 extract. Specific reactivity with the gp160 and gp41 proteins of HIV-1 was verified by Western blot analysis (data not shown).

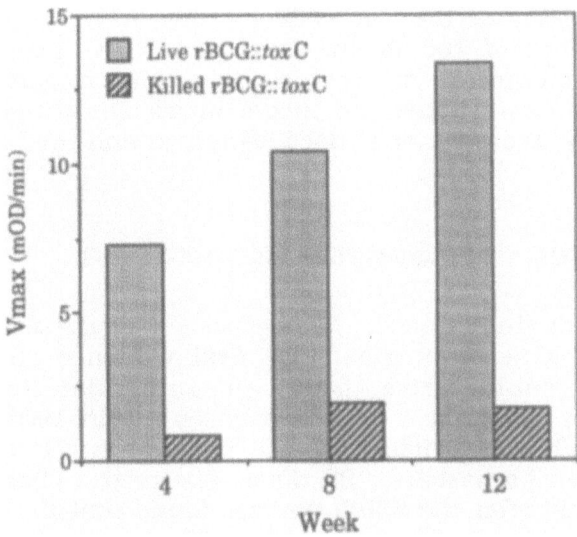

Figure 4. Antibody production to rBCG -expressed tetanus toxin fragment C. and tetanus toxin challenge of mice immunized with rBCG::*tox*C. DNA encoding fragment C of *Clostridium tetani* toxin (*tox*C) was cloned and fused in frame with the sixth codon of BCG *hsp60* on plasmid vector pMV261 to generate rBCG::*tox*C. Expression of the ToxC polypeptide in rBCG::*tox*C was analyzed by Western blot and determined to be 250-fold less than that of β-gal in the rBCG::*lacZ* vaccine used in initial animal studies (data not shown). NIH swiss mice were inoculated with rBCG::*tox*C(10^6 cfu) by ID or intraperitoneal (IP) injection. Parallel inoculations were carried out with identical preparations that had been heat-killed (HK) by incubation for 1 hour at 70^{0}C. Positive control mice received 10 µg toxoid in Freund's complete adjuvant and were boosted at 4 weeks with 10 µg toxoid in Freund's incomplete adjuvant. Sera were obtained monthly, pooled, and analyzed by ELISA with tetanus toxoid.

EFFORTS TO IMPROVE RECOMBINANT ANTIGEN PRESENTATION BY rBCG

All the studies on immune responses engendered by rBCG-delivered model antigens to date have been with recombinant antigens expressed in the cytoplasm of BCG. These studies demonstrate that it is possible to elicit cellular and humoral immune responses, including cytolytic T lymphocytes (CTL) and neutralizing antibody responses to conformational epitopes, to antigens expressed cytoplasmically in rBCG. We have recently constructed vectors which are designed to effect the expression and export of recombinant antigens by rBCG. Two basic types of BCG export vectors have been developed: i) vectors that employ signal sequences derived from mycobacterial genes that encode freely secreted proteins and ii) vectors that employ mycobacterial export signals which effect the export and lipid acylation of surface lipoproteins.[8,9] It is hoped that export (secretion and surface expression) of recombinant antigens by BCG might improve the efficiency of antigen delivery. Furthermore, some protective epitopes of surface proteins which are conformational may require export through the bacterial membrane to fold properly. It is also generally believed that the lipid acylation of peptides or whole proteins can substantially enhance their antigenicity. We have recently confirmed the functionality of the newly developed rBCG expression/export vectors and are currently involved in animal immunogenicity studies to compare immune responses to model protective antigens expressed cytoplasmically, secreted, or surface-expressed and lipid acylated.

PROSPECTS FOR ORAL IMMUNIZATION WITH rBCG

In 1905 in studies on the pathogenesis of tuberculosis in humans, Calmette and Guerin provided the first evidence that pulmonary tuberculosis could arise from an intestinal infection with *Mycobacterium tuberculosis (M. tuberculosis)*.[10] In 1906 Calmette and Guerin, in studies of a highly virulent bovine tubercle bacillus isolate which had been passaged for 38 times, observed a change in colonial morphology, and after the 230th passage found that this organism had lost virulence for mice. They then spent 12 years testing the safety of the attenuated BCG vaccine in mice, cattle and monkeys testing for revertants to virulence. No evidence of reversion was obtained in any species, and in 1921 the child of a mother who died in parturition of tuberculosis was vaccinated for the first time with the new attenuated BCG vaccine by the oral route.[11,12] Since the child's father had also succumbed to tuberculosis and the grandmother had tuberculosis, this child was at enormously high risk. The results of that first experiment were that the child remained free of tuberculosis throughout life and died of other causes at the age of 56. Thus, oral BCG became the first vaccine to be used against tuberculosis.

BCG was given as an oral vaccine until 1976. The reasons for the discontinuation of oral administration of BCG were twofold. First, after intradermal immunization with BCG was established, it became clear that at least two logs more organisms were required for comparable delayed-type hypersensitivity and skin test conversion reactions in individuals who received oral BCG relative to intradermal BCG; thus it was more expensive to give. Second, when given to newborn children up to the age of 6 to 7 months, in whom swallowing of vaccine in food is often not complete, there was relatively high incidence of local lymphadenitis as the high concentration of bacilli drained into regional lymph nodes. While serious consequences seldom ensued, these minor side effects were greater than observed with intradermal injection.

In addition to the objectionable complications, a number of *in vitro* and *in vivo* studies indicated that the need for larger numbers of immunizing organisms was due to the fact that BCG was killed to a significant extent by the acidity of the gastric secretions.[13] For example, it was shown that while there was a major reduction in viable bacilli found in the mesenteric lymph nodes and stool of a given inoculum introduced orally into guinea pigs, bypassing the stomach with a catheter resulted in almost total recovery of organisms introduced into the duodenum.[14] That the major loss of viable BCG by the oral route was the result of killing in the stomach was later confirmed in children by Gaudier and Gernez-Rieux using the freeze-dried BCG vaccine.[15] Therefore, there is a need to apply modern methods of buffering, enteric coating, or microencapsulation to protect recombinant BCG organisms from killing by the acidity of the stomach if rBCG is to be used orally.

Early studies provided evidence that BCG did not just remain localized in the gastrointestinal system after oral vaccination, but BCG could also be found at least 60 to 70 days after vaccination in the mesenteric lymph nodes of vaccinated subjects who had died of accidents or other intercurrent disease[16]. Transitory bacillemia was also observed within 5 hours after oral vaccination.[13] When BCG immunization routes were compared, it was found that oral BCG immunization was as proficient at eliciting tuberculin hypersensitivity and skin test reactivity as parenteral inoculation, confirming that an oral BCG vaccine can systemically sensitize T cells. However, there is essentially no literature on the capacity of oral BCG immunization to elicit systemic or mucosal, cellular or humoral immune responses to mycobacterial antigens. Thus, the extent to which oral BCG immunization would produce systemic or mucosal helper or cytotoxic T cell responses, lymphokine responses or immunoglobulin subclass responses systemically, locally, or in distant secretory compartments remains totally unknown. Therefore, further in-depth studies employing modern immunological techniques are necessary to characterize the nature of the mucosal immune response engendered by oral immunization with BCG to determine the potential for use of rBCG based vaccines as oral immunogens.

REFERENCES

1. C. K. Stover, V. F. de la Cruz, T. R. Fuerst, J. E. Burlein, L. A. Benson, L. T. Bennett, G. P. Bansal, J. F. Young, M. H. Lee, G. F. Hatfull, S. B. Snapper, R. G. Barletta, W. R. Jacobs, Jr., and B. R. Bloom, Induction of antibody and T-cell responses to foreign antigens by recombinant BCG vaccines, *Nature* 351:456 (1991).
2. C. K. Stover, J. E. Burlein, L. T. Bennet, V. F. de la Cruz, G. F. Hatfull, M. H. Lee, W. R. Jacobs Jr., and B. R. Bloom, *in* "Development of BCG as a live recombinant vaccine vehicle", R.M. Chanock, ed., Vaccines 91. Cold Spring Harbor Laboratory Press (1991).
3. W. R. Jacobs, M. Tuckman, and B. R. Bloom, Introduction of foreign DNA into mycobacteria using a shuttle phasmid, *Nature* 327:532 (1987).
4. M. H. Lee, L. Pascopella, W. R. Jacobs, Jr., and G. F. Hatfull, Site-specific integration of mycobacteriophage L5: integration-proficient vectors for Mycobacterium smegmatis, Mycobacterium tuberculosis, and bacille Calmette-Guerin,*Proc. Natl. Acad. Sci. U.S.A.* 88:3111 (1991).
5. V. F. de la Cruz, C. K. Stover, L. A. Benson, S. R. Palasynski, T. R. Fuerst, and J. F. Young, E. Pearce, W. R. Jacobs, and B. R. Bloom, Humoral and cellular immune responses to recombinant mycobacteria (BCG) *in*: Vaccines 91. Cold Spring Harbor Laboratory Press , p. 399 (1991).
6. J. L. Halpern, W. H. Habig, E. A. Neale, S. Stibitz, Molecular cloning and expression of the protective fragment C of tetanus toxin, *Infect. Immun.* 58:1004 (1990).
7. C. K. Stover, J. E. Burlein, W. R. Jacobs, Jr., B. R. Bloom, and V. F. de la Cruz, Protection of mice from tetanus toxin by experimental recombinant BCG vaccine, Manuscript in preparation (1992).
8. M. Hanson, C. L. Vigil, J. E. Burlein, and C. K. Stover, Export and lipid acylation of recombinant antigens to the surface of recombinant BCG. Manuscript in preparation (1992).
9. C. K. Stover, M. Hanson, J. E. Burlein, C. L. Vigil, S. R. Palasynski, A. Barbour, and G. Bansal, Lipid acylation and /or surface expression of OspA by a recombinant BCG vaccine carrier results in greatly enhanced neutralizing antibody responses to Borrelia burgdorferi, Manuscript in preparation, (1992).

10. A. Calmette, and C. Guerin, Origine intestinale de la tuberculose pulmonaire, *Ann. Inst. Pasteur.* 19:601 (1905).
11. A. Calmette, La vaccination preventive contra la tuberculose par BCG, 250 pp. (Masson et cie, Paris). (1927).
12. A. Calmette, C. Guerin, L. Negre, and A. Boquet, Sur la vaccination preventive des enfants, nouvousnes contre la tuberculose par le B.C.G., *Ann. Inst. Pasteur. de Lille* 41:201 (1927).
13. A. Calmette, B. Weill-Halle, A. Saenz, and L. Costil, Demonstration experimentale du passage des bacilles vaccins B.C.G. a traverse la muqueuse intestinale chez l'enfant et chez le singe, *Bull. Acad. Med.* 110:203 (1933).
14. V. M. Schwarting, The action of gastric contents on tubercle bacilli, *Ann. Rev. Tub.* 58:123 (1948).
15. B. Gaudier, and C. Gernez-Rieux, Etude experimentale de la vitalite du B.C.G. au cours de la traversee gastro-intestinale chez des enfants non allergiques vaccines par voie digestive, *Ann. Inst. Pasteur.* 13:77 (1962).
16. J. Zeyland, and E. Piasecka-Zeyland, Etude de 50 autopsies d'enfants vaccines au BCG et morts de maladies non tuberculsosis, *Ann. Inst. Pasteur.* 43:767 (1929).

VACCINIA VIRUS RECOMBINANTS AS POTENTIAL HERPES SIMPLEX

VIRUS VACCINES

James F. Rooney, Charles R. Wohlenberg, and Abner Louis Notkins

Laboratory of Oral Medicine
National Institute of Dental Research
National Institutes of Health
Bethesda, MD 20892

INTRODUCTION

Herpes simplex virus (HSV) is one of the most common infectious agents of the oral cavity. Initial infection, caused primarily by HSV-1, commences when infectious virus particles come in contact with the oral mucosa. Replication of the virus in epithelial cells is followed by uptake of the virus into peripheral sensory nerve endings, retrograde transport of the virus, and establishment of latent infection in the trigeminal ganglion. This latent infection can persist asymptomatically for the life of the infected individual, or may periodically reactivate to release infectious virus particles back into the ganglion, which can then travel down the nerve to the skin surface to cause a typical recurrent herpes lesion.[1] It is estimated that up to 160 million Americans may be infected with HSV type 1, resulting in over 100 million episodes of recurrent disease annually.[2] While the majority of infections are mild and self-limited, infection of the neonate or immunocompromised host can be life threatening. Although antiviral agents have proven effective in the management of acute infections, they cannot eradicate latent infection and may have little effect on the prevalence or transmission of the virus within the community. The development of a vaccine to prevent primary infection and reduce the incidence of recurrent disease would be desirable.

What are the critical properties of a vaccine against HSV? An effective vaccine for HSV must not only limit the symptoms of primary infection, it must protect against the establishment of latent HSV infection, for otherwise it cannot protect against subsequent recurrent disease. In order to protect against the establishment of latent infection, an HSV vaccine must elicit a potent humoral and cellular immune response which is capable of acting locally at the site of primary infection to inhibit viral replication and to restrict viral entry into peripheral sensory nerve endings. An ideal vaccine would provide long-lasting immunity and would protect against both HSV-1 and HSV-2 infection.

A variety of approaches are being taken in the development of a vaccine against HSV, including the use of genetically engineered attenuated live HSV preparations,

Genetically Engineered Vaccines, Edited by
J.E. Ciardi *et al.*, Plenum Press, New York, 1992

purified and recombinantly derived glycoprotein vaccines, synthetic peptide vaccines and recombinant live vector vaccines.[3,4] Although many different HSV antigens have been included in these vaccines, the most commonly utilized antigens have been the HSV glycoproteins, especially glycoprotein D (gD) and glycoprotein B (gB). These two glycoproteins are expressed on the surface of HSV-infected cells and are targets for host cellular and humoral immune responses. Antibodies induced by HSV gD and gB are capable of neutralizing whole virus and exhibit cross-reactivity between HSV-1 and HSV-2.

One approach in the development of a vaccine against HSV utilizes vaccinia virus recombinants which express HSV glycoproteins as potential live virus vaccines.[5-22] We have evaluated three vaccinia virus recombinants which express

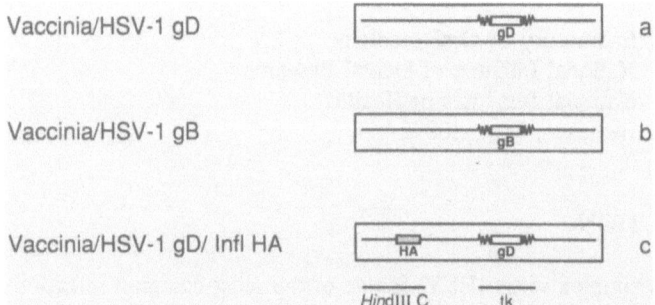

Figure 1. Construction of vaccinia recombinants (open rectangles) which express herpes simplex virus glycoproteins. a. The HSV-1 glycoprotein D gene (strain KOS) was inserted into the thymidine kinase (tk) gene of the vaccinia virus genome (straight line). b. The HSV-1 glycoprotein B gene was inserted into the tk gene of vaccinia. c. A bivalent recombinant was constructed which contained the HSV-1 gD gene in the tk locus of vaccinia and the influenza A hemagglutinin gene in the non-essential HindIII C region of the vaccinia virus genome. For all constructs (a-c) foreign proteins were expressed under the control of vaccinia virus promoters.

herpes simplex virus glycoproteins as potential HSV vaccines. The first expresses the HSV-1 glycoprotein D.[6] The second expresses the HSV-1 glycoprotein B (HSV-1 gD).[7] The third is a bivalent recombinant which expresses the HSV-1 glycoprotein D and the hemagglutinin of influenza A.[12] All three recombinant vaccines are capable of providing a high level of protection against HSV challenge in murine models. For the purpose of this discussion, we would like to focus on our experience with the HSV-1 gD recombinant (vaccinia/gD).

CONSTRUCTION OF THE VACCINIA/HSV-1 gD RECOMBINANT

A plasmid was constructed which contained the active HSV-1 gD coding sequence under the control of a vaccinia virus promoter and flanked by sequences from the vaccinia virus thymidine kinase gene.[6] This plasmid was then transfected into cells which had previously been infected with wild type vaccinia (strain WR). Homologous recombination occurred between the thymidine kinase sequences of the plasmid and those of vaccinia to yield recombinant progeny which contained the HSV-1 gD sequences inserted into the thymidine kinase gene. These progeny were, therefore, thymidine-kinase deficient and could be selected on that basis. One such recombinant was chosen for further evaluation. Cells infected with this recombinant expressed a protein on their surface which cross-reacted with anti-HSV-1 antibody. Polyacrylamide gel electrophoresis of infected cell lysates which had been immunoprecipitated with anti-HSV-1 serum revealed a glycoprotein with molecular weight 60,000, comparable to that of native HSV-1 gD.

PROTECTION AGAINST LETHAL AND LATENT HSV INFECTION

Mice immunized with vaccinia/gD by intradermal inoculation developed typical pox-type lesions which resolved within four weeks. These animals developed antibodies capable of neutralizing HSV *in vitro* and were protected against subsequent lethal intraperitoneal challenge with either HSV-1 or HSV-2.[6]

When vaccinated and control mice were challenged with HSV-1, 81% of mice in the control, unvaccinated group died following HSV-1 challenge, but only 2% of the mice in the vaccinia/gD vaccinated group died. Results following HSV-2 challenge were comparable.

Using a similar mouse model, protection against the development of latent HSV infection was determined.[6,13] Mice were immunized by intradermal inoculation and four weeks later were challenged via lip inoculation with a sublethal dose of HSV-1. Four weeks following HSV-1 challenge the mice were sacrificed and their trigeminal ganglia were explanted onto indicator cell monolayers to observe for cytopathic effects indicative of reactivation of latent HSV infection. The numbers of ganglia positive for HSV in vaccinated and control groups were compared and the percent protection was calculated. Whereas 95% of ganglia in unvaccinated control mice demonstrated evidence of latent infection, only 27% of ganglia from vaccinia/gD-immunized mice showed evidence of HSV infection. This corresponds to a 71% level of protection against latent infection for the vaccinia/gD group. Therefore, a single vaccination with vaccinia/gD can provide substantial protection against both lethal and latent HSV infection.

DURATION OF PROTECTION AND EFFECT OF REVACCINATION

In separate experiments mice were immunized with vaccinia/gD or an alternate vaccinia recombinant expressing a non-HSV gene, the hepatitis B surface antigen.[13] Vaccinated mice and unvaccinated controls were challenged with HSV at various time points after vaccination to determine the duration of protection afforded by a single vaccination with vaccinia/gD. Following lethal intraperitoneal challenge with HSV-1, all unvaccinated controls died, but protection in the vaccinia/gD group was 100% at 18 weeks, 100% at 44 weeks, and 80% at 60 weeks after vaccination. Protection against the development of latent infection was 70% at six weeks, 50% at 41 weeks, and 31% at 60 weeks after vaccination. Therefore, single vaccination with vaccinia/gD can provide protective immunity which persists to greater than one year following immunization.

Additional experiments were conducted to determine the effect of administering a booster dose of the vaccinia/gD vaccine.[13] Mice were vaccinated initially with either vaccinia/gD or a vaccinia recombinant which expressed the hemagglutinin of influenza A (vac/influenza HA). Selected groups of mice were then revaccinated, 14 weeks later, with vaccinia/gD. Vaccinated mice and controls were subsequently challenged with a lethal intraperitoneal injection of HSV-1 and mortality was observed. In the unvaccinated group all mice died following challenge. In the groups of mice that received a single immunization with vaccinia/gD prior to challenge, 80 to 86% of mice survived. In the group which received a second vaccination with vaccinia/gD, antibody titers against HSV increased 10-fold and 100% of animals survived the challenge. However, in the group of animals which were immunized with a non-HSV recombinant (vac/influenza HA) prior to immunization with vaccinia/gD, replication of the vaccinia/gD virus was inhibited. As a result, levels of HSV neutralizing antibody were markedly reduced (as compared to levels observed in vaccinia-naive animals immunized with vaccinia/gD), and only 17% of animals survived lethal challenge. These findings demonstrate that revaccination with vaccinia/gD can increase the immune response to HSV, but that pre-existing immunity to vaccinia (in animals previously immunized with vaccinia or an alternate vaccinia recombinant expressing non-HSV genes) can inhibit the anti-HSV response which occurs following vaccinia/gD vaccination. This latter observation may be an important consideration when devising vaccination strategies utilizing this and other vaccinia recombinants.

MUCOSAL IMMUNIZATION

We conducted additional experiments to determine whether local administration of vaccine to mucosal sites of subsequent HSV challenge could increase levels of protection against latent HSV infection (Table 1).[18] Mice were vaccinated by

Table 1. Efficacy of nasal versus tail vaccination on protection against latent trigeminal ganglionic infection.

Vaccination	Antibody titer	Nasal Challenge	
		No. positive/ no. tested (%)	Percent protection
None	<10	24/40 (60)	-
Nasal	42	7/40 (18)*	71
Tail	178	7/25 (28)+	53

NOTE. Data are from (18). BALB/c mice were vaccinated with 10^8 pfu of vaccinia/gD via tail or nasal routes. At four weeks after vaccination mice were challenged with 4 X 10^6 pfu of HSV-1 (KOS) via nasal route. At six weeks after challenge mice were killed, and trigeminal ganglia were explanted and observed for viral cytopathology. Antibody titers are geometric means.
*$P < 0.001$ compared with nonvaccinated group (two-tailed Fisher's exact test).
+$P = 0.02$ compared with nonvaccinated group (two-tailed Fisher's exact test).

intradermal (tail) or intranasal inoculation with vaccina/gD and were challenged four weeks later with a sublethal dose of HSV by nasal inoculation, an alternate route of HSV challenge which can establish latent infection of the trigeminal ganglion. As compared to controls, animals immunized intradermally with vaccinia/gD demonstrated 53% protection against latent infection, while nasal vaccinated animals demonstrated 71% protection against intranasal HSV challenge. Although nasal-vaccinated animals had higher levels of protection against latent infection than tail-vaccinated animals, the difference between these groups was not statistically significant. We conclude, therefore, that local administration of vaccinia/gD to a mucosal site of subsequent HSV challenge can produce a level of protection at least equivalent to that afforded by intradermal vaccination.

SUMMARY AND FUTURE DIRECTIONS

We have demonstrated that a single vaccination with the vaccinia/gD recombinant can provide substantial protection against both lethal and latent HSV infection. This protection is long-lasting and can be increased by a booster dose of vaccinia/gD. However, preexisting immunity to vaccinia can diminish the response to vaccinia/gD immunization by inhibiting replication of the vector. In murine models, local administration of vaccines to the mucosal site of subsequent challenge provides an effective alternative to intradermal inoculation.

The use of vaccinia recombinants as potential HSV vaccines offers several important advantages. Vaccinia is a live vector and can stimulate both cellular and humoral host immune responses, and this immunity can be long-lasting. Vaccinia recombinants can provide protection against latent infection, but, unlike attenuated live HSV vaccines, they do not establish latent infection themselves. Vaccinia recombinants can be engineered to contain only desired HSV sequences, and can exclude putative oncogenic sequences. Multivalent vaccinia recombinants containing more than one HSV gene can be constructed.

However, vaccinia recombinants have several potential disadvantages as well.[23] The adverse effects of vaccinia immunization in the smallpox eradication campaign are well known; progressive vaccinia, generalized vaccinia, eczema vaccinatum, and post-vaccinial encephalitis have all been documented. In addition, accidental transmission of recombinant vaccine from vaccinated individuals to susceptible contacts remains a possibility. Identification and exclusion of immunocompromised individuals from vaccination is essential, especially with the increasing prevalence of human immunodeficiency virus infection in the general population. Therefore, modification of the vaccinia vector to maximize safety will be necessary before vaccinia/HSV recombinants could be considered suitable for human use. Deletion of vaccinia virus virulence factors, such as thymidine kinase[24] or growth factor,[25] can lead to recombinants which have reduced pathogenicity in animal models. As an example, vaccinia/gD, which is thymidine-kinase deficient, is 2,500 times less potent in murine lethal challenge experiments than the parent WR strain.[24] Deletion of other vaccinia proteins results in virus which has a reduced ability to spread from cell to cell,[26,27] and may be less capable of producing systemic infection. Inclusion of immunomodulators, such as IL-2 or interferon,[28,29] has reduced the pathogenicity of recombinant vectors in immunocompromised murine models. Further studies on the molecular and pathogenic mechanisms of vaccinia virus infection will hopefully lead to additional interventions capable of attenuating recombinant vaccinia virus infection. Clinical trials are currently underway with a vaccinia recombinant which expresses the human immunodeficiency virus gp 160,[30] and the results of this and subsequent trials with improved vectors may help determine the feasibility of using vaccinia recombinants as potential vaccines for HSV.

ACKNOWLEDGMENTS

We wish to acknowledge the important contributions of Kenneth Cremer, Michael Mackett, and Bernard Moss to these studies and of Eloise Mange for editorial assistance.

REFERENCES

1. H. Openshaw, A. Puga, and A.L. Notkins, Herpes simplex virus infection in sensory ganglia: immune control, latency and reactivation, *Fed. Proc.* 38:2660 (1979).
2. J.C. Overall, Jr., Oral herpes simplex: pathogenesis, clinical and virologic course, approach to treatment, *in*: "Viral Infections in Oral Medicine," J. Hooks and G.W. Jordan, eds., Elsevier, New York (1982).
3. K. Cremer and A.L. Notkins, The development of vaccines against herpes simplex virus, *in*: "Herpes Simplex Virus Infection: Biology, Treatment, and Prevention," S.E. Straus, moderator, *Ann. Int. Med.* 103:404 (1985).
4. W.P. Allen, P.J. Hitchcock, eds., Herpes Simplex Virus Vaccine Workshop, *Rev. Inf. Dis.* 13 (Suppl. 11) (1991).
5. B. Moss and C. Flexner, Vaccinia virus expression vectors, *Annu. Rev. Immunol.* 5:305 (1987).
6. K.J. Cremer, M. Mackett, C. Wohlenberg, A.L. Notkins, and B. Moss, Vaccinia virus recombinant expressing herpes simplex virus type 1 glycoprotein D prevents latent herpes in mice, *Science* 228:737 (1985).
7. E.M. Cantin, R. Eberle, J.L. Baldick, B. Moss, D.E. Willey, A.L. Notkins, and H. Openshaw, Expression of herpes simplex virus 1 glycoprotein B by a recombinant vaccinia virus and protection of mice against lethal herpes simplex virus 1 infection, *Proc. Natl. Acad. Sci. USA* 84:5908 (1987).
8. E. Paoletti, B.R. Lipinskas, C. Samsonoff, S. Mercer, and D. Panicali, Construction of live vaccines using genetically engineered poxviruses: biological activity of vaccinia virus recombinants expressing the hepatitis B virus surface antigen and the herpes simplex virus glycoprotein D, *Proc. Natl. Acad. Sci. USA* 81:193 (1984).
9. M. Wachsman, L. Aurelian, C.C. Smith, B.R. Lipinskas, M.E. Perkus, and E. Paoletti, Protection of guinea pigs from primary and recurrent herpes simplex virus (HSV) type 2 cutaneous disease with vaccinia virus recombinants expressing HSV glycoprotein D, *J. Infect. Dis.* 155:1188 (1987).
10. V. Sullivan and G.L. Smith, Expression and characterization of herpes simplex virus type 1 (HSV-1) glycoprotein G (gG) by recombinant vaccinia virus: neutralization of HSV-1 infectivity with anti-gG antibody, *J. Gen. Virol.* 68:2587 (1987).
11. V. Sullivan and G.L. Smith, The herpes simplex virus type 1 US7 gene product is a 66K glycoprotein and is a target for complement-dependent virus neutralization, *J. Gen. Virol.* 69:859 (1988).
12. C. Flexner, B.R. Murphy, J.F. Rooney, C. Wohlenberg, V. Yuferov, A.L. Notkins, and B. Moss, Successful vaccination with a polyvalent live vector despite existing immunity to an expressed antigen, *Nature* 335:259 (1988).
13. J.F. Rooney, C. Wohlenberg, K.J. Cremer, B. Moss, and A.L. Notkins, Immunization with a vaccinia virus recombinant expressing herpes simplex virus type 1 glycoprotein D: long-term protection and effect of revaccination, *J. Virol.* 62:1530 (1988).
14. D.E. Willey, E.M. Cantin, L.R. Hill, B. Moss, A.L. Notkins, and H. Openshaw, Herpes simplex virus type 1-vaccinia virus recombinant expressing glycoprotein B: protection from acute and latent infection, *J. Infect. Dis.* 158:1382 (1988).
15. S. Martin, B. Moss, P.W. Berman, L.A. Laskey, and B.T. Rouse, Mechanisms of antiviral immunity induced by a vaccinia virus recombinant expressing herpes simplex virus type 1 glycoprotein D: cytotoxic T cells, *J. Virol.* 61:726 (1987).
16. S. Martin and B.T. Rouse, The mechanisms of antiviral immunity induced by a vaccinia virus recombinant expressing herpes simplex virus type 1 glycoprotein D: clearance of local infection, *J. Immunol.* 138:3431 (1987).
17. S. Martin, E. Cantin, and B.T. Rouse, Evaluation of antiviral immunity using vaccinia virus recombinants expressing cloned genes for herpes simplex virus type 1 glycoproteins, *J. Gen. Virol.* 70:1359 (1989).
18. J.F. Rooney, C. Wohlenberg, K.J. Cremer, and A.L. Notkins, Immunized mice challenged with herpes simplex virus by the intranasal route show protection against latent infection, *J. Infect. Dis.* 159:974 (1989).
19. M. Wachsman, L. Aurelian, C.C. Smith, M.E. Perkus, and E. Paoletti, Regulation of expression of herpes simplex virus (HSV) glycoprotein D in vaccinia recombinants affects their ability to protect from cutaneous HSV-2 disease, *J. Infect. Dis.* 159:625 (1989).

20. M. Wachsman, J.H. Luo, L. Aurelian, M.E. Perkus, and E. Paoletti, Antigen-presenting capacity of epidermal cells infected with vaccinia virus recombinants containing the herpes simplex virus glycoprotein D, and protective immunity, *J. Gen. Virol.* 70:2513 (1989).

21. J.P. Weir, M. Bennett, E.M. Allen, K.L. Elkins, S. Martin, and B.T. Rouse, Recombinant vaccinia virus expressing the herpes simplex virus type 1 glycoprotein C protects mice against herpes simplex virus challenge, *J. Gen. Virol.* 70:2587 (1989).

22. J.F. Rooney, C.R. Wohlenberg, B. Moss, and A.L. Notkins, Live vaccinia virus recombinants expressing herpes simplex virus genes, *Rev. Infect. Dis.* 13 (Suppl. 11):S898 (1991).

23. I. Arita and F. Fenner, Complications of smallpox vaccination, *in*: "Vaccinia Viruses as Vectors for Vaccine Antigens," G.V. Quinnan, Jr., ed., Elsevier Science Publishing, New York (1985).

24. R.M.L. Buller, G.L. Smith, K. Cremer, A.L. Notkins, and B. Moss, Decreased virulence of recombinant vaccinia virus expression vectors is associated with a thymidine kinase-negative phenotype, *Nature* 317:813 (1985).

25. R.M.L. Buller, S. Chakrabarti, J.A. Cooper, D.R. Twardzik, and B. Moss, Deletion of the vaccinia virus growth factor gene reduces virus virulence, *J. Virol.* 62:866 (1988).

26. J.F. Rodriguez and G.L. Smith, IPTG-dependent vaccinia virus: identification of a virus protein enabling virion envelopment by Golgi membrane and egress, *Nucleic Acids Res.* 18:5347 (1990).

27. R. Blasco and B. Moss, Extracellular vaccinia virus formation and cell-to-cell virus transmission are prevented by deletion of the gene encoding the 37,000-dalton outer envelope protein, *J. Virol.* 65:5910 (1991).

28. C. Flexner, A. Hugin, and B. Moss, Prevention of vaccinia virus infection in immunodeficient mice by vector-directed IL-2 expression, *Nature* 330:259 (1987).

29. M.R.J. Kohonen-Corish, N.J.C. King, C.E. Woodhams, and I.A. Ramshaw, Immunodeficient mice recover from infection with vaccinia virus expressing interferon-γ, *Eur. J. Immunol.* 20:157 (1990).

30. E.L. Cooney, A.C. Collier, P.D. Greenberg, R.W. Coombs, J. Zarling, D.E. Arditti, M.C. Hoffman, S. Hu, and L. Corey, Safety of and immunological response to a recombinant vaccinia virus vaccine expressing HIV envelope glycoprotein, *Lancet* 337:567 (1991).

LIPOSOMES AND CONJUGATE VACCINES FOR ANTIGEN DELIVERY AND INDUCTION OF MUCOSAL IMMUNE RESPONSES

Suzanne M. Michalek, Noel K. Childers, Jannet Katz, Mark Dertzbaugh, Sue Zhang, Michael W. Russell, Frank L. Macrina[1], Susan Jackson, and Jiri Mestecky

Departments of Microbiology and Oral Biology
Schools of Dentistry and Medicine
University of Alabama at Birmingham
Birmingham, AL 35294
[1]Department of Microbiology and Immunology
Virginia Commonwealth University
Richmond, VA 23298

INTRODUCTION

Immunoglobulin A (IgA) is the major immunoglobulin isotype in human external secretions and provides the first line of defense against infections which involve mucosal surfaces.[1-3] Therefore, the development of safe vaccination approaches effective in inducing mucosal IgA responses is of global importance to establish means for protecting the human population against a variety of infectious diseases. In this brief review, we describe our investigations on antigen form, the liposome adjuvant-delivery system and other adjuvants for their use in developing effective and safe oral vaccines for the induction of a mucosal response protective against the infectious oral disease dental caries, in which the principal etiologic agent is *Streptococcus mutans*.[4,5] Results are presented using oral vaccines consisting of purified components of *S. mutans*, anti-idiotypic antibody and cloned *S. mutans* gene products as antigens and liposomes, muramyl dipeptide (MDP) and the B subunit of cholera toxin (CTB) as adjuvants.

ORAL VACCINES AND THE COMMON MUCOSAL IMMUNE SYSTEM

Classic oral vaccines have consisted of killed or attenuated bacteria or viruses.[2,6] These vaccine forms consist of a variety of antigenic determinants and adjuvant molecules and have been shown to induce immune responses. It has been shown that oral administration of *S. mutans* whole cells is effective in inducing salivary IgA immune responses in both humans[2,5,7,8] and experimental rats,[5,9,10] and these responses correlate with protection against *S. mutans* infection. However, a whole-cell vaccine may possess components which could induce a response that may have detrimental effects or one that may not be protective against the infectious organism. Therefore, during the past several years efforts have been directed towards identifying potentially protective antigens for use in vaccine development. Several *S. mutans* antigens have been shown to induce immune responses in experimental animals or humans, and these include the serotype-specific carbohydrate (CHO), glucosyltransferase (GTF) and antigen I/II (Ag I/II).[5] In some of these studies responses were shown to be protective.

The oral route of immunization is a safe way to deliver antigens and to induce a mucosal IgA antibody response in external secretions. The induction of this type of response would be via the common mucosal immune system.[2-4,6] The gut-associated lymphoid tissue (GALT) (e.g., Peyer's patches) are distinct lymphoid follicles located along the small intestine and represent major inductive tissue for IgA responses. The luminal surface of Peyer's patches is covered by a specialized epithelium that consists of cuboidal epithelial cells and unique antigen-sampling cells, termed M cells.[11] These M cells take up antigens present in the gut and deliver them to the underlying lymphoid cells for antigen sensitization of T and precursor IgA B cells. These cells migrate to mucosal tissues where the B cells terminally differentiate and produce IgA antibodies which are secreted into the external secretion.

Several properties of vaccines have been defined with respect to their effectiveness in inducing a response when given by the oral route.[4-6] Oral vaccines consisting of purified soluble antigens have been shown to be less immunogenic than particulate forms. Furthermore, the antigens must survive any potential degradation by enzymes present in the intestine and must reach the appropriate inductive site in adequate quantities and appropriate form for the effective induction of a mucosal IgA immune response. Various approaches are therefore being developed and tested to deliver antigens to and promote antigen uptake by the Peyer's patches and thus augment the induction of mucosal responses.[4,6] These antigen delivery systems include genetically engineered microbial vectors, biodegradable microspheres and liposomes.

LIPOSOME GENERATION AND CHARACTERIZATION FOR ANTIGEN DELIVERY

During the past several years, liposomes have received much attention for use in targeted drug delivery and as carriers/adjuvants for a variety of substances including antigens and antibodies.[12-14] Liposomes are microscopic closed vesicles composed of a bilayered phospholipid membrane. Antigens or other substances (e.g., drugs, adjuvants or DNA) can be incorporated into or associated with these vesicles. The size and composition of liposomes and whether they are multilamellar or unilamellar are among the properties important in determining how effective they will be for effective antigen delivery. We have been especially interested in establishing the usefulness of the liposome as an oral delivery system for inducing mucosal responses. In recent studies, we have used liposomes as an oral delivery system and adjuvant for *S. mutans* antigens, anti-idiotypic antibodies to *S. mutans*, conjugate proteins consisting of *S. mutans* antigens or peptides coupled to the B subunit of cholera toxin (CTB) and conjugate antigens consisting of *S. mutans* carbohydrate coupled to *S. mutans* protein antigen.

The liposomes we have primarily used in our studies consist of dipalmitoyl phosphatidylcholine, cholesterol and dicetylphosphate.[15] We have further standardized our liposome preparations by routinely processing them through a Microfluidizer™ (Microfluidics Corp., Newton, MA). By employing this apparatus, we can reproducibly prepare uniform-sized (~100 nm), unilamellar liposomes that contain the desired amount of antigen for testing. All liposome preparations are then characterized by flow cytometry (FACS). FACS analysis is a convenient and rapid method for characterizing liposome preparations to ensure their uniformity.[15] This procedure involves the use of polystyrene beads (130 nm) which serve as our standard. These particles are used to set the first marker to contain 90% of the beads, and an example of an analysis of different liposomes is presented in Table 1. Forward laser light scatter analysis of a crude sonicated liposome preparation revealed the presence of heterogeneous vesicles with 54% of the vesicles in the

Table 1. FACS analysis of liposome preparations.

Sample	Particle Size Distribution (Percent)	
	Submicron (<130 nm)	Large (>250 nm)
Standard	90	2
Crude Liposomes	54	27
Microfluidized Liposomes	92	1

submicron size range and 27% in the large, aggregated size range. The microfluidized liposomes consisted of more homogeneous vesicles with 92% being in the submicron size range. We have used this method to characterize over 30 liposome preparations for their size and uniformity, and to demonstrate the similarities between liposome preparations from one experiment to another.

An important concept that needed to be addressed to establish the potential usefulness of oral liposome vaccine delivery systems was to obtain evidence that liposomes are taken up by Peyer's patches. In an extensive series of studies, Peyer's patches from rats exposed to microfluidized liposomes were removed and processed for analysis by transmission electron microscopy.[16] Sections of Peyer's patches removed from rats 2 hours after exposure to the liposomes revealed the presence of endosomes within M cells which contained liposomes. These endosomes appeared to be transporting the liposomes through the cell to the underlying lymphoid cells. These findings provided evidence that the homogeneous unilamellar liposomes with a size of ~100 nm are taken up by Peyer's patches and should therefore serve as an oral vaccine delivery system for the induction of IgA antibody responses in external secretion via the common mucosal immune system.

IN VIVO ASSESSMENT OF LIPOSOMES FOR USE AS ORAL ANTIGEN DELIVERY SYSTEMS

In our studies with oral *S. mutans* vaccines, we have primarily employed an experimental rat model which has allowed us to assess the induction of an immune response and to determine its effectiveness in protecting against *S. mutans* infection.[5,10,17] Briefly, germfree rats were given the vaccine by gastric intubation, challenged with a virulent strain of *S. mutans* and then subsequently given the oral vaccine at approximately 10-day intervals. One week after the last immunization, serum and saliva samples were collected and assessed for antibody activity by ELISA, the rats were sacrificed, and the mandibles were removed for microbiological analysis of *S. mutans* plaque levels and for assessment of caries activity. As illustrated in Table 2, rats given purified *S. mutans* GTF in liposomes by gastric intubation had a twofold higher level of salivary IgA antibodies than rats given antigen alone.

Table 2. Effectiveness of oral liposome delivery systems in inducing salivary IgA responses and caries immunity.

Oral Vaccine	Dose (µg)	Salivary IgA Anti-*S. mutans* Activity[*]	*S. mutans* in Plaque (% reduction)	Caries Protection (% reduction)
GTF	50	29	40	39
GTF/liposomes	50	44	85	78
CHO	100	6	0	5
CHO/liposomes	100	20	25	39
CHO/liposomes-GM-53 MDP	100	29	58	57
CHO/liposomes-L18-MDP	100	40	87	73
Anti-Id/liposome	25	53[**]	53	20
NIg/liposome	25	28[**]	0	0
None		<5	0	0

[*]Reciprocol of dilution giving an OD reading of >0.1
[**]OD_{414} x 100 of a 1/80 dilution

This difference in responses was also reflected in the level of protection against *S. mutans* infection seen in these animals. An even more pronounced effect was seen when the CHO of *S. mutans* was tested as an oral vaccine. In rats given an oral vaccine consisting of purified CHO, essentially no salivary IgA response or protection against *S. mutans* infection was observed. When the CHO was incorporated into liposomes and given orally to rats, a salivary IgA response was induced which corresponded with a reduction in the number of *S. mutans* in plaque and with caries protection. Finally, the incorporation of lipophilic MDP into the oral vaccine resulted in an augmentation of the protective response. The L18-MDP derivative was more effective then GM-53 MDP in potentiating the response.

Due to the recent interest that anti-idiotypic (anti-Id) antibodies can induce protective immunity against pathogens, rabbit anti-Id to *S. mutans*-specific antibodies (prepared in rats) were generated, incorporated into liposomes and tested for their ability to induce salivary IgA responses and caries immunity.[18] As indicated in Table 2, rats given liposomes containing anti-Id antibodies to *S. mutans* by gastric intubation exhibited a good salivary immune response which correlated with a 20% reduction in caries activity. Rats given normal IgG (NIg) in liposomes had a slight salivary response, but no protection against caries formation was seen.

Based on the evidence that *S. mutans* Ag I/II is a protective antigen and that conjugated-proteins consisting of Ag I/II and CTB are effective oral vaccines for the induction of mucosal responses in mice,[19,20] and this volume we examined the ability of Ag I/II vaccines to induce caries immunity in our experimental rat model. In an initial experiment we employed liposomes, either crude (c) or microfluidized (m) containing Ag I/II, to establish their effectiveness in inducing protective mucosal immune responses

Table 3. Effect of liposome form and cholera toxin B subunit-conjugate vaccines in augmenting mucosal responses.

Oral Vaccine	Dose (μg)	Salivary IgA Anti-*S. mutans* Activity[*]	*S. mutans* in Plaque (% reduction)	Caries Protection (% reduction)
Ag I/II/c-liposome	10	274	25	30
Ag I/II/m-liposome	10	501	38	40
Ag I/II-CTB	15	325	28	30
GtfB.1-CTB/liposome	50	250	20	15
None		<10	0	0

[*]ELISA Units

(Table 3). Rats given a vaccine consisting of Ag I/II in m-liposomes had a higher salivary IgA response than rats given a vaccine composed of Ag I/II in c-liposomes. These responses correlated with 40% and 30% reduction in caries activity, respectively. These findings show the value of using m-liposomes in oral vaccine development, and provide additional evidence that Ag I/II represents a protective antigen.

In another series of experiments, it was shown that rats given Ag I/II-CTB conjugate by gastric intubation exhibited a salivary anti-Ag I/II response (Table 3). The response induced with the dose of conjugate tested was similar to that obtained in animals given Ag I/II in c-liposomes. Further studies will be required to determine the immunization regimen that is optimal for inducing responses in order to assess the potential of this protein conjugate for use as an oral caries vaccine.

We have also initiated studies to characterize a chimeric protein consisting of an antigenic 15-amino acid peptide sequence (gtfB.1) from the GTF-B of *S. mutans* and

CTB[21,22] and to determine its ability to induce mucosal responses and caries immunity when used as an oral vaccine. As shown in Table 3, when the chimeric protein was incorporated into m-liposomes and given to rats by gastric intubation, a slight salivary anti-GTF response was detected which correlated with protection. Additional dose-response studies are needed to determine the effectiveness of this chimeric protein in inducing protective immune responses against *S. mutans* infection.

Based on our studies in animals, it was also of interest to establish the ability of oral liposome delivery systems containing *S. mutans* antigens to induce mucosal antibody responses in humans. In an extensive series of studies, human volunteers swallowed enteric-coated capsules containing purified CHO of *S. mutans* (500 mg) in liposomes daily for 7 consecutive days.[23] Parotid saliva samples collected from each individual prior to, during and after each immunization were assessed for antibody activity by ELISA. As shown in Table 4, 75% of the subjects responded to the oral vaccine with a peak antibody activity occurring on day 30. A more pronounced salivary response which peaked on day 15 was seen after the second immunization. Further studies in humans will test oral liposome vaccines containing *S. mutans* protein antigens or protein-CHO conjugates for their effectiveness in inducing immune responses protective against *S. mutans* infection. In addition, we are also developing procedures to process liposome vaccines for their practical use in humans.

Table 4. Salivary IgA immune response to CHO in human subjects given a CHO-liposome oral vaccine.

Immunization	Subjects Responding (%)	Mean % Increase*	Mean Peak Response Day
First	75	142	30
Second	75	217	15

*Percent increase in ELISA units/ml over baseline

SUMMARY

In this brief review, emphasis was placed on the effectiveness of liposomes as carriers/vehicles of soluble antigens and as adjuvants for use in oral vaccine development. Evidence was provided that oral administration of antigen in liposomes resulted in a mucosal response which was higher than that obtained when the oral vaccine consisted of antigen alone. Specific mucosal responses were enhanced by incorporating lipophilic MDP into the antigen/liposome vaccines. Antigens shown to be effective in inducing a protective mucosal response when given in an oral liposome vaccine were anti-idiotypic antibodies, purified *S. mutans* GTF, CHO and Ag I/II. Evidence is also provided that CTB may be an effective oral adjuvant when coupled to proteins or peptides by either chemical or genetic methods. Further studies, however, will be required to characterize the effectiveness and safety of CTB in conjugate vaccines for inducing specific mucosal responses and to develop practical means to prepare oral liposome vaccines for use in humans.

ACKNOWLEDGMENTS

We wish to thank Cecily Harmon, Gloria Richardson and Pam White for their expert technical help. The work which has been reviewed was supported by USPHS grants DE 00155, DE 00232, DE 08182, DE 04217, DE 08228, DE 06801, DE 06746, DE 09081 and AI 18745.

REFERENCES

1. N.K. Childers, M.G. Bruce, and J.R. McGhee, Molecular mechanisms of immunoglobulin A defense, *Annu. Rev. Microbiol.* 43:503-536 (1989).
2. J. Mestecky, The common mucosal immune system and current strategies for induction of immune responses in external secretions, *J. Clin. Immunol.* 7:265-276 (1987).
3. J. Mestecky, and J.R. McGhee, Immunoglobulin A (IgA): molecular and cellular interactions involved in IgA biosynthesis and immune response, *Adv. Immunol.* 40:153-245 (1987).
4. J. Mestecky, and J.H. Eldridge, Targeting and controlled release of antigens for the effective induction of secretory antibody responses, *Cur. Opin. Immunol.* 3:492-495 (1991).
5. S.M. Michalek, and N.K. Childers, Development and outlook for a caries vaccine, *Crit. Rev. Oral Biol. Med.* 1:37-54 (1990).
6. J.R. McGhee, and J. Mestecky, In defense of mucosal surfaces. Development of novel vaccines for IgA responses protective at the portals of entry of microbial pathogens, *Infect. Dis. Clin. North Amer.* 4:315-341 (1990).
7. C. Czerkinsky, S.J. Prince, S.M. Michalek, S. Jackson, M.W. Russell, Z. Moldoveanu, J.R. McGhee, and J. Mestecky, IgA antibody producing cells in peripheral blood after antigen ingestion: evidence for a common mucosal immune system in humans, *Proc. Natl. Acad. Sci. (USA)* 84:2449-2453 (1987).
8. J. Mestecky, J.R. McGhee, R.R. Arnold, S.M. Michalek, S.J. Prince, and J.L. Babb, Selective induction of an immune response in human external secretions by ingestion of bacterial antigen, *J. Clin. Invest.* 61:731-737 (1978).
9. J.R. McGhee, and S.M. Michalek, Immunobiology of dental caries: microbial aspects and local immunity, *Annu. Rev. Microbiol.* 35:595-638 (1981).
10. S.M. Michalek, I. Morisaki, C.C. Harmon, S. Hamada, and J.R. McGhee, Effective immunity to dental caries: gastric intubation of *Streptococcus mutans* whole cells or cell walls induces protective immunity in gnotobiotic rats, *Infect. Immun.* 39:645-654 (1983).
11. R.L. Owen, and A.L. Jones, Epithelial cell specialization within human Peyer's patches: an ultrastructural study of intestinal lymphoid follicles, *Gastroenterology* 66:189-203 (1974).

12. M.J. Ostro, Liposomes, *Sci. Amer.* 256:102-111 (1987).

13. B.P. Ram, and P. Tyle, Immunoconjugates: applications in targeted drug delivery for cancer, *Pharm. Res.* 4:181-188 (1987).

14. B.T. Rouse, Liposomes as carriers of antigens as well as other molecules involved in immunity, *J. Am. Vet. Med. Assoc.* 181:988-991 (1982).

15. N.K. Childers, S.M. Michalek, J.H. Eldridge, F.R. Denys, K.A. Berry, and J.R. McGhee, Characterization of liposome suspensions by flow cytometry, *J. Immunol. Methods* 119:135-143 (1989).

16. N.K. Childers, F.R. Denys, N.F. McGee, and S.M. Michalek, Ultrastructural study of liposome uptake by M cells of rat Peyer's patch: an oral vaccine system for delivery of purified antigen, *Regional Immunol.* 3:8-16 (1990).

17. S.M. Michalek, I. Morisaki, S. Kimura, S. Hamada, R. Curtiss III, and J.R. McGhee, Effective immunity to *Streptococcus mutans*-induced dental caries, *in*: "Molecular Microbiology and Immunobiology of *Streptococcus mutans*," S. Hamada, S.M. Michalek, H. Kiyono, L. Menaker, and J.R. McGhee, eds., Elsevier Science Publishers, Amsterdam, pp. 269-278 (1986).

18. S. Jackson, J. Mestecky, N.K. Childers, and S.M. Michalek, Liposomes containing anti-idiotypic antibodies: an oral vaccine to induce protective secretory immune responses specific for pathogens of mucosal surfaces, *Infect. Immun.* 58:1932-1936 (1990).

19. C. Czerkinsky, M.W. Russell, N. Lycke, M. Lindblad, and J. Holmgren, Oral administration of a streptococcal antigen coupled to cholera toxin B subunit evokes strong antibody responses in salivary glands and extramucosal tissues, *Infect. Immun.* 57:1072-1077 (1989).

20. M.W. Russell, and H.-Y. Wu, Distribution, persistence, and recall of serum and salivary antibody responses to peroral immunization with protein antigen I/II of *Streptococcus mutans* coupled to the cholera toxin B subunit, *Infect. Immun.* 59:4061-4070 (1991).

21. M.T. Dertzbaugh, and F.L. Macrina, Inhibition of *Streptococcus mutans* glucosyltransferase activity by antiserum to a sequence peptide, *Infect. Immun.* 58:1509-1513 (1990).

22. M.T. Dertzbaugh, D.L. Peterson, and F.L. Macrina, Cholera toxin B-subunit gene fusion: structural and functional analysis of the chimeric protein, *Infect. Immun.* 58:70-79 (1990).

23. N.K. Childers, S.M. Michalek, D.G. Pritchard, and J.R. McGhee, Mucosal and systemic responses to an oral liposome-*Streptococcus mutans* carbohydrate vaccine in humans, *Regional Immunol.* 3:289-296 (1991).

PERORAL IMMUNIZATION WITH A CHOLERA
TOXIN-LINKED BACTERIAL PROTEIN ANTIGEN
AND SYNTHETIC PEPTIDE

Michael W. Russell,[1] Hong-Yin Wu,[1]
Pamela L. White,[1] Ichiro Takahashi,[2]
Nobuo Okahashi,[2] and Toshihiko Koga[2]

[1]Department of Microbiology
University of Alabama at Birmingham
Birmingham, AL 35294

[2]Department of Dental Research
National Institute of Health
Tokyo 141, Japan

INTRODUCTION

Mucosal surfaces that are exposed to the environment, such as those of the mouth and the alimentary, upper respiratory, and lower genital tracts, cannot be maintained in a sterile condition. The function of the common mucosal immune system (CMIS) that protects these surfaces is therefore quite different from that of the internal or circulatory immune system, which deploys powerful microbicidal effector mechanisms as exemplified by complement and phagocytes that may entail inflammatory consequences. Rather, mucosal immunity must achieve a balance of organisms comprising the normal commensal flora and maintain the integrity of the mucosal membranes. Rational consideration of the attributes of an immune response required for protection against a mucosal infection necessitates attention to the nature of the infection itself, the pathogenic mechanisms involved, and the site of infection where immunity should be expressed. Such adequately detailed information is not often available, but certain principles can be delineated on the basis of current knowledge. Ideally, the response should be targeted towards antigenic molecules that are critically involved in the early stages of the pathogenic process (i.e., before invasion occurs); the organism need not necessarily be killed but should be displaced from its foothold in a manner that does not compromise the mucosal barrier; the response should persist as long as the infection remains a threat, or be rapidly recallable in the event of re-infection; and as mucosal infections are often determined by accessibility or tropism of the organism for particular surfaces, the response should ideally be directed to that site without necessarily involving other sites. The CMIS and its principal effector product, secretory IgA (S-IgA), can be seen as having evolved to meet these needs, and exploitation of this system to elicit protective immunity against a particular infection should take into account what the CMIS and S-IgA can be expected to accomplish. Oral infections such as dental caries afford opportunities to dissect these requirements and to consider how immunity may best be delivered to one site where the causative organism, *Streptococcus mutans* and its allies, resides.

Genetically Engineered Vaccines, Edited by
J.E. Ciardi *et al.*, Plenum Press, New York, 1992

Considerable advances in recent years have led to the identification of the molecular mechanisms whereby *S. mutans* adheres to its substratum, the tooth surface. These include surface adhesins that recognize salivary proteins deposited on the teeth and several enzymes involved in the synthesis of adherent glucans from sucrose. Our work has concentrated on the former, in particular on Ag I/II (also known as antigen B, P1, PAc, spaP), a highly immunogenic surface protein that has structural features in common with other streptococcal surface proteins and which probably acts as an adhesin.[1-5] Protective immunity to dental caries in primates has been achieved by parenteral immunization with Ag I/II which elicits predominantly circulating IgG antibodies that may reach the oral cavity through the gingival crevice,[6] but the exact mechanism of this protection has not been satisfactorily explained. S-IgA, the predominant Ig in saliva, fulfils an anti-adherence function that depends on its structural integrity and probably involves the hydrophilic and negatively charged properties of its Fc region and secretory component.[7] Peroral immunization experiments in rats have demonstrated protective immunity associated with salivary IgA antibodies to *S. mutans* or *S. sobrinus* antigens.[8] Thus, it is believed that immunization designed to induce salivary S-IgA antibodies to Ag I/II would effectively protect against *S. mutans*-induced dental caries. However, most soluble protein antigens, including Ag I/II, are poorly immunogenic by the peroral route, probably because of intestinal digestion and lack of uptake by the gut-associated lymphoid tissues such as Peyer's patches, which are major inductive sites for mucosal immunity. We have overcome this problem by coupling Ag I/II to cholera toxin B subunit (CTB),[9] which is safe, nontoxic, and therefore acceptable for human use, and which is highly immunogenic by the enteric route probably on account of its affinity for cell surface G_{M1} ganglioside.[10] We have monitored the immune responses of mice perorally immunized with small doses of this conjugate by calibrated ELISA of saliva and other secretions and by ELISPOT assay of antibody-secreting cells (ASC) in salivary glands (SG) and other tissues of the CMIS.[11] Furthermore, a 19-amino acid peptide representing part of the sequence of Ag I/II in the A-repeat region[5,12] has been found capable of reacting with antisera to Ag I/II and of inducing protective immune responses in mice.[13] We have therefore investigated immune responses to conjugates of this peptide with CTB. The techniques used allow the distribution, duration, and memory attributes of mucosal immune responses to be analysed quantitatively and at both molecular and cellular levels, and lay the groundwork for studies of protective immunity in rat and primate models of dental caries. The results also suggest that the strategy should be applicable to a number of other infections against which mucosal immunity is desirable.

SEQUENTIAL RESPONSES TO INTRAGASTRIC IMMUNIZATION

Groups of 5 young adult BALB/c mice of either sex were immunized intragastrically (IG) by intubating various doses of Ag I/II-CTB conjugate plus 5 μg of intact cholera toxin (CT) as adjuvant in 0.25 ml of 0.1M NaHCO₃ three times at 10-day intervals.[11] Serum and saliva samples were collected 7 days after each immunization for estimation of antibodies (IgM, IgG, and IgA in serum, and IgA in saliva) to Ag I/II and CT, and of total IgA concentration in saliva by ELISA.[11] Salivary and serum antibody responses to Ag I/II-CTB conjugate were dose-related (Fig. 1), and generally three doses of at least 15 μg were needed to induce strong salivary IgA, and serum IgG and IgA responses to Ag I/II. However, the number of immunizations appeared to be more important than the total dose given, as (for example) three doses of 15 μg were more effective than one or even two doses of 50 μg or 150 μg, especially for generating IgA antibodies to Ag I/II in either saliva or serum. There was a marked tendency for total salivary IgA concentrations to increase over the course of the experiment,[11] but even so, the salivary IgA antibodies relative to total IgA showed a specific increase. Serum IgG and IgA anti-CT responses were less markedly dependent on the dose of Ag I/II-CTB conjugate, but it should be noted that all mice received 5 μg of CT as adjuvant, irrespective of the dose of conjugate. However, it was remarkable that none of these mice developed salivary antibodies to CT, although they all developed strong serum IgG and IgA antibodies to CT. These responses were supported by the finding, in other experiments,[11] that IG immunization with Ag I/II-CTB plus CT adjuvant generated large numbers of IgG and IgA ASC specific for both Ag I/II and CT in the spleen and mesenteric lymph nodes, but only anti-Ag I/II IgA ASC in the SG. Thus the mice responded to both components of the conjugate vaccine in both mucosal and circulatory compartments of the immune system, but the responses were not uniformly distributed within the CMIS. We have postulated that the nonspecific increase in total salivary IgA concentrations seen in this experiment and sustained for a considerably longer period[11] was most likely due to a stimulatory effect of the CT adjuvant which was given to all animals. The magnitude of the effect was similar regardless of the dose of Ag I/II-CTB (0-150 µg), and the increments coincided with the immunization schedule, suggesting that mere aging of the mice was not the cause. A nonspecific increase of intestinal IgA secretion due to CT has been reported previously.[14] Alternatively, it is possible that the cholinergic drug, carbachol, used to stimulate salivary flow, also had some effect. Recent findings that the expression by epithelial cells of secretory component, which transports polymeric IgA into secretions to form S-IgA, is subject to up-regulation by cytokines,[15,16] could afford a mechanism by which either stimulus might operate.

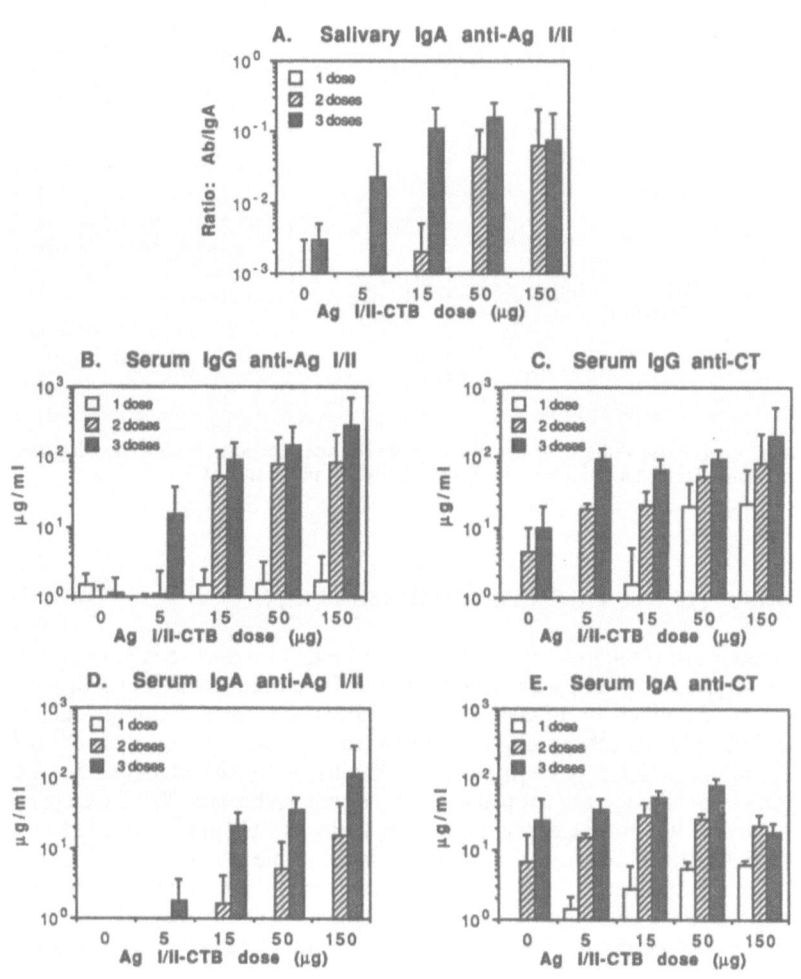

Figure 1. Salivary IgA antibody responses to Ag I/II (A), serum IgG antibody responses to Ag I/II (B) and CT (C), and serum IgA antibody responses to Ag I/II (D) and CT (E) in mice immunized three times IG with various doses of Ag I/II-CTB plus 5 µg of CT.

We postulated that anti-CT IgA ASC were preferentially locating elsewhere in the CMIS, probably the gut lamina propria (LP). To verify this, mice were immunized IG with 30 μg of Ag I/II-CTB plus 5 μg of CT, three times at 10-day intervals. Seven days after the last immunization, mononuclear cells prepared from the intestinal LP,[17] submandibular/sublingual SG,[11] and spleens were analysed by ELISPOT assay[18] for numbers of cells secreting IgM, IgG, and IgA antibodies to Ag I/II and CT, and for total IgM-, IgG-, and IgA-secreting cells.[11] Large numbers of predominantly IgA ASC specific for Ag I/II and for CT were found among LP cells (Fig. 2a), and anti-CT ASC were more numerous than anti-Ag I/II ASC. These results were similar to those obtained by the analysis of specific antibodies in gut wash: high levels of IgA antibodies to both antigen components were found (Fig. 2b). In these mice, SG contained IgA ASC to both antigens (Fig. 2a) and corresponding IgA antibodies in the saliva (Fig. 2b), but anti-Ag I/II responses exceeded anti-CT responses. We noticed on a few other occasions that anti-CT IgA antibodies appeared in the saliva, though always at a substantially lower level than anti-Ag I/II antibodies, and considered that a likely factor was regurgitation of the IG dose into the oral cavity, as sometimes happened, thereby resulting in inadvertant intra-oral exposure to antigen.

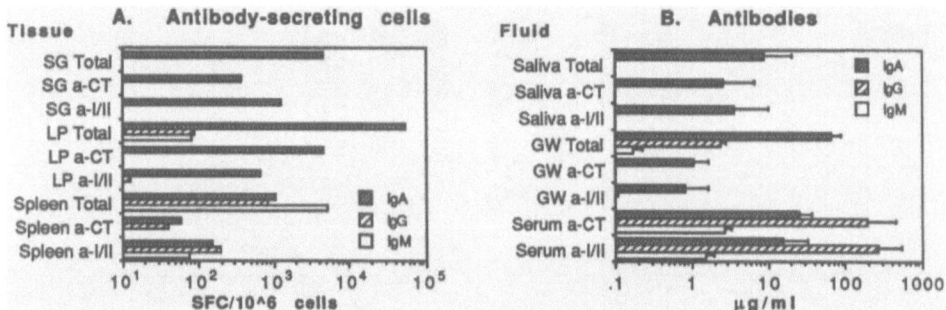

Figure 2. Antibody-secreting cells in various tissues (A) and antibodies in corresponding fluids (B) of mice immunized three times IG with 30 μg of Ag I/II-CTB plus 5 μg of CT.

RESPONSE TO INTRA-ORAL IMMUNIZATION

To determine if the local application of CT could contribute to the localization of an anti-CT response in the SG, we immunized mice intra-orally with 30 μg of Ag I/II-CTB plus 5 μg of CT given in a small volume (20 μl) to limit the amount that would be swallowed directly. Another group was immunized IG with the same dose of Ag I/II-CTB plus CT, and in addition given 5 μg of CT alone intra-orally. Mice immunized intra-orally developed serum and salivary antibody responses comparable with those seen previously in mice immunized IG (Fig. 3), except that salivary IgA anti-CT antibody levels were similar to IgA anti-Ag I/II. Mice immunized IG with conjugate and given oral CT developed similar serum antibodies to those in the intra-oral group: they furthermore developed high salivary IgA anti-CT antibodies to at least the same level as anti-Ag I/II, and substantially higher than ever seen in mice immunized only IG (Fig. 3). Thus it appears that intra-oral exposure to antigen can be instrumental in directing an existing specific mucosal immune response to the saliva. The mechanism involved in this instance remains puzzling. It has previously been reported that local administration of an antigen after primary immunization elsewhere results in an increased number of specific antibody-containing cells in the neighboring submucosa,[19,20] and it was suggested that the locally applied antigen served to stimulate further proliferation of the arriving IgA ASC precursors. Additional, and not necessarily exclusive, explanations postulate a role for locally present antigen in recruiting or retaining specific ASC, either directly, or through the intermediate action of specific T cells and the cytokines that they would produce. In the present case, it is difficult to see how antigen applied to the oral mucosa reaches the major SG, which are located at the end of relatively long ducts, to exert these effects on cells within SG, unless there are pathways of local cell traffic, hitherto unknown, possibly involving minor salivary glands which are amenable to retrograde access by antigen[21] or pharyngeal lymphoid tissues.[22-24]

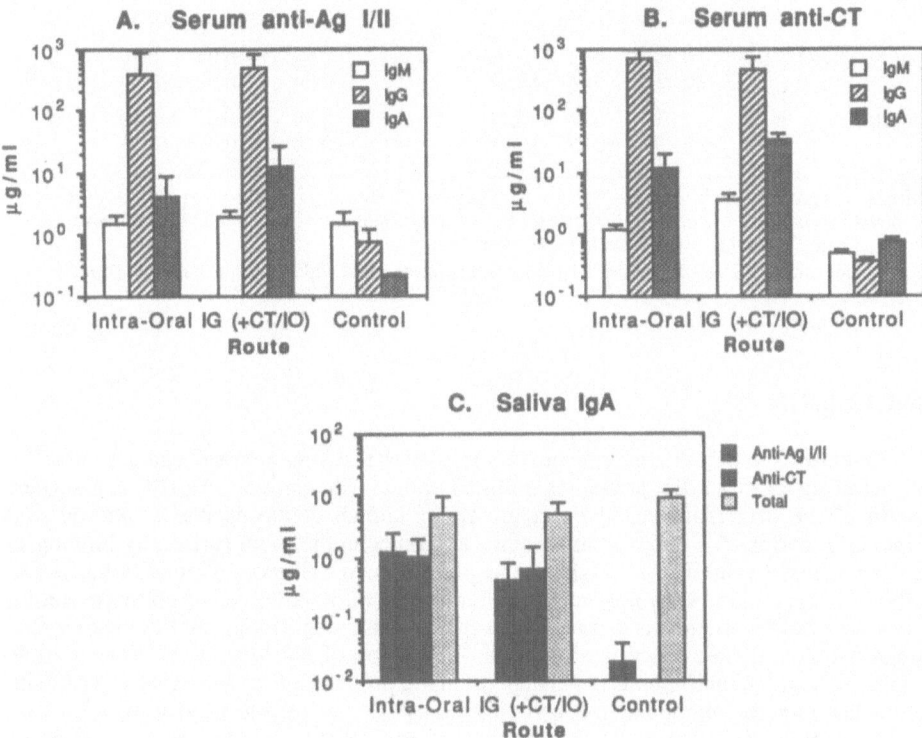

Figure 3. Serum antibody responses to Ag I/II (A) and CT (B), and salivary antibody responses (C) of mice immunized intra-orally with 30 µg of Ag I/II-CTB plus 5 µg of CT, or IG with 30 µg of Ag I/II-CTB plus 5 µg of CT and given 5 µg of CT intra-orally (+CT/IO).

RESPONSE TO PEPTIDE-CTB CONJUGATE

A peptide representing residues 301-319 at the start of the A2 repeat sequence of Ag I/II[13] was synthesized with a C-terminal Cys, coupled to CTB,[11,25] and dialysed to remove excess free peptide. Intraperitoneal (IP) immunization of mice with 25 µg of this conjugate three times at 10-day intervals resulted in modest serum antibody responses that could be detected against both intact Ag I/II and the free peptide (Table 1). IG immunization with 25 µg of peptide-CTB plus 5 µg of CT as adjuvant failed to elicit serum or salivary antibodies (Table 1) as did IG immunization with free peptide alone or mixed with free CT (not shown). IP immunization with free peptide in Freund adjuvant, however, did induce serum antibodies to Ag I/II (IgM, 4.9 ±0.7 µg/ml; IgG, 7.3 ±7.0 µg/ml; N = 4), and IgM and IgG anti-Ag I/II ASC in the spleen (55 and 10 spot-forming cells per 10^6 cells, respectively). These results lend support to those obtained upon intranasal immunization with CTB-peptide containing excess free peptide, which generated serum but not detectable salivary antibodies, and yet induced protection against oral colonization with *S. mutans*.[13]

Table 1. Serum IgM, IgG, IgA, and salivary IgA antibody responses of mice to immunization with Ag I/II peptide 301-319(+Cys) coupled to CTB.

Immunization[a]	Anti-Ag I/II[b]				Anti-peptide[b]			
	IgM	IgG	IgA	Sal. IgA	IgM	IgG	IgA	Sal. IgA
Peptide-CTB, IP	2.11 ±0.15	8.29 ±12.05	0.44 ±0.28	nd[c]	1.80 ±0.37	8.23 ±11.81	0.30 ±0.31	nd
Peptide-CTB, IG	0.78 ±0.23	0.11 ±0.04	0.08 ±0.07	0	0.93 ±0.06	0.05 ±0.02	0.18 ±0.07	0
Controls	1.33 ±0.40	0.14 ±0.07	0.12 ±0.08	0	0.98 ±0.18	0.04 ±0.01	0.19 ±0.06	0

[a] Mice (3 per group) were immunized with 25 µg of peptide-CTB conjugate three times at 10-day intervals and tested 7 days later. IP immunization was given with magnesium/aluminum oxide; IG immunization was given with 5 µg of CT as adjuvant in 0.1M NaHCO3.
[b] Antibodies (mean ±SD, µg/ml) assayed by ELISA on plates coated with Ag I/II or with free peptide 301-319(+Cys).
[c] Not determined.

DISCUSSION

Our efforts to induce mucosal antibody responses that would protect against a specific oral organism illustrate the principles outlined above. The antigen, Ag I/II, is a surface protein of the principal causative organisms of human dental caries, *S. mutans* and *S. sobrinus*, and is now believed to mediate adherence to the tooth surface by binding to salivary pellicle proteins.[4,26] S-IgA, the predominant salivary immunoglobulin and antibody isotype induced by peroral immunization, is well known for its ability to inhibit adherence of microorganisms to mucosal surfaces. Our recent findings[27] have shown that human S-IgA antibodies specifically inhibit interaction of Ag I/II with salivary pellicle protein and that the inhibition is dependent upon the presence of the Fc region of IgA with its attached secretory component: S-IgA cleaved by IgA protease to yield antigen-binding Fab fragment is less effective.[28] This finding suggests that S-IgA may be more effective than serum-type IgA or IgG antibodies, which lack secretory component and, in the case of IgG, have a different, more hydrophobic and less charged Fc region.[7]

However, the CT-linked system of peroral immunization that we have adopted also induces strong serum IgA and IgG antibodies, as does peroral immunization with CT alone.[29] Although it may be argued that such responses are irrelevant for intra-oral protection, parenteral immunization which generates serum IgG antibodies in the apparent absence of salivary antibody has been found effective in reducing the incidence of dental caries,[6] and although the mechanism of action remains uncertain, small quantities of circulating immunoglobulins undoubtedly reach the oral cavity.[30] For protection against a potentially invasive mucosal pathogen, the generation of both mucosal IgA and circulating IgG (and IgA) antibodies could have a particular advantage. The former would inhibit colonization of the mucosal surface: the latter would act against organisms that had evaded the mucosal defenses and invaded the tissues. IgG antibodies having the capacity to activate complement and facilitate phagocytosis would be especially effective against invading bacteria. IgA is less able to recruit these two powerful antibacterial defense mechanisms, which have inflammatory side effects. Indeed, it may be a particular advantage of IgA for mucosal protection that it appears to be non-inflammatory and possibly anti-inflammatory,[31] since inflammation compromises the integrity of the mucosal barrier. The generation of both mucosal IgA, and circulating IgG and IgA antibodies may therefore provide not only immune defense, but also damage-limiting capability.

The memory attributes of the CMIS have been debated.[32] Most data show that S-IgA antibody responses decline to background levels after withdrawal of the stimulating antigen, and it now seems clear that the long-term survival of B memory cells requires the persistence of antigen. However, features characteristic of anamnestic responses have been observed upon secondary immunization of the CMIS.[33,34] We have reported elsewhere[11] that the salivary and serum antibody responses of mice immunized IG with Ag I/II-CTB plus CT adjuvant decline after about 3 months, and that a single repeated immunization generates a substantial anamnestic response but only in serum IgG and IgA isotypes: a minor increase in salivary IgA antibody could be largely accounted for by a similar increase in total salivary IgA. Thus, stimulation of the CMIS is capable of generating memory, although its expression in response to recall immunization may be in the circulatory rather than the mucosal compartment. Clearly much remains to be learned about the memory functions within the CMIS, and how they might be exploited.

The concept of the CMIS at its simplest implies that antigen-sensitized IgA-committed cells generated in the central inductive sites are disseminated uniformly throughout the various mucosal surfaces, and that consequently, S-IgA antibodies should be equally expressed in all secretions. However, that does not seem to be the case, and there is increasing evidence for compartmentalization within the CMIS (J. Mestecky et al.; this volume). Our data illustrate one facet of this and suggest that it may be possible, by local application of antigen after central induction of a mucosal immune response, to direct the response to a particular tissue site where protection is required. Compartmentalized networks of cell traffic could also afford the opportunity to immunize subcomponents of the CMIS selectively. This could have practical physiological advantage, since a strong immune response, measured, for example, in terms of the fraction of cells devoted to synthesizing antibodies to a single antigen (see Fig. 2a), involves the commitment of a substantial proportion of the lymphoid cell population. If the total number of lymphoid cells remains roughly constant within any one tissue, this implies that an immune response to one antigen must be at the expense of other responses to other antigens. Thus, for example, generating a strong salivary antibody response intended to combat an infection restricted to the oral cavity might also generate a similar response in other mucosal tissues where it would be not only unnecessary and hence physiologically wasteful, but also possibly detrimental to immune protection against other pathogens. Indeed, compartmentalization within the CMIS and an ability to respond to locally present antigens may be an adaptation to overcome this problem. If so, it should be possible to exploit these features of the CMIS to induce specific immune responses at particular mucosal sites. Success in achieving such desirable attributes of a mucosal immune response can be quantitatively monitored at cellular and humoral levels by techniques such as those employed in this study.

ACKNOWLEDGMENTS

These studies were begun in collaboration with Drs. Cecil Czerkinsky, Nils Lycke, and Jan Holmgren, University of Göteborg, Sweden, whose encouragement and advice are gratefully acknowledged. We also acknowledge collaborative efforts with Drs. Suzanne Michalek, Zina Moldoveanu, and Jiri Mestecky on related aspects of these studies. The work was supported by US PHS grants DE-06746, DE-08182, and DE-08228.

REFERENCES

1. M.W. Russell, L.A. Bergmeier, E.D. Zanders, and T. Lehner, Protein antigens of *Streptococcus mutans*: Purification and properties of a double antigen and its protease-resistant component, *Infect. Immun.* 28:486 (1980).

2. A.S. Bleiweis, S.F. Lee, L.J. Brady, A. Progulske-Fox, and P.J. Crowley, Cloning and inactivation of the gene responsible for a major surface antigen of *Streptococcus mutans*, *Arch. Oral Biol.* 35 (Supplement):15 (1990).

3. J.K.-C. Ma, C.G. Kelly, R.A. Whiley, and T. Lehner, Conservation of the gene encoding streptococcal antigen I/II in oral streptococci, *Infect. Immun.* 59:2686 (1991).

4. D.R. Demuth, M.S. Lammey, M. Huck, E.T. Lally, and D. Malamud, Comparison of *Streptococcus mutans* and *Streptococcus sanguis* receptors for human salivary agglutinin, *Microb. Pathog.* 9:199 (1990).

5. N. Okahashi, C. Sasakawa, M. Yoshikawa, S. Hamada, and T. Koga, Molecular characterization of a surface protein antigen gene from serotype *c Streptococcus mutans*, implicated in dental caries, *Mol. Microbiol.* 3:673 (1989).

6. T. Lehner, M.W. Russell, J. Caldwell, and R. Smith, Immunization with purified protein antigens from *Streptococcus mutans* against dental caries in rhesus monkeys, *Infect. Immun.* 34:407 (1981).

7. M. Kilian, J. Mestecky, and M.W. Russell, Defense mechanisms involving Fc-dependent functions of immunoglobulin A and their subversion by bacterial immunoglobulin A proteases, *Microbiol. Rev.* 52:296 (1988).

8. S.M. Michalek and N.K. Childers, Development and outlook for a caries vaccine, *Crit. Rev. Oral Biol. Med.* 1:37 (1990).

9. C. Czerkinsky, M.W. Russell, N. Lycke, M. Lindblad, and J. Holmgren, Oral administration of a streptococcal antigen coupled to cholera toxin B subunit evokes strong antibody responses in salivary glands and extramucosal tissues, *Infect. Immun.* 57:1072 (1989).

10. J. Holmgren, Action of cholera toxin and the prevention and treatment of cholera, *Nature* 257:797 (1981).

11. M.W. Russell and H.-Y. Wu, Distribution, persistence, and recall of serum and salivary antibody responses to peroral immunization with protein antigen I/II of *Streptococcus mutans* coupled to the cholera toxin B subunit, *Infect. Immun.* 59:4061 (1991).

12. C. Kelly, P. Evans, L. Bergmeier, S.F. Lee, A. Progulske-Fox, A.C. Harris, A. Aitken, A.S. Bleiweis, and T. Lehner, Sequence analysis of the cloned streptococcal cell surface antigen I/II, *FEBS Lett.* 258:127 (1989).

13. I. Takahashi, N. Okahashi, K. Matsushita, M. Tokuda, T. Kanamoto, E. Munekata, M.W. Russell, and T. Koga, Immunogenicity and protective effect against oral colonization by *Streptococcus mutans* of synthetic peptides of a streptococcal surface protein antigen, *J. Immunol.* 146:332 (1991).

14. S.R. Hamilton, D.F. Keren, J.K. Boitnott, S.M. Robertson, and J.H. Yardley, Enhancement by cholera toxin of IgA secretion from intestinal crypt epithelium, *Gut* 21:365 (1980).

15. L.M. Sollid, D. Kvale, P. Brandtzaeg, G. Markussen, and E. Thorsby, Interferon-γ enhances expression of secretory component, the epithelial receptor for polymeric immunoglobulins, *J. Immunol.* 138:4303 (1987).

16. J.O. Phillips, M.P. Everson, Z. Moldoveanu, C. Lue, and J. Mestecky, Synergistic effect of IL-4 and IFN-γ on the expression of polymeric Ig receptor (secretory component) and IgA binding by human epithelial cells, *J. Immunol.* 145:1740 (1990).

17. N. Lycke, A sensitive method for the detection of specific antibody production in different isotypes from single lamina propria cells, *Scand. J. Immunol.* 24:393 (1986).

18. C. Czerkinsky, Z. Moldoveanu, J. Mestecky, L.-Å. Nilsson, and Ö. Ouchterlony, A novel two colour ELISPOT assay I. Simultaneous detection of distinct types of antibody-secreting cells, *J. Immunol. Methods* 115:31 (1988).

19. A.J. Husband and J.L. Gowans, The origin and antigen-dependent distribution of IgA-containing cells in the intestine, *J. Exp. Med.* 148:1146 (1978).

20. P. Weisz-Carrington, S.R. Grimes, and M.E. Lamm, Gut-associated lymphoid tissues as a source of an IgA immune response in respiratory tissues after oral immunization and intrabronchial challenge, *Cell Immunol.* 106:132 (1987).

21. P.N.R. Nair and H.E. Schroeder, Duct-associated lymphoid tissue (DALT) of minor salivary glands and mucosal immunity, *Immunology* 57:171 (1986).

22. C. Bolduc, J.D. Waterfield, and N. Deslauriers, Tissue distribution and cytofluorometric analysis of oral mucosal T cells in the BALB/c mouse, *Res. Immunol.* 141:461 (1990).

23. C.F. Kuper, D.M.H. Hameleers, J.P. Bruijntjes, I. van der Ven, J. Biewenga, and T. Sminia, Lymphoid and non-lymphoid cells in nasal-associated lymphoid tissue (NALT) in the rat, *Cell Tissue Res.* 259:371 (1990).

24. P.J. Koornstra, F.I.C.R.S. De Jong, L.F.M. Vlek, E.H.M.A. Marres, and P.J.C. Van Breda Vriesman, The Waldeyer ring equivalent in the rat. A model for analysis of oronasopharyngeal immune responses, *Acta Otolaryngol.* 111:591 (1991).

25. D. Bessen and V.A. Fischetti, Influence of intranasal immunization with synthetic peptides corresponding to conserved epitopes of M protein on mucosal colonization by group A streptococci, *Infect. Immun.* 56:2666 (1988).

26. M.W. Russell and B. Mansson-Rahemtulla, Interaction between surface protein antigens of *Streptococcus mutans* and human salivary components, *Oral Microbiol. Immunol.* 4:106 (1989).

27. G. Hajishengallis and M.W. Russell, Inhibition of *Streptococcus mutans* Ag I/II binding to saliva-coated hydroxyapatite by S-IgA antibodies, *J. Dent. Res.* 71:147, Abstr. 330 (1992).

28. J. Reinholdt and M. Kilian, Interference of IgA protease with the effect of secretory IgA on adherence of oral streptococci to saliva-coated hydroxyapatite, *J. Dent. Res.* 66:492 (1987).

29. N. Lycke, L. Lindholm, and J. Holmgren, IgA isotype restriction in the mucosal but not in the extramucosal immune response after oral immunizations with cholera toxin or cholera B subunit, *Int. Arch. Allergy Appl. Immunol.* 72:119 (1983).

30. S.J. Challacombe, M.W. Russell, J.E. Hawkes, L.A. Bergmeier, and T. Lehner, Passage of immunoglobulins from plasma to the oral cavity in rhesus monkeys, *Immunology* 35:923 (1978).

31. M.W. Russell, J. Reinholdt, and M. Kilian, Anti-inflammatory activity of human IgA antibodies and their Fabα fragments: inhibition of IgG-mediated complement activation, *Eur. J. Immunol.* 19:2243 (1989).

32. L.-Å. Hanson and P. Brandtzaeg, The mucosal defense system, *in*: "Immunological Disorders in Infants and Children", E.R. Stiehm, ed., W. B. Saunders Co., Philadelphia, PA (1988).

33. J. Mestecky, J.R. McGhee, R.R. Arnold, S.M. Michalek, S.J. Prince, and J.L. Babb, Selective induction of an immune response in human external secretions by ingestion of bacterial antigen, *J. Clin. Invest.* 61:731 (1978).

34. M. Quiding, I. Nordström, A. Kilander, G. Anderson, L.Å. Hanson, J. Holmgren, and C. Czerkinsky, Intestinal immune responses in humans. Oral cholera vaccination induces strong intestinal antibody responses and interferon-γ production and evokes local immunological memory, *J. Clin. Invest.* 88:143 (1991).

PEPTOMERS AS VACCINE CANDIDATES

Frank A. Robey, Tracy A. Harris, Anh Ky Nguyen,
Niels H.H. Heegaard and Drago Batinić

Peptide and Immunochemistry Unit
Laboratory of Cellular Development and Oncology
National Institute of Dental Research
National Institutes of Health
Bethesda, MD 20892

INTRODUCTION

This presentation is about an approach we are taking to address several issues that have safe and effective vaccines as the endpoint. Our increasing ability to control immune responses, especially in targeting specific epitopes, makes new approaches to vaccine development conceptually feasible. Because it is well known that the outer-most components of pathogens are most important for infection, envelope proteins and capsular polysaccharides present the pathogen's "weak spots" (i.e., regions of the pathogen where immune intervention will block infection and escort the pathogen away to be destroyed).

Although bacteria may thrive quite well on components of the host outside of any cell, viruses must enter cells in order to replicate and cause destruction. To enter a cell, the virus first must bind to a receptor which exists on the cell surface. Of course, the receptor relates to the function of the cell and the host of which it is a component; it is not present for the purpose of binding viruses. However, for specific viruses and cells, these shared recognition points attract much attention for therapeutic and vaccine-related intervention.

The key point is that viruses and many bacteria often contain surface components that are compatible with the host and, although the immune response is intended to protect the host, in many cases, immune responses against invading pathogens result in cross-reactions against certain host components. The concept of antigenic mimicry is not new but it is becoming well recognized as a cause of disease by an increasing number of pathogens. Examples of antigenic mimicry involved in autoimmunization are listed in Table 1.[1,2]

The growing number of diseases resulting from antigenic mimicry adds urgency to the approach that we are trying to develop--that of using new forms of subunits derived from a pathogen in order to induce immune responses that are specific only for a particular pathogen. This is in addition to the basic research question which asks, what is the minimum structural requirement necessary for an immunogen to stimulate protective immunity? By addressing the issues of minimum requirements

Table 1. Examples of antigenic mimicry involved in autoimmunization.

Exogenous Antigen	Autoantigen	Disease
Streptococcal M antigen	Cardiac myosin	Rheumatic carditis
M. tuberculosis	Cartilage Proteoglycan	Rheumatic arthritis
Klebsiella nitrogenase	HLA-B27	Reiter's syndrome
Yersenia 19 kDa protein	HLA-B27	Reiter's syndrome
Yersenia protein	TSH receptor	Graves' disease
Adenovirus 72-kD protein	La (SS-B)	Sjögren's syndrome
Polio virus VP2	Acetylcholine receptor	Myasthenia gravis
Measles virus P3	Myelin basic protein	Encephalitis (EAE)
Papilloma virus E2	Insulin receptor	Type I diabetes mellitus
Visna polymerase	Myelin basic protein	Encephalitis (EAE)
Hepatitis B polymerase	Myelin basic protein	Encephalitis (EAE)
Retroviral p30 protein	U1RNA	Systemic Lupus Erythematosus
Bacterial phospholipid	DNA	Systemic Lupus Erythematosus
Microbial tRNA	Jo	Polymyositis
Trypanozoma cruzi	Cardiac muscle	Chagas' disease
Mycoplasma	I-antigen	Hemolytic anemia
Adenovirus E1B protein	A-gliadin	Celiac disease

for a vaccine we also are attempting to minimize possible cross-reactions that could challenge the long-term safety of a vaccine.

The primary target of our research efforts and, indeed, an excellent model for retroviral vaccine development is the human immunodeficiency virus type 1 (HIV-1), the virus that initiates the disease known as AIDS. With regard to the subject of HIV-1 and autoimmunity, it was reported in 1988 that regions of gp160, the envelope glycoprotein on the surface of HIV-1, had an amino acid sequence (N-G-T-E-R-V-R) that was homologous to the beta chain of HLA-DR (E-G-T-D-R-V-I).[3] HLA-DR is believed to play a major role in the immune system and it is certainly possible for anti-HLA-DR antibodies to block HLA function and lead to the types of immune-related problems that are present in patients suffering from AIDS. Indeed, antibodies against this peptide were found in the sera of several patients suffering from AIDS[3].

Further work along the lines of HIV-induced autoimmunity were recently reported by Zaitseva et al.,[4] who demonstrated that 75 percent of sera tested from HIV-1-infected individuals contained antibodies that bound to human B-lymphoma cells bearing the major histocompatibility class II molecule. These authors reported that three highly conserved amino acid sequences in α- and β-chains of the class II molecule and three homologous fragments of HIV-1 gp120 and gp41 were identified.

The findings reported in Refs. 3 and 4 encourage a redirection of the thinking about a strategy for developing HIV-1 vaccines and, as a result, now we are involved with making subunit immunogens derived from HIV-1 that do not share homology with HLA-DR or any other host protein.

Based on the current understanding of the mechanisms by which vaccines provide protection, the most effective vaccine which could provide protection against HIV-1 would contain components obtained from HIV-1 or would be composed of the entire attenuated virus itself. However, by designing synthetic materials that could be used as immunogens in place of the naturally occurring, virus-derived components, we would have the latitude to dictate which epitopes on the surface of HIV-1 are recognized by the host's immune system. With this strategy, we can circumvent possibilities of inducing autoimmune phenomena by avoiding the use of viral immunogens that share homology with the host, although, admittedly, we may well be sacrificing efficacy.

For HIV-1, a thorough understanding of the mechanism of infection is important in order to design rational infection-blocking agents such as targeted antibodies produced by the host. One model of the mechanism by which HIV-1 binds to and starts infecting CD4-bearing cells is illustrated in Figure 1 where part of gp120 from the virus is shown interacting with cell surface CD4 in an initial tight binding of the virus to the cell.

A region of gp120, called the principle neutralizing determinant (PND) of the V3 region of gp120, may actually function in the fusion/infection processes by interacting with a section of CD4 that is not involved with the initial binding or with cell surface sulfated polysaccharides (SP).[5] Because antibodies to the PND block HIV-1 infection and HIV-1-induced cell fusion,[6] our hypothesis (Figure 1) shows that the mechanism by which anti-PND antibodies block infection and fusion involves the prevention of PND binding to CD4 (amino acids 78-94) or SP or both in order to interfere with this part of the infection or fusion mechanisms.

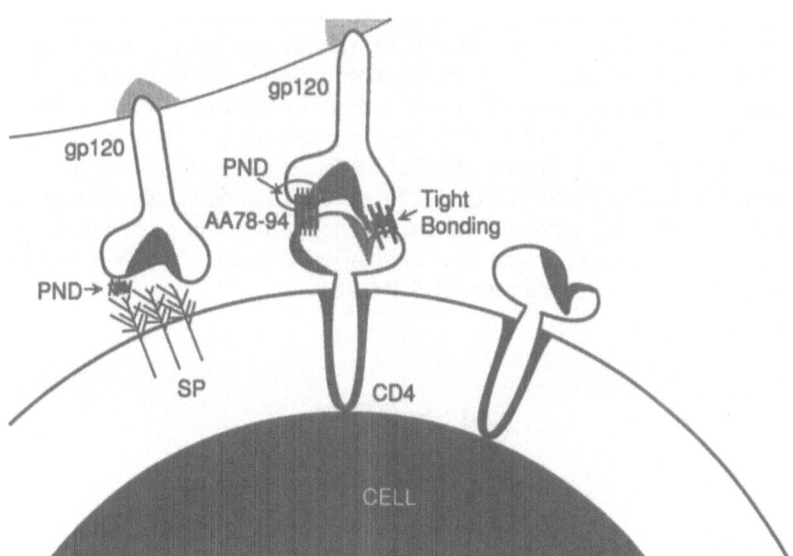

Figure 1. Hypothetical model to explain the binding of the envelope protein of HIV-1, gp120, to cells that express the CD4 receptor. A region of gp120 must first bind to the cell and this is the tight binding region. Second, parts of the V3 loop from gp120, referred to as the principle neutralizing determinant (PND) here, could bind to another region of CD4 (amino acids 78-94) or to sulfated polysaccharides (SP) on the cell surface. Antibodies against the PND block HIV-1 infection and, most likely, block the ability of the PND region in gp120 to bind to amino acids 78-94 of CD4 and/or to cell surface SP.

An attempt is being made to synthesize materials that mimic naturally occurring regions of gp120, the viral envelope protein of HIV-1, not the whole protein. Since we agree with the theory that a linear peptide often will not fill the desired requisites for an effective immunogen,[7] we are designing and evaluating alternative peptide forms which might be more representative of the peptide in a protein than would a peptide in solution or conjugated to a carrier. As we learn more about the molecular mechanisms of infection by a specific pathogen, some of which are shown in Figure 1 above for HIV-1, molecular approaches to immunizing with subunits should become a reality. As for as we know, at this time there are no mechanisms of viral infection that are fully understood. That alone may explain why there are no approved vaccines designed to provide protection to the host specifically by interfering with a known mechanism.

PEPTOMERS

"Peptomer" is a new term to define a peptide polymer that is formed by the specific cross-linking of a peptide to itself. Specifically cross-linking synthetic peptides is a new procedure that was developed in this laboratory over the past three years[8-10]. Most importantly, the chemistry to form peptomers is very simple and can be performed by anyone who is operating an automated peptide synthesizer. Peptomers are compounds which are synthesized under controlled conditions and this is important for reproducibility of biological activity, quality control and lot-to-lot consistency.

The general reaction scheme for forming a peptomer from a synthetic peptide containing a bromoacetylated amino acid at the amino terminus is given in Figure 2. In the example shown, the peptomer is formed in a configuration that links the N-terminus to the C-terminus of the synthetic peptide by way of a sulfhydryl group on a C-terminal cysteine which reacts with the bromoacetyl moiety at the other end. The resulting linkage is a very stable thioether and the by-product of the reaction is simply NaBr, an innocuous salt.[9] Some of the experimental details include adding the bromoacetylated peptide to a slightly alkaline buffer, such as phosphate buffered saline, at a concentration of at least 10 mg/ml. The polymerization occurs spontaneously and the reaction mix is allowed to stir for several hours until the pattern of the polymers formed appears constant as judged by analytical gel filtration or sodium dodecyl sulfate polyacrylamide gel electrophoresis (SDS-PAGE).

BrAc-Peptide-Cys

Ph = 7-8

BrAc-Peptide-(-Peptide-)$_n$-Peptide-Cys
+ NaBr

Figure 2. Peptomer formation with an N-bromoacetyl-derivatized synthetic peptide used as the starting material. The peptomer forms spontaneously when the bromoacetylated peptide containing cysteine is placed into aqueous buffer at neutral or slightly alkaline pH. The only byproduct of the polymerization is NaBr, an innocuous salt.

In addition to the peptomers in which the monomers are linked "head-to-tail" as shown on Figure 2, we have designed a technique that allows the specific linking of peptides together at any desired position in the peptide chain. BBAL[11] is a tBOC amino acid derivative of lysine that contains an extended bromoacetyl moiety at the epsilon amine position. Bromoacetyl moieties are stable under most of the reaction conditions used in tBOC peptide synthesis and BBAL provides the capability to place a bromoacetyl moiety at any desired position in a peptide. The peptide chemist can now link peptides to themselves in many configurations. For a peptide that contains both cysteine and BBAL, the positioning of the two allows the creation of several new peptide derivatives from a single amino acid sequence simply by changing the positions of the BBAL and the cysteine. In addition to peptomers, BBAL will allow the synthesis of numerous new cyclic peptides.

An example of an SDS-PAGE of a sample peptomer is shown on the next page in Figure 3. The SDS-PAGE can be used to evaluate the consistency of the synthesis of a peptomer; specific patterns on the gel could act as a fingerprint for a particular peptomer.

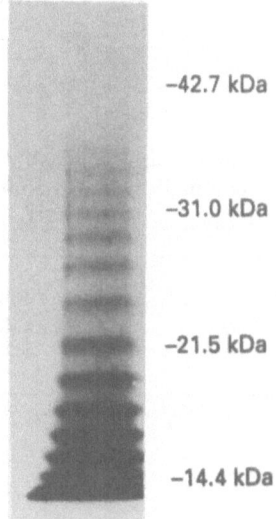

-42.7 kDa

-31.0 kDa

-21.5 kDa

-14.4 kDa

Figure 3. Sodium dodecylsulfate polyacrylamide gel (SDS-PAGE) of a sample peptomer. The molecular weights are given to the right of the gel.

Peptomer formation and immunogenicity are presently very active topics of research in this laboratory; choice of solvent, proton scavengers and temperature as well as the composition of the monomeric peptide will influence the sizes of the final peptomer mixture. It is possible to make peptomers with molecular weights in excess of 20,000 but larger molecular weights may prove to be optimal for immunization purposes.

It has been shown in the literature that peptide polymers enhance immunogenicity of peptides[12] and a peptomer has been shown to elicit antibody responses in nude mice[13] indicating that peptomers may be useful as immunostimulants in immunocompromised individuals. Peptomers could offer certain conformational properties to a specific peptide that are not found in the corresponding monomer or protein-conjugated peptide. We have found in many instances that peptomers are more immunogenic than the monomeric conjugated synthetic peptide, whereas in a few others a peptomer may actually decrease the immunogenicity (unpublished data). This is a further indication of the fact that the peptomer offers properties to a peptide that are not attainable from the free or conjugated peptide.

The ELISA results obtained with a single peptide in our AIDS work are given in Table 2. The peptide T-R-K-I-S-I-Q-R-G-P-G-R-C is derived from the PND of the IIIB isolate of HIV-1, and the cysteine at the C-terminus is added for conjugation or polymerization. It is clear that the titer of the anti-peptomer antibodies exceeds that of the free peptides by several orders of magnitude. Antibodies to 131GRC react strongly and specifically with recombinant (Genentech, San Francisco, CA) and native gp120 from HIV-1 IIIB in ELISAs and Western blots. As expected, the antibodies shown here do not block *in vitro* HIV-1 infection because, even though the peptide sequence comes from the V3 region of gp120, it falls short of the Gly-Pro-Gly surrounding regions that appear to be necessary for inhibition. The peptomer labeled T-T is the peptomer specifically linked through the carboxy terminus using BBAL (*vide supra*). The BBAL was placed in front of the cysteine in the C-terminus of the monomeric peptide.

Table 2. ELISA titers of rabbit anti gp120 using V3-based immunogens after 6 weeks. All injections initially were in complete Freund's adjuvant.

Free Peptide (131GR IIIB)	1:5000
Peptomer H-T (131GR IIIB)	1:250,000
Peptomer T-T (131GR IIIB)	1:5000

CONCLUDING REMARKS

This brief introduction to the concept of peptomers is intended to introduce the scientific community working on mucosal immunity to an alternative approach for developing synthetic vaccines. The ease of syntheses and the future needs for such vaccines makes this approach attractive but there are drawbacks that are being understood as time permits.

Often, peptomers that we have synthesized are poorly soluble. The property of poor solubility in aqueous media may contribute to the stability of the peptomer *in vivo* and this is in contrast to most immunogenic synthetic peptides. However, as a peptomer represents a magnification of the properties of a peptide, peptomer complexes with carriers may be quite useful in mucosal immunity because the peptomer is likely to attach more strongly to certain carrier beads or alum than the monomeric peptide.

Several factors contribute to inducing an immune response and the antibodies formed are often against discontinuous epitopes on a protein. Because it is very clear that conformation of a peptide would be involved in the cross-reaction of an anti-peptide antibody with a native protein, any design that would address the facts that peptides have very little structure in solution and exhibit a multitude of interconverting structures in solution should be closely investigated. Peptomers were designed to bridge the gap between the conformations found in small peptides and those found in the parent protein, and as more data become available we may be able to evaluate them critically as viable vaccine candidates.

There are no synthetic peptides used in vaccines today, but with the explosion of information related to immunogenicity, adjuvants, epitopes, mimetopes, autoimmunity, etc., the minimum requirements for total protection of a host from a particular pathogen will soon be realized.

REFERENCES

1. R.S. Schwartz and S.K. Datta, Autoimmunity and autoimmune disease, *in*: "Fundamental Immunology," 2nd ed., W.E. Paul, ed., Raven Press, New York (1989).
2. I. Roitt, "Essential Immunology," Blackwell Scientific Publications, Oxford (1980).
3. H. Golding, F.A. Robey, F. T. Gates, W. Lindner, T. Hoffman, and B. Golding, Identification of homologous regions in human immunodeficiency virus 1 gp41 and human MHC class II domain, *J. Exp. Med.* 167:914 (1988).
4. M.B. Zaitseva, S.A. Moshnikov, A.T. Kozhich, H.A. Frolova, O.D. Makarova, S.P. Pavlikov, I.G. Sidorovich, and B.B. Brondz, Antibodies to MHC Class II peptides are present in HIV-1-positive sera, *Scand. J. Immunol.* 35:267 (1992).
5. D. Batinić and F.A. Robey, The V3 region of the envelope glycoprotein of human immunodeficiency virus type 1 binds sulfated polysaccharides and CD4-derived synthetic peptides, *J. Biol. Chem.* 267:6664 (1992).
6. J.R. Rusche, K. Javaherian, C. McDanal, J. Petro, D.L. Lynn, R. Grimaila, R. A. Langlois, R.C. Gallo, L.O. Arthur, P.L. Fischinger, D.P. Bolognesi, S.D. Putney, and T.J. Matthews, Antibodies that inhibit fusion of human immunodeficiency virus-infected cells bind a 24-amino acid sequence of the viral envelope, gp120, *Proc. Natl. Acad. Sci. USA* 85:3198 (1988).
7. W.G. Laver, G.M. Air, R.G. Webster and S.J. Smith-Gill, Epitopes on protein antigens: misconceptions and realities, *Cell* 61:553 (1990).
8. W. Lindner and F.A. Robey, Automated synthesis of N-chloroacetyl-modified peptides for the preparation of synthetic peptide polymers and peptide-protein immunogens, *Int. J. Pept. Protein Res.* 30:794 (1987).
9. F.A. Robey and R.L. Fields, Automated synthesis of N-bromoacetyl-modified peptides for the preparation of synthetic peptide polymers, peptide-protein conjugates and cyclic peptides, *Anal. Biochem.* 177:373 (1989).
10. N. Kolodny and F.A. Robey, Conjugation of synthetic peptides to proteins: quantitation from S-carboxymethylcysteine released upon acid hydrolysis, *Anal. Biochem.* 187:136 (1990).
11. J.K. Inman, P.F.Highet, N. Kolodny, and F.A. Robey, Synthesis of N^α-(tert-butoxycarbonyl)-N^E-[N-(bromoacetyl)-β-alanyl]-L-lysine: its use in peptide synthesis for placing a bromoacetyl crosslinking function at any desired position, *Bioconjugate Chem.* 2:458 (1991).
12. F. Borras-Cuesta, Y. Fedon and A. Petit-Camurdan, Enhancement of peptide immunogenicity by linear polymerization, *Eur. J. Immunol.* 18:199 (1988).
13. K. Hillman, O. Shapiro-Nahor, R. Blackburn, D. Hernandez and H. Golding, A polymer containing a repeating peptide sequence can stimulate T-cell-independent IgG antibody production *in vivo*, *Cell. Immunol.* 134:1 (1991).

STRUCTURAL AND FUNCTIONAL STUDIES OF
HERPES SIMPLEX VIRUS GLYCOPROTEIN D

Gary H. Cohen,[1,2] Martin I. Muggeridge,[1,2] Deborah Long,[1,2]
Donald A. Sodora,[1,2] and Roselyn J. Eisenberg[2,31]

[1]Department of Microbiology and [2]Center for Oral Health Research
School of Dental Medicine
[3]Department of Pathobiology, School of Veterinary Medicine
University of Pennsylvania
Philadelphia, PA 19104

INTRODUCTION

Herpes simplex virus (HSV) is a major human pathogen which causes a variety of diseases including cold sores, eye and genital infections, neonatal infections and encephalitis.[1-3] Almost 80% of American adults have antibodies against the oral form of the virus, HSV-1,[4] and 16% of adults[5] or 22% of women[6] are infected with the genital form, HSV-2. Both HSV-1 and HSV-2 establish lifelong latent infections and reactivate periodically to produce recurrent infections.[1,2,3] Thus, these viruses represent significant morbidity problems in general, and HSV-1 is of fundamental importance in oral health. The goal of a large number of laboratories is to produce a vaccine that will prevent infection or ameliorate the severity of the recurrent form of the disease. At present, two human viral glycoprotein subunit vaccines are in clinical trials in the United States. In both cases, the vaccine contains the essential glycoprotein gD. One vaccine uses the full-length form of this protein isolated from HSV-1–infected cells (gD-1),[7] and the other uses a genetically engineered truncated form of gD from HSV-2 (gD-2t).[8,9]

In this article we will outline studies in our laboratory which address basic questions about the structure of gD. The results of these studies have important and general implications for the successful use of gD as a vaccine, regardless of the serotype used or the source of the protein. We emphasize that this article is not a literature review of HSV glycoproteins or vaccines, and direct the reader to three excellent reviews on HSV vaccines[8,10,11] and other reviews on HSV glycoproteins.[12-15] We consider four questions here: 1) What is the antigenic profile of gD especially relating to epitopes involved in virus neutralization? 2) Which sites on the molecule are important for its function and how do these sites relate to its antigenic structure? 3) What is the contribution of the N-linked oligosaccharides on gD to its structure and function? 4) How important are the cysteine residues to structure and function of gD?

Genetically Engineered Vaccines, Edited by
J.E. Ciardi *et al.*, Plenum Press, New York, 1992

STRUCTURE OF HSV AND SOME GENERAL PROPERTIES OF gD

The HSV virion (Fig. 1) has a double-stranded DNA genome of 150 kilobases. There are at least 72 open reading frames, of which at least 10 code for glycoproteins.[16,17] The glycoproteins are referred to by alphabetical designations, gB, gC, gD, gE, gG, gH, gI, gJ, gK, and gL. The DNA genome is encased in an icosahedral protein capsid which is surrounded by an amorphous protein layer called the tegument, and then by a lipid envelope containing the 10 virus glycoproteins. At time of infection, these virion glycoproteins act singly or in concert to trigger a fusion event at the plasma membrane of a susceptible cell. After infection, they are expressed on all cellular membranes, including the plasma membrane. Because they are present in the virion envelope and on infected cell plasma membranes, the glycoproteins are major targets of the host's immune response. gD is an essential protein which is required for virus penetration into susceptible cells.[18,19] It is strongly implicated in receptor binding,[20,21] cell fusion,[22-24] and neuro-invasiveness.[25] gD, when injected into test animals, induces viral neutralizing antibodies.[26-28] The vaccinated animals are protected from HSV infection and establishment of latency.[7-9,29-35]

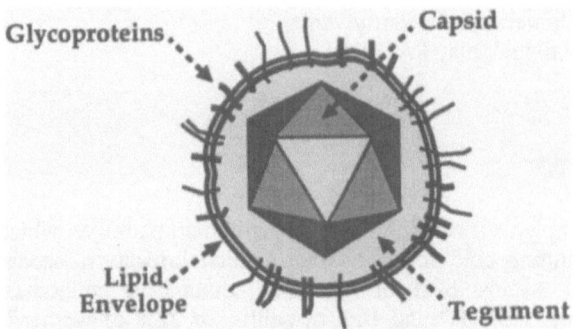

Figure 1. Diagram of a herpes simplex virion.

The major historical events which focused attention on gD as a subunit vaccine are shown in Figure 2. Briefly: (i) monoclonal antibodies (MAbs) to gD, when passively administered to animals, protected them against lethal virus challenge;[36] (ii) purified gD[28] was shown to protect animals against lethal infection by either the oral[29,31] or genital serotype of HSV[31]; (iii) the glycoprotein also was shown to prevent the establishment of HSV latency;[37] (iv) gD was shown to protect against recurrent HSV-2 infections in guinea pigs;[38] and (v) human clinical trials were begun in 1990 with some promising results for stimulation of an antibody response.[7,9] Some other important events in this history, not detailed here, are also summarized on Figure 2.

STRUCTURAL PROPERTIES OF gD

Figure 3 details some structural properties of gD. After removal of the 25 amino acid signal sequence,[39] gD-1 has 369 amino acids[40] and gD-2 has 368.[41,42] The two proteins are 85% homologous, and both have three N-linked

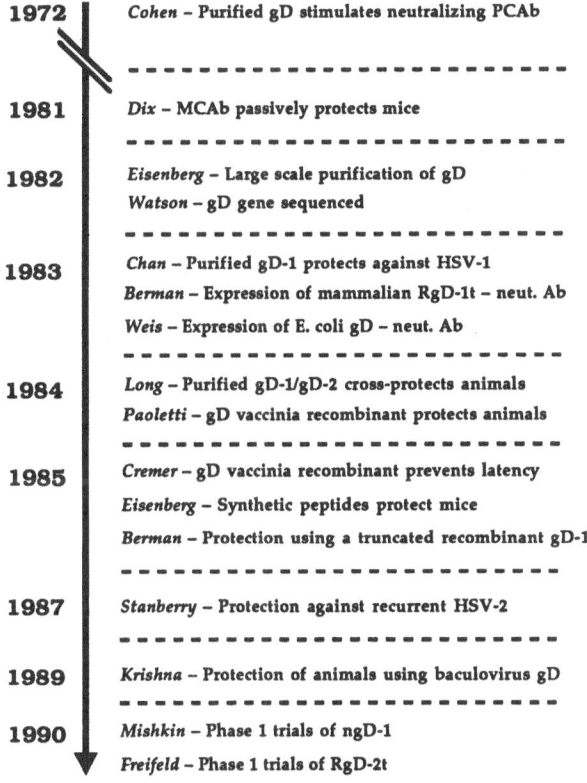

Figure 2. Time-line for progress towards a gD subunit vaccine.

oligosaccharides[43] (balloons), two or three O-linked oligosaccharides,[44] and six cysteine residues[40-42] in their extracellular domain. A seventh cysteine residue is located in the hydrophobic membrane-anchoring domain of gD-1 near the carboxyl terminus of the protein.[40] The first six cysteines are strictly conserved in both number and spacing in gD-1 and gD-2.[40-42] A very similar pattern of cysteine spacing is also found in gD homologs of several animal herpesviruses.[45-49]

Prior work has provided a detailed picture of the major antigenic sites of gD and the epitopes within them, based on the properties of a collection of MAbs.[14] The antigenic sites are divided into those that are continuous versus those that are discontinuous.[14,50-53] Continuous epitopes consist of a linear array of amino acids, whereas discontinuous epitopes are those which are brought together as a result of protein folding. MAbs which recognize discontinuous epitopes are particularly important since they are exquisitely sensitive probes to detect subtle conformational

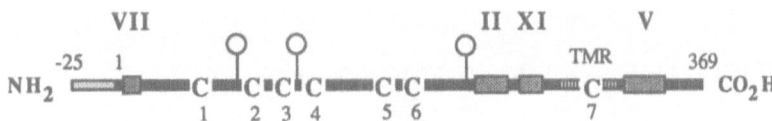

Figure 3. Stick model of gD-1.

changes in the protein brought about by either mutational alterations or variations in physicochemical conditions. Localization of these sites led to the model of gD-1 structure, shown in Figure 4. Four of the continuous antigenic sites are shown: site VII (residues 11 to 19),[54,55] site II (residues 264 to 279),[52] site XI (residues 284-301)[52] and site V (residues 340 to 356)[56]. Three discontinuous sites are also

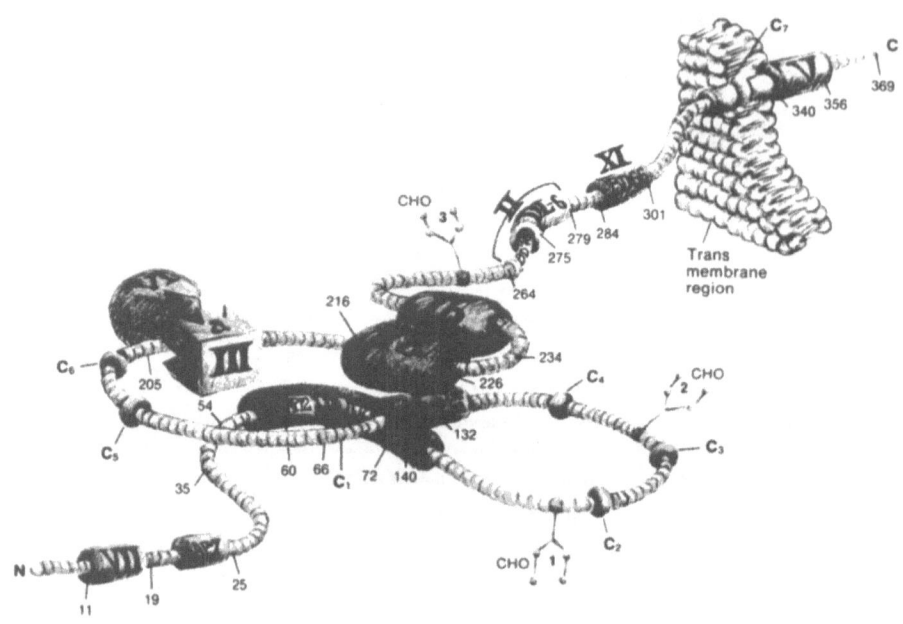

Figure 4. 3D model for gD-1.

shown (I, III and VI)[57] In some cases, we have indicated the position of an amino acid in an epitope corresponding to a specific (and sometimes ungrouped) MAb, such as AP7, for which a monoclonal antibody resistant mutation was identified at position 25[23]. To position other antigenic sites, we examined the properties of mutant forms of gD that were unable to bind MAbs within a particular group. For example, site III and VI include residues downstream as well as upstream of residue 205. A mutation at residue 54 oblates the binding of two groups III MAbs, but binding of group VI MAbs is unaffected (D. Sodora, G. Cohen and R. Eisenberg, unpublished data). This placed residue 54 in proximity with residues just downstream of aa205 so that group III MAbs could contact both parts of gD.

DETAILED MAPPING OF ANTIGENIC SITE I OF gD

Antigenic site I is recognized by the group I MAbs, of which seven have currently been identified[53]. Each MAb blocks the binding of the others within the group, and all seven bind to discontinuous epitopes that are conserved between gD-1 and gD-2[53]. Binding of group I MAbs results in neutralization of virus infectivity, presumably due to interference with the function of gD in virus penetration.[23,28,58-61] In addition, these MAbs block cell-to-cell spread of virus and also cell fusion by syncytial strains.[18,19,24] It therefore seemed likely that site I overlaps a functional site on the protein. We have used a variety of truncation, deletion and virus antibody escape (*mar*) mutants to characterize this site.

Truncation and deletion mutants were created in a gD gene within a eukaryotic expression vector, and the corresponding proteins were then produced in transiently transfected cells. Reaction of group I MAbs with cell extracts on native Western blots suggested that they could be subdivided into the Ia's (MAbs HD1 and LP2) and the Ib's (MAbs DL11, 4S, D2, III-114 and III-174).[53] The shortest fragment with which Ia MAbs react is 1-226, whereas the Ib MAbs require aa1-258. Analysis of deletion mutants confirmed this division, since only the Ia MAbs reacted with a mutant lacking amino acids 234-244 (Δ234-244).[53]

Five of the group I MAbs were used to select *mar* mutant viruses, and cross-neutralization experiments demonstrated that mutants selected with Ia MAbs were still neutralized by Ib MAbs, and vice versa. DNA sequencing showed that both Ia MAbs selected a Ser to Asn substitution at aa216, consistent with results from the analysis of truncation mutants.[23,53] Substitutions selected by the Ib MAbs were Gln to Leu at aa132 (D2) and Ser to Asn at aa140 (DL11 and 4S), identifying an upstream portion of site I. Moreover, the change at 140 created a new site for attachment of an N-linked oligosaccharide, which is used.[53]

Evidence to support the hypothesis that site I overlaps a functional site on the protein was obtained by using oligonucleotide-directed mutagenesis to create a gD gene containing all three *mar* mutations within site I; despite the fact that the corresponding protein folded correctly when expressed in transfected cells, the gene could not be recombined intact into the virus genome. Instead, only the change at aa216 was retained, implying that the combination of mutations was lethal.[62] Further characterization of the functional activity of gD mutants relied on a complementation assay.

This assay takes advantage of a gD-minus virus called F-gDβ, in which the gD gene has been replaced with a β-galactosidase gene.[20,21] Because gD is essential for virus replication, the virus can be propagated only in cells that synthesize gD, such as the VD60 cell line. When stocks of virus produced in VD60 cells are used to infect non-gD-expressing cells, virions are produced but they lack gD protein and are non-infectious.[20,21] However, infectious virus can be obtained if the cells are first transfected with a plasmid expressing gD and then infected with F-gβ.[62] Progeny virus still lacks a gD gene, but it incorporates the plasmid-encoded gD protein and is therefore able to enter VD60 cells and subsequently form plaques. Comparison of the number of plaques resulting from transfection with a mutated gD gene with the number after transfection with the wild-type gene gives the percentage complementation, which is related to the functional activity of the mutant protein. Analysis of genes containing *mar* mutations agreed with earlier results in that individual mutations had little effect on complementation, whereas the 140/216 double mutant and the 132/140/216 triple mutant had very low activity.[62]

Combining the complementation assay with an antigenic analysis of mutant proteins has been a powerful approach to mapping a gD functional site. Of several in-frame deletion mutants, those lacking discontinuous epitopes had no complementation activity.[63] However, the remainder, with only local conformational changes, were split between those with activity and those without. Thus, the N-terminus of the protein (aa7-21) and the cytoplasmic tail (aa338-369) were dispensable, whereas aa234-244 were essential.[63] Since absence of these latter residues also abolished binding of group Ib MAbs, this result suggests that site I does indeed overlap a functional site on gD.

WHAT IS THE IMPORTANCE OF N-LINKED OLIGOSACCHARIDES (N-CHO) ON gD FOR ANTIGENIC STRUCTURE AND FUNCTION?

gD contains three predicted recognition sites for the addition of N-CHO, and all are utilized. We used site-directed mutagenesis to alter each site individually, and a fourth mutant was altered at all three sites.[64,65] The mutated genes were inserted into an expression vector and analyzed. We found that the N-CHO moieties individually and in combination are not essential for the proper processing or surface expression of gD in transfected cells. No significant change in protein structure could be detected in the proteins carrying mutations (threonine or serine to alanine), which eliminated N-CHO site 2 or site 3. However, substitution of alanine for threonine (residue 96) at N-CHO site 1 eliminated the first N-CHO addition signal, but the change in the amino acid in itself adversely affected protein structure. We tested several other mutations which would eliminate the N-CHO signal and found that substitution of glutamine (Q) for asparagine at residue 94 was least deleterious to protein structure. We also prepared a triple mutant (QAA) which eliminated all three signals and tested the effect of each change on gD function. Despite changes in antigenic conformation, each mutant form of gD complemented the infectivity of FgDβ. Thus, marked changes in gD structure do not necessarily lead to a loss of gD function. These studies predicted that viruses containing gD without N-CHO would be able to replicate normally in cell culture. We carried out recombination experiments[66] and isolated a virus, called F-gD(QAA), which contained the mutated form of gD lacking N-CHO. FgD(QAA) and the wild-type virus F-gD(WT) were remarkably similar in properties, except that F-gD(QAA) produced smaller plaques (50% of the size of the wild type). In addition, the gD protein that was detergent extracted from cells infected with the mutant was antigenically altered. However the gD protein that was transported to the cell surface during infection appeared to be antigenically intact. Flow cytometry analysis using MAbs to sites VI and Ib showed that the amount and structure of gD found on infected cell surfaces was unaffected by the presence or absence of N-CHO. We concluded from this study that the N-CHO present on gD are not required for its biological role in HSV infection but contribute to some extent to the structure of gD and to efficient plaque formation and cell-to-cell spread. Although this protein lacks N-CHO, we found that it contains O-linked oligosaccharides. Further studies will be necessary to assess the contribution of these modifications to gD structure and function. Our results predict that the QAA-gD protein isolated and purified from infected cells would not function efficiently as a subunit vaccine due to its altered antigenic structure. However, the protein might be more stable if it is secreted from the cell and recovered in the absence of detergents.

WHAT IS THE CONTRIBUTION OF CYSTEINE RESIDUES TO THE STRUCTURE AND FUNCTION OF gD?

One of our goals has been to localize those epitopes on gD which are necessary for stimulation of a protective immune response to virus infection. In early animal studies we found that the efficacy of gD as a vaccine was strictly dependent on intact disulfide bonds. The fully reduced and alkylated protein stimulates production of non-neutralizing antibodies and does not afford protection (our unpublished data). This molecule no longer reacts with any MAbs to discontinuous epitopes, including those in Group I. In contrast, boiling gD in sodium dodecyl sulfate under non-reducing conditions does not destroy discontinuous epitopes. We constructed internal deletion mutations of gD, and those which eliminated any one of the first six cysteines resulted in a protein with profoundly altered structure.[67] Several deletions outside of this grouping of cysteine residues were less deleterious and resulted in an infectious mutant virus.[68]

To systematically analyze the importance of each cysteine residue for the antigenic structure and function of gD, we constructed seven different mutations in the gD gene by site-directed mutagenesis, specifically altering each cysteine residue to a serine.[69] The mutated genes were cloned into an expression vector and transiently expressed in mammalian cells. Replacement of the cysteine of gD-1 at position 333 by serine (Cys-7 mutant) had a minimal effect on carbohydrate processing or the reactivity of MAbs that recognized discontinuous epitopes. In contrast, replacement of any one of the other six cysteine residues (Cys-1 through Cys-6) resulted in a major reduction or complete loss of MAb binding. In addition, mutations of any of these six cysteines had profound effects on oligosaccharide processing, and transport of gD. On the basis of these results, we postulated that the first six cysteines of gD form three disulfide bonds, all of which are required for proper folding, and that impaired processing is due to misfolding of the protein. Furthermore, the experimental variation that we detected in the amount of antigenic activity and processing suggested that some of the mutations might be temperature sensitive (ts). Five of the mutants, Cys-1, Cys-2, Cys-4, Cys-5, and Cys-6 are in fact ts for one or more of the following properties: antigenic conformation, formation of aggregates, processing and transport to the cell surface.[70] When we analyzed each of the mutants for virus function, using the complementation assay, only two mutant proteins, Cys-2 and Cys-4, could complement F-gDβ, and the complementation activity was ts. Moreover, the efficiency of complementation by these mutants was low even at permissive temperature. Although both mutants functioned normally in virus penetration, they were more susceptible to thermal inactivation than was the wild-type virus. Thus the protein, once formed at permissive temperatures can be inserted into the membrane of the virus and function normally. Two pairs of mutants (Cys1/Cys-5 and Cys-2/Cys-4) had similar phenotypes, and these results led us to predict that mutants with similar phenotypes might each represent one partner in a disulfide pair. On this basis, we postulated that Cys-1 might be bonded to Cys-5 and Cys-2 to Cys-4. By default we paired Cys-3 with Cys-6, even though these two mutants were not very similar to each other. A model depicting this possible disulfide bond structure is shown in Figure 5. At present we are continuing to pursue the disulfide assignment problem in two ways. First, we are now studying the properties of additional mutants, each of which lacks two cysteine residues. The argument here is that if both partners in a disulfide pair are missing, the protein might still be able to fold into a structure close to that of the native protein with two disulfide bonds intact. However, if the double mutation eliminates one partner in each of two different disulfide bonds, the protein will be much less stable, because only one correct disulfide bond could form. Thus, creation of the 15 possible double Cys mutants should yield only 3 proteins whose properties resemble those of the native molecule and 12 which should serve as excellent negative controls. Our second approach is to carry out direct biochemical analysis of the purified protein. The strategy is to fragment the protein by chemical and enzymatic means under non-reducing conditions, and then to analyze the structure of the individual disulfide-bonded peptides. Knowledge of the actual disulfide bond structure of gD will allow us to re-model the protein in a more folded fashion. Clearly, however, the ultimate goal is to solve the structure of gD by X-ray crystallographic means. Once this is done, we will have a much better understanding of the relationship of the actual structure of gD to its antigenic and functional properties.

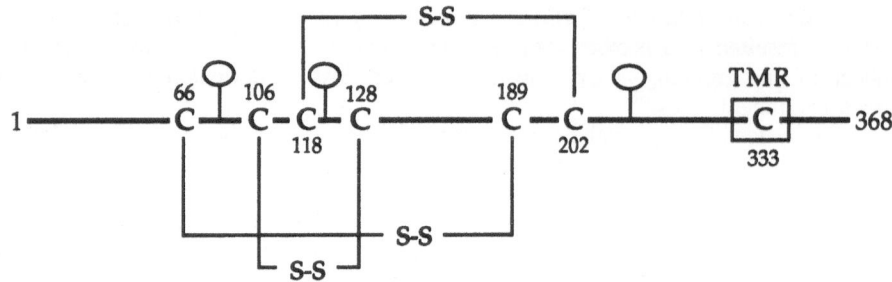

Figure 5. Predicted disulfide bond pattern for gD.

FINAL CONCLUSIONS

We conclude from these studies that the three N-linked oligosaccharides in gD are not necessary for HSV infection but are critical for the maintenance of the antigenic structure of the molecule when isolated from the cell. Clearly, carbohydrates are important in HSV vaccine development. Secondly, the three disulfide bonds in the extracellular portion of gD are critical for antigenic structure of the protein and function of the virus. Alteration of any one of the three pairs results in global destruction of structure and function. Vaccines which incorporate gD must be designed to maintain all three disulfide bonds. Knowledge of disulfide bond structure will contribute substantially to our understanding of the antigenic structure of gD. Finally, a functional site on gD has been mapped. In the folded structure of gD, it is close to, but distinct from, the major antigenic sites Ia and Ib. Although group I MAbs have powerful virus neutralizing activity, they are not the only MAbs to gD with this property. Whether other parts of gD will be found to contribute to its function will require further analysis of gD mutants and further localization of discontinuous epitopes.

ACKNOWLEDGMENTS

This investigation was supported by Public Health Service grants DE-08239 from the National Institute of Dental Research and AI-18289 from the National Institute of Allergy and Infectious Diseases and by a grant to G.H.C. and R.J.E. from the American Cyanamid Co. D.L. is a predoctoral trainee supported by Public Health Service grant NS-07180 from the National Institute of Neurological Diseases and Stroke.

REFERENCES

1. T.J. Hill, Herpes simplex virus latency, *in*: "The Herpesviruses," B. Roizman, ed., Plenum Press, New York (1985).
2. B. Roizman and A.E. Sears, An inquiry into the mechanisms of herpes simplex virus latency, *Ann. Rev. Microbiol.* 41:543 (1987).
3. J.G. Stevens, Human herpesviruses: a consideration of the latent state, *Microbiol. Rev.* 53:318 (1989).
4. L.C. Goldstein and R.C. Nowinski, Diagnosis of herpesviruses with monoclonal antibodies, *in*: "Human Herpesvirus Infections," C. Lopez and B. Roizman, eds., Raven Press, New York (1986).

5. R.E. Johnson, A.J. Nahmias, L.S. Magder, F.K. Lee, C.A. Brooks, and C.B. Snowden, A seroepidemiologic survey of the prevalence of herpes simplex virus type 2 infection in the United States, *N. Engl. J. Med.* 321:7 (1989).

6. M.K. Breinig, L.A. Kingsley, J.A. Armstrong, D.J. Freeman, and M. Ho, Epidemiology of genital herpes in Pittsburgh: serologic, sexual, and racial correlates of apparent and inapparent herpes simplex infections, *J. Infect. Dis.* 162:299 (1990).

7. E.M. Mishkin, J.R. Fahey, Y. Kino, R.J. Klein, A.S. Abramovitz, and S.J. Mento, Native herpes simplex virus glycoprotein D vaccine: immunogenicity and protection in animal models, *Vaccine* 9:147 (1991).

8. R.L. Burke et al., Development of herpes simplex virus subunit vaccine, in: "Vaccines 89. Modern Approaches to New Vaccines Including Prevention of AIDS," R.A. Lerner, H. Ginsberg, R.M. Chanock, and F. Brown, eds., Cold Spring Harbor Laboratory (1989).

9. A.G. Freifeld et al., Phase 1 testing of a recombinant herpes simplex type 2 glycoprotein D vaccine, 13th Interscience Conference on Antimicrobial Agents and Chemotherapy (1990).

10. B. Meignier, Vaccination against herpes simplex virus infections, in: "The Herpesviruses. Immunobiology and Prophylaxis of Human Herpesvirus Infections," B. Roizman and C. Lopez, eds., Plenum Press, New York (1985).

11. R.D. Dix, Prospects for a vaccine against herpes simplex virus types 1 and 2, *Prog. Med. Virol.* 34:89 (1987).

12. P.G. Spear, Antigenic structure of herpes simplex viruses, in: "Immunochemistry of Viruses. The Basis for Serodiagnosis and Vaccines," M.V.H. van Regenmortel and A.R. Neurath, eds., Elsevier Science Publishers B. V., Amsterdam (1985).

13. P.G. Spear, Glycoproteins specified by herpes simplex viruses, in: "The Herpesviruses," B. Roizman, ed., Plenum Press, New York (1985).

14. M.I. Muggeridge, S.R. Roberts, V.J. Isola, G.H. Cohen, and R.J. Eisenberg, Herpes simplex virus, in: "Immunochemistry of Viruses, II. The Basis for Serodiagnosis and Vaccines," M.H.V. Van Regenmortel and A.R. Neurath, eds., Elsevier Biochemical Press, Amsterdam (1990).

15. H.S. Marsden, Herpes simplex virus glycoproteins and pathogenesis, in: "Molecular basis of Virus Disease. Fortieth Symposium of the Society for General Microbiology," W.C. Russell and J.W. Almond, eds., Cambridge University Press (1987).

16. D.J. McGeoch, A. Dolan, S. Donald, and F.J. Rixon, Sequence determination and genetic content of the short unique region in the genome of herpes simplex virus type 1, *J. Mol. Biol.* 181:1 (1985).

17. D.J. McGeoch et al., The complete DNA sequence of the long unique region in the genome of herpes simplex virus type 1, *J. Gen. Virol.* 69:1531 (1988).

18. A.O. Fuller and P.G. Spear, Anti-glycoprotein D antibodies that permit adsorption but block infection by herpes simplex virus 1 prevent virion-cell fusion at the cell surface, *Proc. Natl. Acad. Sci. USA* 84:5454 (1987).

19. S.L. Highlander, S.L. Sutherland, P.J. Gage, D.C. Johnson, M. Levine, and J.C. Glorioso, Neutralizing monoclonal antibodies specific for herpes simplex virus glycoprotein D inhibit virus penetration, *J. Virol.* 61:3356 (1987).

20. D.C. Johnson and M.W. Ligas, Herpes simplex viruses lacking glycoprotein D are unable to inhibit virus penetration: quantitative evidence for virus-specific cell surface receptors, *J. Virol.* 62:4605 (1988).

21. M.W. Ligas and D.C. Johnson, A herpes simplex virus mutant in which glycoprotein D sequences are replaced by β-galactosidase sequences binds to but is unable to penetrate into cells, *J. Virol.* 62:1486 (1988).

22. A.J. Forrester, V. Sullivan, A. Simmons, B.A. Blacklaws, G.L. Smith, A.A. Nash, and A.C. Minson, Induction of protective immunity with antibody to herpes simplex virus type 1 glycoprotein H (gH) and analysis of the immune response to gH expressed in recombinant vaccinia virus, *J. Gen. Virol.* 72:369 (1991).

23. A.C. Minson, T.C. Hodgman, P. Digard, D.C. Hancock, S.E. Bell, and E.A. Buckmaster, An analysis of the biological properties of monoclonal antibodies against glycoprotein D of herpes simplex virus and identification of amino acid substitutions that confer resistance to neutralization, *J. Gen. Virol.* 67:1001 (1986).

24. A.G. Noble, G.T.-Y. Lee, R. Sprague, M.L. Parish, and P.G. Spear, Anti-gD monoclonal antibodies inhibit cell fusion induced by herpes simplex virus type 1, *Virology* 129:218 (1983).

25. K.M. Izumi and J.G. Stevens, Molecular and biological characterization of a herpes simplex virus type 1 (HSV-1) neuroinvasiveness gene, *J. Exp. Med.* 172:487 (1990).

26. G. Cohen, M. Ponce de Leon, and C. Nichols, Isolation of a herpes simplex virus-specific antigenic fraction which stimulates the production of neutralizing antibody, *J. Virol.* 10:1021 (1972).

27. G.H. Cohen, M. Katze, C. Hydrean-Stern, and R.J. Eisenberg, Type-common CP-1 antigen of herpes simplex virus is associated with a 59,000-molecular-weight envelope glycoprotein, *J. Virol.* 27:172 (1978).

28. R.J. Eisenberg, M. Ponce de Leon, L. Pereira, D. Long, and G.H. Cohen, Purification of glycoprotein gD of herpes simplex virus types 1 and 2 by use of monoclonal antibody, *J. Virol.* 41:1099 (1982).

29. W.-L. Chan, Protective immunization of mice with specific HSV-1 glycoproteins, *Immunology* 49:343 (1983).

30. R.J. Eisenberg et al., Synthetic glycoprotein D-related peptides protect mice against herpes simplex virus challenge, *J. Virol.* 56:1014 (1985).

31. D. Long, T.J. Madara, M. Ponce de Leon, G.H. Cohen, P.C. Montgomery, and R.J. Eisenberg, Glycoprotein D protects mice against lethal challenge with herpes simplex virus types 1 and 2, *Infect. Immun.* 37:761 (1984).

32. E. Paoletti, B.R. Lipinskas, C. Samsonoff, S. Mercer, and D. Panicali, Construction of live vaccines using genetically engineered poxviruses: biological activity of vaccinia virus recombinants expressing the hepatitis B virus surface antigen and the herpes simplex virus glycoprotein D, *Proc. Natl. Acad. Sci. USA* 81:193 (1984).

33. R.D. Dix and J. Mills, Acute and latent herpes simplex virus neurological disease in mice immunized with purified virus-specific glycoproteins gB or gD, *J. Med. Virol.* 17:9 (1985).

34. L.A. Lasky, D. Dowbenko, C.C. Simonsen, and P.W. Berman, Protection of mice from lethal herpes simplex virus infection by vaccination with a secreted form of cloned glycoprotein D, *Bio/Technology* 2:527 (1984).

35. J.F. Rooney, C. Wohlenberg, K.J. Cremer, and A.L. Notkins, Immunized mice challenged with herpes simplex virus by the intranasal route show protection against latent infection, *J. Infect. Dis.* 159:974 (1989).

36. R.D. Dix, L. Pereira, and J.R. Baringer, Use of monoclonal antibody directed against herpes simplex virus glycoproteins to protect mice against acute virus-induced neurological disease, *Infect. Immun.* 34:192 (1981).

37. K.J. Cremer, M. Macket, C. Wohlenberg, A.L. Notkins, and B. Moss, Vaccinia virus recombinant expressing herpes simplex virus type 1 glycoprotein D prevents latent herpes in mice, *Science* 228:737 (1985).

38. L.R. Stanberry, D.I. Bernstein, R.L. Burke, C. Pachl, and M.G. Myers, Vaccination with recombinant herpes simplex virus glycoproteins: protection against initial and recurrent genital herpes, *J. Infect. Dis.* 155:914 (1987).

39. R.J. Eisenberg, D. Long, R. Hogue-Angeletti, and G.H. Cohen, Amino-terminal sequence of glycoprotein D of herpes simplex virus types 1 and 2, *J. Virol.* 49:265 (1984).

40. R.J. Watson, J.H. Weis, J.S. Salstrom, and L.W. Enquist, Herpes simplex virus type-1 glycoprotein D gene: nucleotide sequence and expression in *Escherichia coli*, *Science* 218:381 (1982).

41. L.A. Lasky and D.J. Dowbenko, DNA sequence analysis of the type-common glycoprotein D genes of herpes simplex virus types 1 and 2, *DNA* 3:23 (1984).

42. R.J. Watson, DNA sequence of the herpes simplex virus type 2 glycoprotein D gene, *Gene* 26:307 (1983).

43. G. Cohen, D. Long, J. Matthews, M. May, and R. Eisenberg, Glycopeptides of the type-common glycoprotein gD of herpes simplex virus types 1 and 2, *J. Virol.* 46: 679 (1983).

44. F. Serafini-Cessi, F. Dall'Olio, N. Malagolini, L. Pereira, and G. Campadelli-Fiume, Comparative study on O-linked oligosaccharides of glycoprotein D of herpes simplex virus types 1 and 2, *J. Gen. Virol.* 69:869 (1988).

45. E.A. Petrovskis, J.G. Timmins, M.A. Armentrout, C.C. Marchioli, R.J.J. Yancey, and L.E. Post, DNA sequence of the gene for pseudorabies virus gp50, a glycoprotein without N-linked glycosylation, *J. Virol.* 59:216 (1986).

46. S.K. Tikoo, D.R. Fitzpatrick, L.A. Babiuk, and T.J. Zamb, Molecular cloning, sequencing, and expression of functional bovine herpesvirus 1 glycoprotein gIV in transfected bovine cells, *J. Virol.* 64:5132 (1990).

47. C.C. Flowers, E.M. Eastman, and D.J. O'Callaghan, Sequence analysis of a glycoprotein D gene homolog within the unique short segment of the EHV-1 genome, *Virology* 180:175 (1991).

48. L.J.N. Ross, M.M. Binns, and J. Pastorek, DNA sequence and organization of genes in a 5.5 kbp EcoRI fragment mapping in the short unique segment of Marek's disease virus (strain RB1B), *J. Gen. Virol.* 72:949 (1991).

49. L.J.N. Ross and M.M. Binns, Properties and evolutionary relationships of the Marek's disease virus homologues of protein kinase, glycoprotein D and glycoprotein I of herpes simplex virus, *J. Gen. Virol.* 72:939 (1991).

50. R.J. Eisenberg, D. Long, L. Pereira, B. Hampar, M. Zweig, and G.H. Cohen, Effect of monoclonal antibodies on limited proteolysis of native glycoprotein gD of herpes simplex virus type 1, *J. Virol.* 41:478 (1982).

51. R.J. Eisenberg et al., Localization of epitopes of herpes simplex virus type 1 glycoprotein D, *J. Virol.* 53:634 (1985).

52. V.J. Isola, R.J. Eisenberg, G.R. Siebert, C.J. Heilman, W.C. Wilcox, and G.H. Cohen, Fine mapping of antigenic site II of herpes simplex virus glycoprotein D, *J. Virol.* 63:2325 (1989).

53. M.I. Muggeridge et al., Antigenic analysis of a major neutralization site of herpes simplex virus glycoprotein D, using deletion mutants and monoclonal antibody-resistant mutants, *J. Virol.* 62:3274 (1988).

54. G.H. Cohen et al., Localization and synthesis of an antigenic determinant of herpes simplex virus glycoprotein D that stimulates production of neutralizing antibody, *J. Virol.* 49:102 (1984).

55. B. Dietzschold, R.J. Eisenberg, M. Ponce de Leon, E. Golub, F. Hudecz, A. Varrichio, and G.H. Cohen, Fine structure analysis of type-specific and type-common antigenic sites of herpes simplex virus glycoprotein D, *J. Virol.* 52:431 (1984).

56. J.T. Matthews, G.H. Cohen, and R.J. Eisenberg, Synthesis and processing of glycoprotein D of herpes simplex virus types 1 and 2 in an in vitro system, *J. Virol.* 48:521 (1983).

57. G.H. Cohen, V.J. Isola, J. Kuhns, P.W. Berman, and R.J. Eisenberg, Localization of discontinuous epitopes of herpes simplex virus glycoprotein D: use of a nondenaturing ("native" gel) system of polyacrylamide gel electrophoresis coupled with Western blotting, *J. Virol.* 60:157 (1986).

58. A.O. Fuller and P.G. Spear, Specificities of monoclonal and polyclonal antibodies that inhibit adsorption of herpes simplex virus to cells and lack of inhibition by potent neutralizing antibodies, *J. Virol.* 55:475 (1985).

59. G. Kümel, H.C. Kaerner, M. Levine, C.H. Schröder, and J.C. Glorioso, Passive immune protection by herpes simplex virus-specific monoclonal antibodies and monoclonal antibody-resistant mutants altered in pathogenicity, *J. Virol.* 56:930 (1985).

60. L. Pereira, T. Klassen, and J.R. Baringer, Type-common and type-specific monoclonal antibody to herpes simplex virus type 1, *Infect. Immun.* 29:724 (1980).

61. S.D. Showalter, M. Zweig, and B. Hampar, Monoclonal antibodies to herpes simplex virus type 1 proteins, including the immediate-early protein ICP 4, *Infect. Immun.* 34:684 (1981).

62. M.I. Muggeridge, T.-T. Wu, D.C. Johnson, J.C. Glorioso, R.J. Eisenberg, and G.H. Cohen, Antigenic and functional analysis of a neutralization site of HSV-1 glycoprotein D, *Virology* 174:375 (1990).

63. M.I. Muggeridge, W.C. Wilcox, G.H. Cohen, and R.J. Eisenberg, Identification of a site on herpes simplex virus type 1 glycoprotein D that is essential for infectivity, *J. Virol.* 64:3617 (1990).

64. D.L. Sodora, G.H. Cohen, and R.J. Eisenberg, Influence of asparagine-linked oligosaccharides on antigenicity, processing, and cell surface expression of herpes simplex virus type 1 glycoprotein D, *J. Virol.* 63:5184 (1989).

65. D.L. Sodora, G.H. Cohen, M.I. Muggeridge, and R.J. Eisenberg, Absence of asparagine-linked oligosaccharides from glycoprotein D of herpes simplex virus type 1 results in a structurally altered but biologically active protein, *J. Virol.* 65:4424 (1991).

66. D.L. Sodora, R.J. Eisenberg, and G.H. Cohen, Characterization of a recombinant herpes simplex virus which expresses a glycoprotein D lacking asparagine-linked oligosaccharides, *J. Virol.* 65:4432 (1991).

67. G.H. Cohen, W.C. Wilcox, D.L. Sodora, D. Long, J.Z. Levin, and R.J. Eisenberg, Expression of herpes simplex virus type 1 glycoprotein D deletion mutants in mammalian cells, *J. Virol.* 62:1932 (1988).

68. V. Feenstra, M. Hodaie, and D.C. Johnson, Deletions in herpes simplex virus glycoprotein D define nonessential and essential domains, *J. Virol.* 64:2096 (1990).

69. W.C. Wilcox, D. Long, D.L. Sodora, R.J. Eisenberg, and G.H. Cohen, The contribution of cysteine residues to antigenicity and extent of processing of herpes simplex virus type 1 glycoprotein D, *J. Virol.* 62:1941 (1988).

70. D. Long, G.H. Cohen, M.I. Muggeridge, and R.J. Eisenberg, Cysteine mutants of herpes simplex virus type 1 glycoprotein D exhibit temperature-sensitive properties in structure and function, *J. Virol.* 64:5542 (1990).

MOLECULAR, IMMUNOLOGICAL AND FUNCTIONAL CHARACTERIZATION OF THE MAJOR SURFACE ADHESIN OF *STREPTOCOCCUS MUTANS*

A.S. Bleiweis, P.C.F. Oyston, and L.J. Brady

University of Florida
Gainesville, FL 32610

INTRODUCTION

At a conference held at the National Institutes of Health in 1976 to discuss aspects of caries vaccine development, we defined the ideal immunogen: non-toxic, inductive of protective antibodies that are specific for *Streptococcus mutans*, able to induce mainly salivary antibodies and noncross-reactive with human tissues (Bleiweis and Chiu, 1976). Potential immunogens recognized at that time included: group-specific cell wall polysaccharides of *S. mutans*; glucosyltransferases, and lipoteichoic acid (Bleiweis and Chiu, 1976). Lipoteichoic acid was dismissed from consideration as it was found to be a heterophilic antigen present in most Gram-positive microorganisms (Wicken and Knox, 1981). The serotype-specific wall polysaccharides have been well characterized as to molecular structure and compose a significant part of the outer wall of many *S. mutans* strains. Unfortunately, they have been found to be non-immunogenic in test animals in their purified states. [For a review of this class of polysaccharides see Hamada and Slade, 1980.] Glucosyltransferases, however, still remain viable as candidate immunogens and will be discussed later in this workshop.

Since that NIH conference, several oral streptococci including *S. mutans*, *S. sobrinus*, *S. gordonii* (formerly *S. sanguis*) and others have been found to possess a major surface protein of $M_r \sim 185,000\text{-}200,000$, believed to play a key role in adhesion to dental or other oral surfaces. First described by Russell and Lehner (1978) as a $M_r \sim 185,000$ moiety that is cleaved by trypsin to two distinct antigens (I and II), the protein was designated surface antigen (SA) I/II. Analogous proteins have since been identified on a variety of strains from a range of serotypes, as shown in Table 1. In this paper we will refer to the protein as P1.

With a view to developing an effective, safe caries vaccine based on this major surface protein, we undertook a characterization of P1. This paper will give an overview of the work we and others have done so far on the immunological, molecular and functional aspects of this *S. mutans* protein.

Genetically Engineered Vaccines, Edited by
J.E. Ciardi *et al.*, Plenum Press, New York, 1992

Table 1. P1-Like proteins of the streptococci.

Protein	Gene	Species	Serotype	Strain	Reference
P1		*S. mutans*	c	Ingbritt	Forester et al., 1983
	spaP	*S. mutans*	c	NG5	Kelly et al., 1989
I/II		*S. mutans*	c	Ingbritt	Russell & Lehner, 1978
Antigen B		*S. mutans*	c	Ingbritt	Russell, 1979
IF		*S. mutans*	c	Ingbritt	Hughes et al., 1980
PAc	*pac*	*S. mutans*	c	MT8148	Okahashi et al., 1989a
					Okahashi et al., 1989b
MSL-1	*msl-1*	*S. mutans*	c	KPSK2	Demuth et al., 1990b
SpaA	*spaA*	*S. sobrinus*	d	B13N	Abiko et al., 1989
		S. sobrinus	g	6715	Holt et al., 1982
					LaPolla et al., 1991
PAg	*pag*	*S. sobrinus*	g	6715	Okahashi et al., 1986
				MT3791	Tokuda et al., 1991
SR	*sr*	*S. mutans*	f	OMZ175	Ackermans et al., 1985
					Ogier et al., 1990
SSP-5	*ssp-5*	*S. gordonii* (formerly *sanguis*)		M5	Demuth et al., 1988
					Demuth et al., 1990a

IMMUNOLOGICAL CONSIDERATIONS

Since antigen P1 was identified as a major surface antigen of *S. mutans* there has been much attention paid to it as a potential vaccine. Several groups have found antigen P1 to be an effective caries vaccine in monkeys (Lehner et al., 1981) and mice (Czerkinsky et al., 1989; Iwaki et al., 1990; Takahashi et al., 1990), and anti-P1 antibodies have been shown to protect against colonization by *S. mutans* in passive immunization studies in rats (Otake et al., 1991), monkeys (Lehner et al., 1985) and humans (Ma et al., 1987; Ma and Lehner, 1990). Protection against caries in primates following active immunization was associated with the production of IgG antibodies in serum and gingival crevicular fluid, the titer of which was maintained at a high level over many weeks following only a single round of immunization (Lehner et al., 1981). This suggested that P1 could be an effective caries vaccine. However, before such a vaccine could be developed the problem of apparent heart cross-reactivity had first to be addressed.

In 1976 we reported that antibodies from rabbits hyperimmunized with whole cells of several mutans streptococcal serotypes cross-reacted with epitopes in human myocardial tissue (van de Rijn et al., 1976). Several investigators, in confirming our findings, subsequently reported that a cell wall-localized protein, possibly P1, was responsible for inducing heart cross-reactive antibodies (reviewed by Russell and Wu, 1990). However, van de Rijn et al. (1976) had observed that *S. rattus* could elicit heart cross-reactive antibodies even though it was found to lack antigen P1 (Russell, 1980; Ayakawa et al., 1987). Therefore, our laboratory produced a panel of monoclonal antibodies (MAbs) raised against P1 in an attempt to resolve the issue. None of the MAbs cross-reacted with heart tissue (Ayakawa et al., 1987), confirming the observation of Smith et al. (1984) who also failed to detect cross-reactivity with their monoclonals. Also, none of the polyclonal rabbit antisera we prepared against native or recombinant P1 displayed heart cross-reactivity (Bleiweis et al., 1990). Indeed, rather than sharing epitopes with heart tissue, Ogier et al. (1990) found the SR protein of *S. mutans* OMZ175 to cross-react with epitopes on IgG molecules.   As heart preparations contain significant amounts of serum proteins (Russell and Wu, 1990) including IgG, the previously described cross-reactivity of P1 may be due to epitopes shared with IgG rather than heart tissue.   In addition we have identified a group of membrane-localized proteins that were cross-reactive with anti-human heart serum (Ayakawa et al., 1985).   These antigens were present in *S. mutans, S. sanguis* and *S. pyogenes* and possessed myosin-like epitopes (Ayakawa et al., 1988).   Collectively, these observations have cast doubt on the claims that antigen P1 is a tissue cross-reactive antigen.

Proteins antigenically related to P1 were found on the surface of *S. mutans* serotypes c, e and f, *S. sobrinus* serotypes d and g and *S. cricetus* serotype a when probed with monospecific antiserum (Moro and Russell, 1983) or with MAbs produced in our laboratory (Ayakawa et al., 1987) and by Smith and Lehner (1989). These streptococcal surface proteins are closely related, both immunologically and biochemically (Forester et al., 1983; Holt et al., 1982; Ogier et al., 1990), and a P1-like protein, SSP-5, is also expressed by *S. gordonii*   (formerly *S. sanguis*) (Demuth et al, 1988). In fact, we have found that although overlapping, but not identical, subsets of anti-P1 MAbs are cross-reactive with SpaA and SSP-5 by Western immunoblot analysis and with *S. sobrinus* and *S. gordonii* whole cells by radioimmunoassay, in general, the degree of cross-reactivity of these MAbs with SSP-5 appears to be higher than that with SpaA (Brady et al., 1991b). This occurs despite the fact that the P1 and SSP-5 deduced amino acid sequences are 56% homologous while the P1 and SpaA deduced sequences demonstrate 66% homology (LaPolla et al., 1991).   Therefore, in designing a vaccine based on P1 to prevent colonization by *S. mutans*, care must be taken not to unbalance the oral ecology by eliminating noncariogenic species, such as *S. gordonii*.

Anti-P1 antisera and MAbs have been used to detect significant quantities of P1 in the culture supernatant of some *S. mutans* strains, suggesting that in these strains P1 was being shed from the bacterial surface, while in other strains the protein was not predominantly released from the surface (Smith et al., 1984). Electron microscopy of immunogold-labelled bacteria has shown the P1 protein to be associated with the layer of peritrichous fibrils surrounding the cell in "retainer" strains, whereas "nonretainer" strains released P1 into the culture supernatant and did not possess a layer of fibrils on the cell surface (Ayakawa et al., 1987), as shown in Figure 1.   The reason for some strains being nonretainers is not known, nor has it yet been established how common this phenomenon is *in vivo*.   Obviously, a vaccine utilizing this antigen may not protect against colonization by cariogenic nonretainer strains that may be able to adhere to surfaces within the oral cavity by other mechanisms.

MOLECULAR CHARACTERIZATION

The genes encoding P1-like proteins have been cloned and sequenced from a number of strains and species, as summarized in Table 1.   Our laboratory cloned and sequenced the *spaP* gene from the *S. mutans* serotype c strain NG5 into *E. coli* JM109 by a shotgun method using pUC18 as the vector (Lee et al., 1988) to give the clone *E. coli* SM2949. The gene was contained in a 5.2 kb *HindIII* fragment. The expression of P1 was independent of the *lac* inducer and orientation of the insert indicating that the fragment contained its own promoter.   Ouchterlony double-diffusion tests revealed that the cloned P1 product and the native P1 were almost butnot entirely identical. This partial identity occurred because the 5.2 kb fragment did not contain the whole *spaP* sequence. As we later discovered, the clone only contained 4,019 of the 4,782 bp which comprise the *spaP* gene, and some 763 bp from the 3' terminus of the gene had not been included in the clone (Kelly et al., 1989). Subsequent collaboration between our laboratory and that of Dr. Lehner has produced the 3' sequence by PCR and this has been ligated to the pSM2949 sequence to give a full clone, *E. coli* SMI/II (unpublished data).

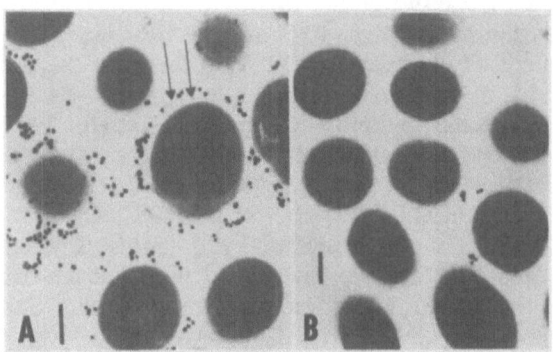

Figure 1. Immunoelectron microscopy of thin sections of a retainer, *S. mutans* Ingbritt 175 (A), and a nonretainer strain, Ingbritt 162 (B), reacted with MAbs raised against P1. Note the association of the gold label with the layer of peritrichous fibrils on the retainer strain. Bars, 0.2 μm. From Ayakawa et al. (1987).

Southern hybridization studies in our laboratory detected DNA sequences highly homologous to *spaP* in *S. mutans* serotypes c, e and f, *S. sobrinus* serotype d and *S. cricetus* serotype a (Lee et al., 1988). Using limited probes generated by PCR, Ma et al. (1991) detected hybridization of regions of *spaP* with DNA from both mutans streptococci and non-mutans alpha-haemolytic streptococci, indicating that some regions of the gene are highly conserved and P1 may be part of a family of related proteins. The genes encoding P1-like proteins from several strains and species of oral streptococci have been sequenced. Comparison of the deduced amino acid sequences revealed considerable but varying degrees of homology between proteins, with many of the amino acid substitutions being conservative (LaPolla et al., 1991). Irrespective of the degree of primary sequence homology, the predicted proteins all possessed highly conserved common structural features, as illustrated in Figure 2. All the proteins were preceded by an amino-terminal leader sequence which would be removed by post-translational processing. The signal sequences of both *S. mutans* *spaP* and *pac* gene products were 38 residues in length (Kelly et al., 1989; Okahashi et al., 1989b), whereas that of the *S. sobrinus* *spaA* gene product was 50 residues (LaPolla et al., 1991). The amino-terminal third of each protein was shown to include an alanine-rich (A) region, comprising three 82-residue tandem repeats encompassing residues 219 to 464 of P1 (Kelly et al., 1989). The central portion of each molecule was shown to include a proline-rich (P) region comprising three 39-residue tandem repeats, from residue 847 to 963 of P1. The P-region of the SpaA protein of *S. sobrinus* serotype g strain 6715 only possessed two tandem repeats (LaPolla et al., 1991); however, there is some strain-to-strain variation in SpaA, as the protein of the serotype g strain MT3791 possessed a typical P-region

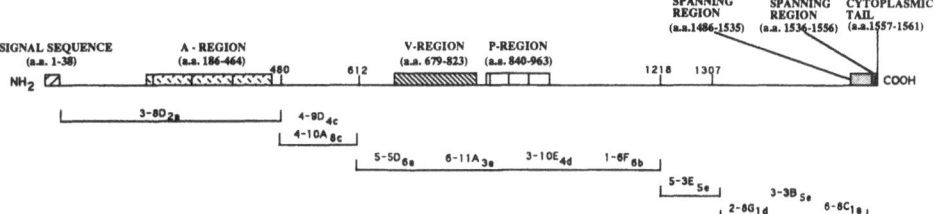

Figure 2. Schematic representation of the P1 molecule from *S. mutans* serotype c strain NG5. Numbers refer to the amino acid predictions deduced from the nucleotide sequence of the cloned *spaP* gene (Kelly et al., 1989). The A-region and P-region designations identify the alanine-rich and proline-rich repeats respectively. The V-region indicates the variable region identified by restriction fragment length polymorphism analysis of *spaP* (Brady et al., 1991a). The bars below the map indicate the approximate binding domains of the anti-P1 MAbs, deduced from Western immunoblotting experiments (Brady et al., 1991b). The order of the antibodies within each section is arbitrary and does not reflect the location of binding on the P1 molecule.

with three tandem repeats (Tokuda et al., 1991). Variable numbers of tandem repeats have been observed for other streptococcal proteins, for example group A streptococcal M protein (Fischetti, 1989) and streptococcal immunoglobulin-binding proteins (Fahnestock et al., 1990). The carboxy-termini of P1 and related proteins were all shown to possess a proline-rich sequence, similar to that described for other streptococcal proteins. This structure is believed to span the cell wall (Fischetti, 1989). Lastly was found a hydrophobic transmembrane domain. A charged cytoplasmic tail was observed for the proteins from serotype c isolates (Kelly et al., 1989; Okahashi et al., 1989b).

When the *spaP* and *pac* genes, both derived from serotype c strains, were compared a total of 36 predicted amino acid substitutions were found (Kelly et al., 1989). However, rather than being scattered, 20 of these substitutions were clustered within a variable (V) region under 150 amino acids long encompassing residues 679 to 823 of P1 and 679 to 827 of PAc. Only two amino acid changes occurred within the A-region and one within the P-region. The existence of this variable region among serotype c isolates was confirmed by a restriction fragment length polymorphism (RFLP) study undertaken in our laboratory in collaboration with Dr. Lehner's group (Brady et al., 1991a). Two families of V-region were identified after sequencing PCR-generated DNA of this region from a number of strains. Isolates were either *pac*-like or *spaP*-like. The significance of these two V-regions is not known, but their occurrence was found to be independent of geographic origin of the streptococcal strains. We have been unable to identify any immunological differences between P1 expressed by *pac*-like and *spaP*-like strains using our panel of anti-P1 MAbs (Brady et al., 1991b).

Secondary structure predictions have been performed on the predicted amino acid sequences of P1 (Figure 3), SSP-5, PAc and SpaA (Kelly et al., 1989; Demuth et al., 1990a; Okahashi et al., 1989b; LaPolla et al., 1991). The A-region contains a high proportion of alanine residues and other amino acids arranged in a heptad periodicity which would be predicted to form an alpha-helical structure typical of coiled-coil proteins. This is a stable structure and one found commonly in fibrous and surface proteins (Cohen and Parry, 1986). It has also been predicted for such

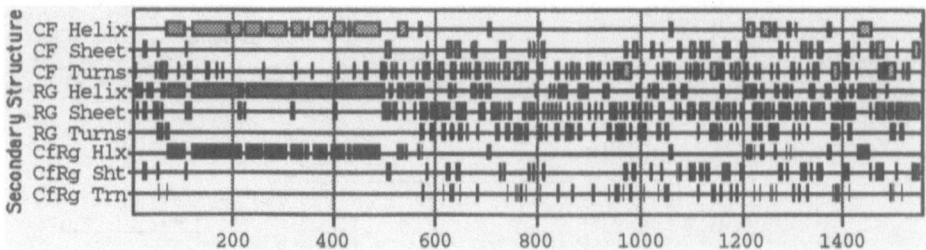

Figure 3. Secondary structure prediction of the P1 protein deduced from the sequence of the cloned *spaP* gene (Kelly et al., 1989) using the IBI MacVector software package. Predictions of regions likely to form an α helix, ß sheet or ß turn are based on the algorithms of Chou-Fasman (CF) (1978) and Robson-Garnier (RG) (Garnier et al., 1978) and also on a composite of these two methods (CfRg).

streptococcal proteins as the M protein of group A streptococci (Fischetti, 1989) and the PspA protein of *S. pneumoniae* (Talkington et al., 1991). In the P-region, the relative abundance of proline residues would preclude compact folding and would probably result in an extended, pleated structure with many beta turns. The proteins have a conserved C-terminal sequence found in all the P1-like proteins except SSP-5 (Demuth et al., 1990a). The consensus sequence, L-P-X-T-G, is common to a number of streptococcal wall proteins and was first identified for group A streptococcal M protein (Fischetti, 1989). This sequence is believed to represent a signal for a thiol-dependent membrane anchor-cleaving enzyme (MACE) which post-translationally modifies proteins allowing subsequent attachment to a putative membrane anchor protein (Pancholi and Fischetti, 1989).

This laboratory has become interested in why some strains do not retain P1 on the bacterial cell surface. Nonretention does not appear to be due to loss of any carboxy-terminal sequence encoding anchoring regions of the protein as four oligonucleotide probes complementary to regions within the 3' terminus of *spaP* hybridized with the gene from nonretainers (Brady et al., 1991a). Also, when the 3' sequence of the retainer strain MT8148 *pac* (Okahashi et al., 1989b) and the nonretainer NG5 *spaP* (Kelly et al., 1989) genes were compared they were found to be identical. Another nonretainer, strain GS5, possessed the 3' sequence as shown by the oligonucleotide probes, but only produced a truncated protein of $M_r \sim 155,000$ compared to the $M_r \sim 185,000$ protein released by NG5. When the *pac* gene encoding the $M_r \sim 185,000$ product was inserted into GS5 the intact protein was found on the cell surface (Koga et al., 1990). This would suggest that it is the gene itself which is different in GS5 and that the protein is not truncated by post-transcriptional or post-translational modifications. Although this would suggest that nonretention by GS5 may be the result of a point mutation resulting in premature termination, sequencing of the *spaP* gene from GS5 did not identify any such premature stop codons (Crowley and Bleiweis, unpublished observation). Therefore, a mutation elsewhere in the gene may be responsible for the production of the truncated protein. This has been shown to be the case for streptokinase where correct post-translational carboxy-terminal cleavage is dependent on an intact leader sequence (Park et al., 1991). However, a different mechanism than that responsible for the nonretention of P1 from strain GS5 must be involved for nonretainers such as NG5 and Ingbritt 162, which release full-length P1. In these cases the mutation may have occurred outside the *spaP* gene, for example in the putative MACE or membrane anchor proteins.

To study the *spaP* gene further, this laboratory produced mutants defective in the expression of P1 (Lee et al., 1989). The *spaP* gene, encoding the P1 protein which had been cloned into pUC18, was insertionally inactivated with the plasmid pVA981, which carried a tetracycline resistance marker (Lee et al., 1989). This gave a plasmid of 11.7 kb in which the ampicillin gene was rendered defective. The linearized plasmid was introduced into the serotype c parent strain NG8 by electroporation, where homologous recombination integrated the defective gene into the chromosome at the *spaP* locus. The transformants were able to grow in the presence of tetracycline but not ampicillin. These P1-deficient mutants were designated 807, 858 and 834, and strain 834 was used in functional studies as shall be discussed later. Western immunoblotting using polyclonal antiserum was unable to detect P1 in mutanolysin extracts of strain 834 and the mutant no longer expressed fibrils on the bacterial cell surface (Lee et al., 1989). However, a truncated polypeptide of 612 amino acids is expressed by this mutant, which represents the P1 molecule up to the site of insertional inactivation (Brady et al., 1991a).

FUNCTIONAL CONSIDERATIONS

P1 and related molecules are believed to function as adhesins enabling organisms which express them on their surfaces to interact with components of the salivary pellicle coating oral surfaces (Lee et al., 1989; Okahashi et al., 1989a; Koga et al., 1990; Brady et al., 1991c) as well as with salivary components coating other plaque microorganisms (Lamont, et al. 1991). Their exact biological function(s), however, remain somewhat unclear. Experimentally, it has been shown that retainer strains of *S. mutans* serotype c will adhere to salivary agglutinin-coated hydroxyapatite (HA) beads (Lee et al., 1989; Brady et al., 1991c) and saliva-coated HA (Koga et al., 1990) in a calcium-dependent fashion. We have shown also that P1 is directly responsible for adherence of these bacteria to agglutinin-coated HA since pure P1 (both recombinant and native) acts as an efficient fluid-phase competitive inhibitor of this phenomenon. The agglutinin is a high molecular weight $(M_r \sim 300,000)$ glycoprotein which can be purified from human saliva (Ericson and Rundegren, 1983) and identified using specific monoclonal antibodies (Takano et al., 1991). Antigen SR appears to be similarly involved in the adherence of *S. mutans* serotype f strains. Ogier, et al. (1990) report that 38% of the binding of salivary glycoproteins to a serotype f strain (OMZ175) is due to this protein. Less is known about the contribution of the SpaA protein of *S. sobrinus* strains to their ability to adhere to components of the salivary pellicle. We have found substantial variation in the ability of *S. sobrinus* serotype g isolates to adhere to agglutinin-coated HA beads despite similar levels of expression of cell surface SpaA (Brady et al., 1991c). We have also found that polyclonal anti-P1 rabbit antiserum is a much more efficient inhibitor of *S. mutans* serotype c adherence than is polyclonal anti-SpaA rabbit antiserum (Figure 4), even though these two cross-reactive antisera demonstrate similar Western immunoblot and ELISA titers against both P1 and SpaA.

In addition to adherence, another phenomenon which appears to involve the interaction of these bacterial cell surface proteins with salivary components is aggregation. Aggregation can be monitored by measuring the decrease in optical density (700 nm) over time of a bacterial suspension in the presence of saliva or salivary components. Formation of bacterial aggregates, either single species or co-aggregates, would represent a non-immune mechanism of clearance of organisms from the oral cavity as the aggregates are physically removed by swallowing.

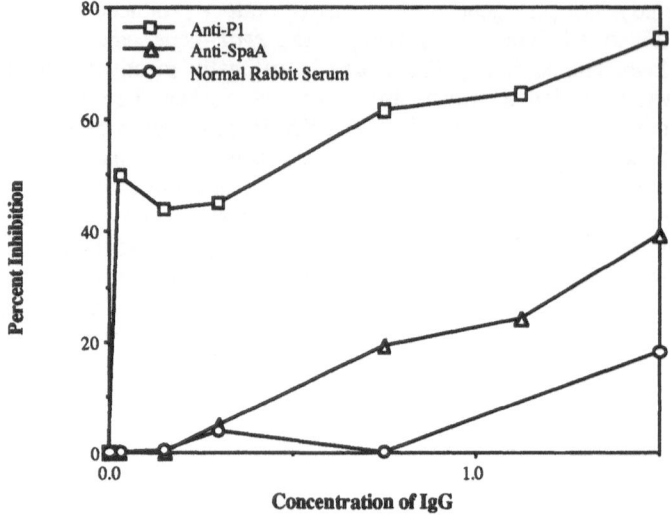

Figure 4. Inhibition of adherence of *Streptococcus mutans* serotype c P1 retainer strain NG8 to agglutinin coated-hydroxyapatite beads by polyclonal antisera. Immunoglobulin was purified by affinity chromatography using a column of immobilized protein A and the adherence inhibition was performed as described by Brady et al. (1991c).

Demuth et al. (1989) have shown that transformation of a nonaggregating *Enterococcus faecalis* strain with the *SSP-5* gene from the *S. gordonii* strain M5 (formerly *S. sanguis*) conferred an aggregation-positive phenotype in the presence of salivary agglutinin. Like adherence, the aggregation was also calcium-dependent. P1 retainer strains of *S. mutans* serotype c have also been reported to aggregate in the presence of whole saliva (Lee et al., 1989; Koga et al., 1990) and purified agglutinin (Demuth et al., 1990a; Brady et al., 1991c). Demuth et al. (1990b) reported that agglutinin-mediated aggregation of *S. gordonii* was inhibitable by sialic acid, whereas that of *S. mutans* serotype *c* was not, suggesting that the binding sites on agglutinin which interact with P1 and SSP-5 may differ.

The interaction of P1 with salivary components appears to involve hydrophobic interactions. P1-deficient mutants and P1 nonretainer strains are more hydophilic than are wild-type retainer strains (Lee et al., 1989; Okahashi et al., 1989a; Ohta et al., 1989; Koga et al., 1990). Lee et al. (1989) demonstrated that the hydrophilic P1-deficient mutants were unable to adhere to agglutinin-coated HA, unlike the parent NG8 strain, and that agglutinin-mediated aggregation was significantly diminished also. Koga et al. (1990) reported that transformation of the *pac* gene into the nonretainer hydrophilic strain GS5, which resulted in expression of PAc on the cell surface, also resulted in an increase in cell surface hydrophobicity and conferred the ability to adhere to saliva-coated HA and to aggregate in the presence of whole saliva.

Recent research in our laboratory has involved the use of our anti-P1 monoclonal antibodies to understand the interaction of P1 with agglutinin at the molecular level. The approximate binding domains of these MAbs have been determined by Western immunoblotting of truncated and full-length P1 polypeptides (Brady et al., 1991b). These binding sites are illustrated in the context of the schematic diagram of P1 shown in Figure 2. It should be noted that at present it is unclear whether the three MAbs which map closest to the carboxy-terminus of P1 are actually binding directly within that region or are merely dependent on that segment of the protein for the formation of their cognate epitopes. All 11 anti-P1

MAbs were tested for their ability to inhibit adherence of the P1 retainer strain NG8 to agglutinin-coated HA and to inhibit the aggregation of the same strain in the presence of fluid-phase agglutinin (Brady et al., 1991c). The results of these experiments are summarized in Table 2. The MAbs are listed in descending order of their relative inhibitory activity. Several MAbs demonstrated marked differences in their ability to inhibit adherence versus aggregation. This suggests that the way in which P1 interacts with agglutinin when it is immobilized (adherence) is different from the way it interacts with fluid-phase agglutinin (aggregation). This is reminiscent of the phenomenon of "cryptitopes" involved in bacterial adhesion described by Gibbons (1989). Cryptitopes are hidden receptors for bacterial adhesins which may be generated due to conformational changes or enzymatic modifications of the receptor molecules. As can be seen by locating the binding domains of MAbs on Figure 2, those exhibiting moderate to strong inhibition of adherence or aggregation do not map to discrete regions based on the linear structure of P1. Since virtually nothing is known regarding the tertiary structure of P1 and related molecules, more work is necessary to understand whether the MAbs exert their inhibitory effects by direct interaction with a functional domain, by steric hindrance, or by inducing conformational changes which may affect regions of P1 other than their binding sites.

Table 2. Relative inhibitory activities of anti-P1 monoclonal antibodies.[1]

Degree of Inhibition	Anti-Adherence	Anti-Aggregation
Strong	$4\text{-}10A_{8c}$	$4\text{-}9D_{4c}$
	$1\text{-}6F_{6b}$	$6\text{-}8C_{1a}$
	$6\text{-}8C_{1a}$	$4\text{-}10A_{8c}$
	$4\text{-}9D_{4c}$	**$2\text{-}8G_{1d}$**
	$5\text{-}5D_{6a}$	**$3\text{-}8D_{2a}$**
Moderate	$5\text{-}3E_{5e}$	**$3\text{-}10E_{4d}$**
	$3\text{-}3B_{5e}$	$5\text{-}5D_{6a}$
Weak	**$3\text{-}8D_{2a}$**	$3\text{-}3B_{5e}$
	$6\text{-}11A_{3a}$	**$1\text{-}6F_{6b}$** [2]
		$5\text{-}3E_{5e}$
		$6\text{-}11A_{3a}$ [2]
None	$3\text{-}10E_{4d}$	$1\text{-}5F_{2a}$ [3]
	$2\text{-}8G_{1d}$	
	$1\text{-}5F_{2a}$ [3]	

[1] Information compiled from Brady et al. (1991c). Both anti-adherence and anti-aggregation assays were performed using the *S. mutans* serotype c P1 retainer strain, NG8, as the test organism. Antibodies are listed in descending order of inhibitory activity. Antibodies shown in bold print demonstrated substantial differences in anti-adherence versus anti-aggregation behavior.

[2] These antibodies demonstrated a concentration dependent effect in that inhibition was only observed at intermediate, but not high or low, concentrations.

[3] Anti-*Actinobacillus actinomycetemcomitans* negative control antibody.

One MAb, 3-8D$_{2a}$, reacts with the amino-terminal region of P1. This MAb and two others react with mutant 834 generated by insertional inactivation of *spaP* (Lee et al., 1989). By comparison of molecular and immunologic data, we have deduced that 834 produces a truncated polypeptide which corresponds to amino acid residues 1-612 of P1 (Brady et al., 1991b). We now know, by subcloning of PCR generated DNA, that MAb 3-8D$_{2a}$ reacts directly within the alanine-rich repeat region of P1 (Crowley et al., unpublished data). We have also shown that although 834 is completely unable to adhere to agglutinin-coated HA, it does demonstrate a limited, albeit significantly decreased, capacity to aggregate in the presence of fluid-phase agglutinin (Brady et al., 1991c). MAb 3-8D$_{2a}$ is a much better inhibitor of aggregation than of adherence (see Table 2). Therefore, the combined results of limited aggregation of 834 and the anti-aggregation inhibitory activity of 3-8D$_{2a}$ suggest that the A-region may be involved in aggregation but not adherence. The ability to differentiate, and selectively inhibit, the two related but distinct phenomena of adherence and aggregation would be clinically significant since adherence would result in colonization by the organism and be detrimental to the host, while aggregation would result in clearance of the organism and be beneficial to the host.

SUMMARY

In the 15 years since the last major NIH conference that dealt with anti-caries vaccines, we have learned much. Certainly, whole bacteria or bacterial fractions may not be proper immunogens due to the possibility of inducing tissue cross-reactivity. Our own experience (van de Rijn et al., 1976) illustrates that pitfall. But even in the era of genetically engineered vaccines, we first must understand the biological functions of our chosen immunogen before employing that pure protein in a vaccine. Our recent work (Brady et al., 1991c) indicates that antigen P1, a ubiquitous protein found on several oral streptococci, may possess different, but possibly overlapping, functional domains influencing reactions with fluid-phase salivary agglutinin (aggregation) versus fixed agglutinin (adherence). A proper vaccine would induce antibodies against the latter domain(s) thereby retarding colonization. An improper vaccine that induces antibodies against aggregation-related domains on P1 would lessen the host's ability to clear those bacteria from the oral cavity. After carefully identifying appropriate functional domains and obtaining sub-clones of the larger gene that yield truncated polypeptides typical of adherence-specific regions that are also immunogenic, we may be in a position to create the most effective vaccine. In studies employing the polymerase chain reaction (PCR) and standard cloning procedures, we have already begun to produce such polypeptides. Once a library of polypeptides is assembled, they may be tested for functional activity and for lack of induction of cross-reactivity with nonpathogenic streptococci (i.e., *S. gordonii*). Certain of these recombinant-specified polypeptides could serve as the basis for an anti-caries vaccine. Alternatively, peptides may be synthesized that resemble these sub-molecular regions for use in a vaccine or as competitive inhibitors of adherence but not aggregation. Clearly, a vaccine against dental caries remains a real possibility for the future.

ACKNOWLEDGEMENTS

Original research reviewed above was supported by Public Health Service grant no. R37 DE-08007 to A.S.B.

REFERENCES

Abiko, Y., Hayakawa, M., Alki H., Saito, S., and Takiguchi, H., 1989, Cloning of the gene for cell surface protein antigen A from *Streptococcus sobrinus* (serotype d), *Archs Oral Biol* 34 (7): 571.

Ackermans, F., Klein, J.P., Ogier, J.A., Bazin, H., Cormont, F., and Frank, R.M., 1985, Purification and characterization of a saliva-interacting cell wall protein from *Streptococcus mutans* serotype f by using monoclonal antibody immunoaffinity chromatography, *Biochem J* 228: 211.

Ayakawa, G.Y., Siegel, J.L., Crowley, P.J., and Bleiweis, A.S., 1985, Immunochemistry of the *Streptococcus mutans* BHT cell membrane: detection of determinants cross-reactive with human heart tissue, *Infect Immun* 48: 280.

Ayakawa, G.Y., Boushell, L.W., Crowley, P.J., Erdos, G.W., McArthur, W.P., and Bleiweis, A.S., 1987, Isolation and characterization of monoclonal antibodies specific for antigen P1, a major surface protein of mutans streptococci, *Infect Immun*, 55: 2759.

Ayakawa, G.Y., Bleiweis, A.S., Crowley, P.J., and Cunningham, M.W., 1988, Heart cross-reactive antigens of mutans streptococci share epitopes with group A streptococci and myosin, *J Immunol* 140: 253.

Bleiweis, A.S. and Chiu, T., 1976, Candidate vaccines as immunogens for a caries vaccine, *in*: "Immunological Aspects of Dental Caries," W. Bowen, R.J. Genco, T.C. and O'Brien, eds., IRL Press, Washington D.C.

Bleiweis, A.S., Lee, S.F., Brady, L.J., Progulske-Fox, A., and Crowley, P.J., 1990, Cloning and inactivation of the gene responsible for a major surface antigen on *Streptococcus mutans*, *Arch Oral Biol* 35 (Suppl): 15S.

Brady, L.J., Crowley, P.J., Ma, J.K-C., Kelly, C., Lee, S., Lehner, T., and Bleiweis, A.S., 1991a, Restriction fragment length polymorphisms and sequence variation within the *spaP* gene of *Streptococcus mutans* serotype c isolates, *Infect Immun* 59: 1803.

Brady, L.J., Piacentini, D.A., Crowley, P.J., and Bleiweis, A.S., 1991b, Identification of monoclonal antibody-binding domains within antigen P1 of *Streptococcus mutans* and cross-reactivity with related surface antigens of oral streptococci, *Infect Immun* 59: 4425.

Brady, L.J., Piacentini, D.A., Crowley, P.J., Oyston, P.C.F., and Bleiweis, A.S., 1991c, The role of antigen P1 in salivary agglutinin-mediated adherence and aggregation of oral streptococci and differentiation by monoclonal antibodies. Submitted for publication.

Chou, P.Y. and Fasman, G.D., 1978, Prediction of the secondary structure of proteins from their amino acid sequence, *Adv Enzymol Relat Areas Mol Biol* 47: 45.

Cohen, C. and Parry, A.D., 1986, α-Helical coiled coils- a widespread motif in proteins, *Trends Biochem Sci* 6: 245.

Czerkinsky, C., Russell, M.W., Lycke, N., Lindbald, M., and Holmgren, J., 1989, Oral administration of a streptococcal antigen coupled to cholera toxin B subunit evokes strong antibody responses in salivary glands and extramucosal tissues, *Infect Immun* 57: 1072.

Demuth, D.R., Davis, C.A., Corner, A.M., Lamont, R.J., Leboy, P.S., and Malamud, D., 1988, Cloning and expression of a *Streptococcus sanguis* surface antigen that interacts with a human salivary agglutinin, *Infect Immun* 56: 2484.

Demuth, D.R., Berthold, P., Leboy, P.S., Golub, E.E., Davis, C.A., and Malamud, D., 1989, Saliva-mediated aggregation of *Enterococcus faecalis* transformed with a *Streptococcus sanguis* gene encoding the SSP-5 surface antigen, *Infect Immun* 57: 1470.

Demuth, D.R., Golub, E.E., and Malamud, D., 1990a, Streptococcal-host interactions: structural and functional analysis of a *Streptococcus sanguis* receptor for a human salivary glycoprotein, *J Biol Chem* 265: 7120.

Demuth, D.R., Lammey, M.S., Huck, M., Lally, E.T., and Malamud, D., 1990b, Comparison of *Streptococcus mutans* and *Streptococcus sanguis* receptors for human salivary agglutinin, *Microbial Pathogen* 9: 199.

Ericson, T. and Rundegren, J., 1983, Characterization of a salivary agglutinin reacting with a serotype c strain of *Streptococcus mutans*, *Eur J Biochem* 133: 255.

Fahnestock, S.R., Alexander, P., Filpula, D., and Nagle, J., 1990, Structure and evolution of the streptococcal genes encoding protein G, *In*: M.D.P. Boyle (ed.) Bacterial Immunoglobulin-Binding Proteins, I. Academic Press, San Diego: 133.

Fischetti, V.A., 1989, Streptococcal M protein: molecular design and biological behavior, *Clin Microbiol Rev* 2: 285.

Forester, H., Hunter, N., and Knox, K.W., 1983, Characteristics of a high molecular weight extracellular protein of *Streptococcus mutans*, *J Gen Microbiol* 129: 2779.

Garnier, J., Osguthorpe, D.J., and Robson, B., 1978, Analysis of the accuracy and implications of simple methods for predicting the secondary structure of globular proteins, *J Mol Biol* 120: 97.

Gibbons, R.J., 1989, Bacterial adhesion to oral tissues: a model for infectious diseases, *J. Dent Res* 68: 750.

Hamada, S. and Slade, H.D., 1980, Biology, immunology, and cariogenicity of *Streptococcus mutans*, *Microbiol Rev* 44: 331.

Holt, R.C., Abiko, Y., Saito, S., Smorawinska, M., Hansen, J.B., and Curtiss, R., 1982, *Streptococcus mutans* genes that code for extracellular proteins in *Escherichia coli* K12, *Infect Immun* 38: 147.

Hughes, M., MacHardy, S.M., Sheppard, A.J., and Woods, N.C., 1980, Evidence for an immunological relationship between *Streptococcus mutans* and human cardiac tissue, *Infect Immun* 27: 576.

Iwake, M., Okahashi, N., Takahashi, I., Kanamoto, T., Sugita-Konishi, Y., Aibara, K., and Koga, T., 1990, Oral immunization with recombinant *Streptococcus lactis* carrying the *Streptococcus mutans* surface protein antigen gene, *Infect Immun* 58: 2929.

Kelly, C., Evans, P., Bergmeier, L., Lee, S.F., Progulske-Fox, A., Harris, A.C., Aitken, A., Bleiweis, A.S., and Lehner, T., 1989, Sequence analysis of the cloned streptococcal cell surface antigen I/II, *FEBS Letts* 258: 127.

Koga, T., Okahashi, I., Takahashi, T, Kanamoto, H., Asakawa, H., and Iwaki, M., 1990, Surface hydrophobicity, adherence, and aggregation of cell surface protein antigen mutants of *Streptococcus mutans* serotype c, *Infect Immun* 58: 289.

Lamont, R.J., Demuth, D.R., Davis, C.A., Malamud, D., and Rosan, B., 1991, Salivary-agglutinin-mediated adherence of *Streptococcus mutans* to early plaque bacteria, *Infect Immun* 59: 3446.

LaPolla, R.J., Haron, J.A., Kelly, C.G., Taylor, W.R., Bohart, C., Hendricks, M., Pyati, J., Graff, R.T., Ma, J.K-C., and Lehner, T., 1991, Sequence and structural analysis of surface protein antigen I/II (SpaA) of *Streptococcus sobrinus*, *Infect Immun* 59: 2677.

Lee, S.F., Progulske-Fox, A., and Bleiweis, A.S., 1988, Molecular cloning and expression of a *Streptococcus mutans* major surface protein antigen, P1 (I/II), in *Escherichia coli*, *Infect Immun* 56: 2114.

Lee, S.F., Progulske-Fox, A., Erdos, G.W., Piacentini, D.A., Ayakawa, G.Y., Crowley, P.J., and Bleiweis, A.S., 1989, Construction and characterization of isogenic mutants of *Streptococcus mutans* deficient in major surface protein antigen P1 (I/II), *Infect Immun* 57: 3306.

Lehner, T., Russell, M.W., Caldwell, J., and Smith, R., 1981, Immunization with purified protein antigens from *Streptococcus mutans* against dental caries in rhesus monkeys, *Infect Immun* 34: 407.

Lehner, T., Caldwell, J., and Smith, R., 1985, Local passive immunization by monoclonal antibodies against streptococcal antigen I/II in the prevention of dental caries, *Infect Immun* 50: 796.

Ma, J.K-C., Smith, R., and Lehner, T., 1987, Use of monoclonal antibodies in local passive immunization to prevent colonization of human teeth by *Streptococcus mutans*, *Infect Immun* 55: 1274.

Ma, J.K-C. and Lehner, T., 1990, Prevention of colonization of *Streptococcus mutans* by topical application of monoclonal antibodies in human subjects, *Archs Oral Biol* 35 (Suppl): 115S.

Ma, J.K-C., Kelly, C.G., Munro, G., Whiley, R.A., and Lehner, T., 1991, Conservation of the gene encoding streptococcal antigen I/II in oral streptococci, *Infect Immun* 59: 2686.

Moro, I. and Russell, M.W., 1983, Ultrastructural localization of protein antigens I/II and III in *Streptococcus mutans*, *Infect Immun* 41: 410.

Ogier, J.A., Wachsmann, D., Scholler, M., Lepoivre, Y., and Klein, J.P., 1990, Molecular characterization of the gene *sr* of the saliva interacting protein from *Streptococcus mutans* OMZ175, *Archs Oral Biol* 35 (Suppl): 25S.

Ohta, H., Kato, H., Okahashi, N., Takahashi, I., Hamada, S., and Koga, T., 1989, Characterization of a cell-surface protein antigen of hydrophilic *Streptococcus mutans* strain GS-5, *J Gen Microbiol* 135: 981.

Okahashi, N., Koga, T., and Hamada, S., 1986, Purification and immunochemical properties of a protein antigen from serotype g *Streptococcus mutans*, *Microbiol Immunol* 30: 34.

Okahashi, N., Sasakawa, C., Yoshikawa, M., Hamada, S., and Koga, T., 1989a, Cloning of a surface protein antigen gene from serotype c *Streptococcus mutans*, *Mol Microbiol* 3: 221.

Okahashi, N., Sasakawa, C., Yoshikawa, M., Hamada, S., and Koga, T., 1989b, Molecular characterization of a surface protein antigen from serotype c *Streptococcus mutans* implicated in dental caries, *Mol Microbiol* 3: 673.

Otake, S., Nishihara, Y., Makimura, M., Hatta, H., Kim, M., Yamamoto, T., and Hirasawa, M., 1991, Protection of rats against dental caries by passive immunization with hen-egg-yolk antibody (IgY), *J Dent Res* 70: 162.

Pancholi, V. and Fischetti, V.A., 1989, Identification of an endogenous membrane anchor-cleaving enzyme for group A streptococcal M protein, *J Exp Med* 170: 2119.

Park, S.K., Lee, B.R., and Byun, S.M., 1991, The leader sequence of streptokinase is responsible for its post-translational carboxy-terminal cleavage, *Biochem Biophys Res Comm* 174: 282.

Russell, M.W. and Lehner, T., 1978, Characterization of antigens extracted from cells and culture fluids of *Streptococcus mutans* serotype c, *Arch Oral Biol* 23: 7.

Russell, M.W. and Wu, H., 1990, *Streptococcus mutans* and the problem of heart cross-reactivity, *Crit Revs Oral Biol Med* 1: 191.

Russell, R.R.B., 1979, Wall-associated antigens of *Streptococcus mutans*, *J Gen Microbiol* 114: 109.

Russell, R.R.B., 1980, Distribution of cross-reactive antigens A and B in *Streptococcus mutans* and other oral streptococci, *J Gen Microbiol* 118: 383.

Smith, R., Lehner, T., and Beverley, P.C.L., 1984, Characterization of monoclonal antibodies to *Streptococcus mutans* antigenic determinants I/II, I, II and III and their serotype specificities, *Infect Immun* 46: 168.

Smith, R. and Lehner, T., 1989, Characterization of monoclonal antibodies to common surface protein epitopes on the cell surface of *Streptococcus mutans* and *Streptococcus sobrinus*, *Oral Microbiol Immunol* 4: 153.

Takahashi, I., Okahashi, N., Kanamoto, T., Asakawa, H., and Koga, T., 1990, Intranasal immunization of mice with recombinant protein antigen of serotype c *Streptococcus mutans* and cholera toxin B subunit, *Archs Oral Biol* 35: 475.

Takano, K., Bogert, M., Malamud, D., Lally, E., and Hand, A.R., 1991, Differential distribution of salivary agglutinin and amylase in the golgi apparatus and secretory granules of human salivary gland acinar cells, *Anatom Rec* 230: 307.

Talkington, D.F., Crimmins, D.L., Voellinger, D.C., Yother, J., and Briles, D.E. , 1991, A 43-kilodalton pneumococcal surface protein, PspA: isolation, protective abilities and structural analysis of the amino terminal sequence, *Infect Immun* 59: 1285.

Tokuda, M., Okahashi, N., Takahashi, I., Nakai, M., Nagaoka, S., Kawagoe, M., and Koga, T., 1991, Complete nucleotide sequence of the gene for a surface protein antigen of *Streptococcus sobrinus*, *Infect Immun* 59: 3309.

Van de Rjn, I., Bleiweis, A.S., and Zabriskie, J.B., 1976, Antigens in *Streptococcus mutans* cross-reactive with human heart muscle, *J Dent Res* 55C: 59.

Wicken, A.J. and Knox, K.W., 1981, Chemical composition and properties of amphiphiles, *in*: "Chemistry and Biological Activities of Bacterial Surface Amphiphiles," G.D. Shockman and A.J. Wicken, eds., Academic Press, New York: 1.

REACTIVE ANTIGENS OF THE PERIODONTOPATHIC BACTERIUM

Actinobacillus actinomycetemcomitans

R.C. Page,[1,2,3] T.J. Sims,[1] B.J. Moncla,[2] R.P. Darveau,[2,4]
B. Bainbridge, and L.D. Engel[1,2]

Research Center in Oral Biology[1]
Departments of Periodontics[2] & Pathology [3]
University of Washington
Seattle, WA 98195

Bristol Myers Squibb[4]
Seattle, WA 98121

INTRODUCTION

Vaccines may be useful in the prevention and control of periodontal diseases. The field of periodontal vaccine development, however, is only in its infancy relative to developments in the fields of caries and oral viral infections. The specific microbial species on which to focus is unresolved, and the biologically and quantitatively significant antigens have not been identified.

Actinobacillus actinomycetemcomitans serotype b (Aa) has been strongly implicated in the etiology of early-onset forms of periodontitis, especially juvenile periodontitis.[1,2]

The purpose of our study was to identify and biochemically characterize the quantitatively major antigen(s) of Aa Y-4 that are immunologically recognized by patients with early-onset periodontitis. Our strategy has been to fractionate sonicates of Aa and use high-titer patient sera to assess the relative immune reactivity of the fractions by enzyme-linked immunoassay (ELISA), Western blot, and dotblot. Components manifesting the greatest immune reactivity were further purified chromatographically and their chemical composition determined by gas-liquid chromatography (GLC). Our studies show that the serotype-specific antigen constitutes the bulk of the antibody binding activity from high-titer sera, and that the reactivity is localized in the sugars of the O antigen sidechains of the high molecular mass lipopolysaccharide (LPS).

MATERIALS AND METHODS

Materials and methods used for the experiments reported here have been described in detail[3-5] and are only briefly summarized here because of space limitations.

Genetically Engineered Vaccines, Edited by
J.E. Ciardi *et al.*, Plenum Press, New York, 1992

Sera were from patients ranging in age from 10 to 30 years and were diagnosed as having localized juvenile periodontitis by standard criteria. Control sera were from periodontally normal subjects ranging in age from 5 to 38 years. *A. actinomycetemcomitans* cultures obtained from the American Type Culture Collection were ATCC 43717, ATCC 43718, and ATCC 43719, which are serotypes a, b, and c, respectively. In some experiments, strains 75, Y-4, and 67 obtained from J.J. Zambon (State University of New York, Buffalo) were used.

LPS was prepared by the phenol-water method of Westphal and Jann,[6] protein as described by Sims et al.,[3] and polysaccharide by the method of Zambon et al.[7] as well as by the phenol-water method described by Sims et al.[4] Polysaccharide fractions were further purified by passage over a column of polymyxin B-agarose by the method of Issekutz.[8] Antigen prepared as described by Amano et al.[9] was kindly provided by Dr. T. Koga, National Institute of Health, Tokyo, Japan. All fractions were derivatized as described by Bryn and Jantzen,[10] and analyzed by GLC. Total protein and DNA-RNA content were measured as described by Sims et al.,[3] and the presence and location of LPS and absence of protein in isolated LPS fractions separated by SDS-PAGE was demonstrated by silver[11] and Coomassie blue staining, respectively. LPS was separated on a column of Sephacryl S-300.[5] High pressure liquid chromatography (HPLC) was performed using a Hewlett Packard 5890 series II chromatograph, using a Bio-Sil TSK-250 column eluted with 0.25 M ammonium acetate buffer, pH 7.0. Column effluent was monitored by phenol-sulfuric acid assay for carbohydrate. ELISA and ELISA inhibition assays were performed as described by Page et al.[5] and Western blots as described by Towbin et al.[12]

RESULTS

Sera obtained from 19 patients with juvenile periodontitis and sera from 19 periodontally normal control subjects were surveyed by ELISA using whole sonic extracts of the three Aa serotypes as plate antigens.[4] Sera with endpoint titers of 50-fold or more above background were designated as high-titer and those below 50-fold were designated low-titer.

Sonicates of all three serotypes of Aa were subjected to SDS-PAGE and the resulting patterns stained with Coomassie blue to demonstrate protein components. As seen in Figure 1, distinct bands of components ranging in molecular mass from around 18 kDa to near 200 kDa were observed. The band patterns of the three serotypes were very similar, although minor differences were noted, indicating a high degree of similarity in the protein components of the three serotypes. Using 11 different patient and control sera, Western blots were prepared and representative patterns are shown in Figure 1. All of the sera studied detected a component of approximately 28 kDa. Patterns for all of the sera known to be high-titer by ELISA were highly positive for antibody binding to high molecular mass components (>28 kDa) for one or more serotypes. Notably, the immunopositive material did not migrate in discrete bands, as observed for proteins in gels stained with Coomassie blue, but rather as a smear. Patient sample P1, known by ELISA to be immunodominant for serotypes a and b, intensely stained high molecular mass material in lanes containing serotype a and b but not c sonicate. Likewise, patient serum P0 specific only for serotype c, and P18 known to be specific for only serotype b, intensely stained high molecular mass material only in the homologous lanes. In contrast, patient sample P4 and control sample C1, known to be low-titer and not immunodominant for any serotype by ELISA, only faintly stained discrete bands in the high molecular mass region, and detected these only faintly. Thus the serotype-specific antibody appeared to account for most antibody binding under the conditions used, and to bind specifically to material that migrated not as discrete bands as did the protein components, but rather as a smear in the high molecular mass region of the gels.

To further delineate the nature of the high molecular mass, serotype-associated material, and to determine which antigens were located on the cell surface, we prepared Western blots using whole-cell sonicates digested or not digested with proteinase K, and incubated them in sera that had been adsorbed or not adsorbed with intact Aa cells. To assure detection of all antigenic components, the SDS-PAGE sample load was increased to 2 mg/ml instead of 1 mg/ml, and the blots were developed at a serum dilution of 1:500 rather than 1:1000, as was the case in the experiment described above. Patterns containing all three serotypes and developed with serum P18, known to be immunodominant for serotype b, are shown in Figure 2. As shown in lane set 1 containing undigested sonicates of serotypes a, c, and b and developed using unadsorbed serum immunodominant for serotype b, protein bands are antibody-positive for all three serotypes, and the lane containing the serotype b material exhibits an intensely stained smear in the high molecular mass portion of the patterns. In contrast, proteinase K-digested material separated and developed in the same way (lane set 2) revealed no antibody binding in the position of protein bands, but persistent staining was observed for the diffuse type-specific material. Lane set 4 containing digested serotype a, c,

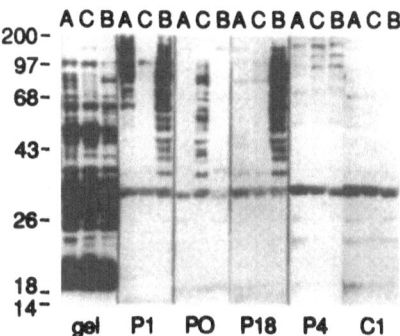

Figure 1. Blot patterns for representative serum samples from patients and controls. Triplicate lanes labeled "gel" are SDS-PAGE patterns of the three *A. actinomycetemcomitans* serotypes stained with Coomassie blue to demonstrate protein components. The remaining lane sets are Western blots of the three serotypes developed with sera from four patients and one control subject, each sample diluted 1:1000. Molecular mass markers are displayed to the left of the gel, and the serotype is indicated at the top of each lane (from Sims et al., 1991, ref. 4, with permission).

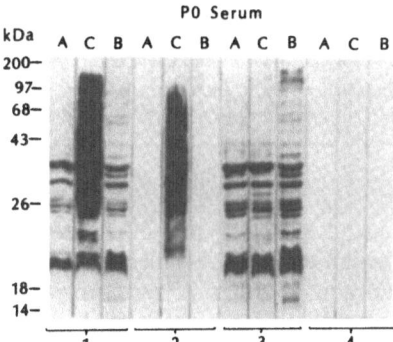

Figure 2. Western blot of PAGE-separated components of *A. actinomycetemcomitans* serotypes a, c, and b (labeled A, C, and B), either digested with proteinase K for 60 min (panels 2,4) or undigested (panels 1,3), and stained to show IgG binding to bands after incubation with *A. actinomycetemcomitans* serotype b-positive serum at a dilution of 1:500, either adsorbed with saturating amounts of intact serotype b cells (panels 3,4), or unadsorbed (panels 1,2) (from Sims et al., 1991, ref. 4, with permission).

and b material, respectively, and developed with the same b-adsorbed serum, revealed no staining. In contrast, lane set 3, representing the same serotypes not digested but also developed with b-adsorbed serum, revealed staining of proteins for all three serotypes but absence of binding against the serotype-specific material. Thus the serotype-specific material was not protein and was exposed on the cell surface. Furthermore, the material migrates in SDS-PAGE in the position of LPS.

Whole-cell adsorption experiments were done to determine the proportions of serotype-specific antibodies and crossreactive antibodies in five high-titer sera, using ELISA plates coated with whole-cell sonicates (Table 1). Samples of each serum were adsorbed or not adsorbed with either serotype a, b, or c cells. The relative degree of removal of IgG against each serotype was then determined by comparing the A_{405} in

Table 1. Relationship between magnitude of IgG titer against *Actinobacillus actinomycetemcomitans* sonicate antigen and degree of inhibition by adsorption with whole cells from homologous and heterologous serotypes.*

Serum	Plate Antigen	Baseline Titer Ratio[a]	PI[b] Homologous	Heterologous a	b	c
P1	b	832	90.3	11.1	-	14.7
P18	b	437	84.1	8.3	-	1.9
P1	a	187	79.1	-	10.7	14.2
P9	c	113	90.2	28.1	1.2	-
P9	a	94	90.5	-	6.3	43.4
C0	c	53	72.2	1.8	8.2	-
P0	c	52	55.5	15.9	7.2	-
P0	b	50	72.4	18.2	-	35.0
P18	a	24	72.4	-	70.9	68.6
P18	c	18	86.1	84.2	80.9	-
P1	c	12	92.4	80.3	79.0	-
P9	b	16	90.0	90.0	-	89.7
P0	a	11	60.0	-	67.9	58.6
C0	a	6	86.0	-	85.2	83.3
C0	b	6	85.5	87.9	-	90.4
Pearson's correlation:			*0.26*	*-0.50*	*-0.76*	*-0.71*

[a] Baseline titer ratio = titer to indicated serotype/mean of three lowest titers for the same serotype for normal control sera.

[b] PI = 100 x [1 - (ELISA A_{405} for serum diluted 1:800 preadsorbed with intact cells from the indicated serotype/absorbance for the same serum without adsorption)].

*From Sims et al., 1991, ref. 4, with permission.

ELISA in wells containing adsorbed samples (diluted 1:800) with that in unadsorbed controls. The resulting data, expressed as percent inhibition (PI), are ranked in Table 1 on the basis of decreasing titer-to-control baseline ratio. Results of adsorbing each sample with heterologous cells (those from a serotype other than that of the plate sonicate) were compared with those for cells that were homologous relative to the plate antigen. There was a significant negative correlation between baseline titer ratios and heterologous PI values ($P < 0.01$), but there was no correlation between titer and homologous inhibition levels. For high-titer sera with titer baseline ratios of 53 to 611, only one PI value was above the range of 1.2% to 28%. In contrast, most of the heterologous PI values for low-titer sera (with ratios less than 24) were above 80%. Furthermore, for all low-titer sera tested, both homologous and heterologous PI values were high and approximately the same for a given plate antigen, regardless of the serotype used for adsorption, indicating that the activity was mainly against crossreactive cell components. In contrast, most of the antibody in high-titer sera reflect a high degree of serotype specificity, as indicated by the difference between homologous and heterologous inhibition levels.

Additional ELISA inhibition experiments were done to determine what proportion of the antibody binding activity could be accounted for by LPS. Wells were coated with sonicate of serotype b and developed with a serum with a high-titer to serotype b in the presence or absence of potential inhibitors of binding. Antibody binding was inhibited in a dose-dependent manner up to 50% by adding an ethanol precipitate of the culture supernatant and up to 90% by 50 μg of whole-cell serotype b sonicate per ml (Fig. 3). In contrast, LPS isolated either from *A. actinomycetemcomitans* culture medium or from the cells by the phenol water method of Westphal and Jann[6] inhibited antibody binding greater than 90% at approximately 5 μg/ml.

To confirm the above observation and determine whether the inhibitory activity of LPS was serotype-specific, additional inhibition experiments were performed using LPS isolated from all three serotypes and representative sera with high and low titers against sonicate antigens. For three sera with a very high titer against b sonicate and a Western blot pattern dominant for b (P14,P18,P13), homologous LPS strongly inhibited antibody binding against b sonicate, giving PI values of 89.4%, 90.6%, and 84.4%, respectively (Table 2). In contrast, inhibition due to heterologous LPS at the

Table 2. Relationship between magnitude of IgG titer against Aa sonicate antigen *Actinobacillus actinomycetemcomitans* sonicate antigen and the degree of inhibition by LPS isolated from homologous vs. heterologous serotypes.*

Serum	Sonicate Plate Antigen	Smear On Western Blot	Baseline Titer Ratio	Homologous	LPS Inhibition Heterologous		
					a	b	c
P14	b	+	478	89.4	8.4		21.9
P18	b	+	437	90.6	9.2		19.8
P13	b	+	323	84.4	3.1		7.1
C8	c	+	111	57.0	35.2	4.9	
C8	a	+	54	47.5		9.4	38.4
C8	b	−	26	19.8	9.2		8.5
C1	b	−	6	7.6	1.4		6.1
P13	c	−	5	6.8	6.0	2.8	
C1	c	−	5	4.5	2.5	4.9	
P4	c	−	2	4.5	4.8	4.9	
P4	b	-	2	7.4	4.1		4.2

* From Sims et al., 1991, ref. 4, with permission.

same concentration was much lower: 3.1% to 9.2% for a LPS, and 7.1% to 21.9% for c LPS, clearly indicating that LPS was the major sonicate component recognized, and that the epitopes recognized by these sera were highly serotype-specific. Using serum C8, immunodominant for both serotypes a and c, homologous inhibition on c and a plates was 57.0% and 47.5%, respectively. Heterologous inhibition by b LPS was only 4.9% and 9.4%, but for a and c LPS, it was 35.2% and 38.4% for c and a plates,

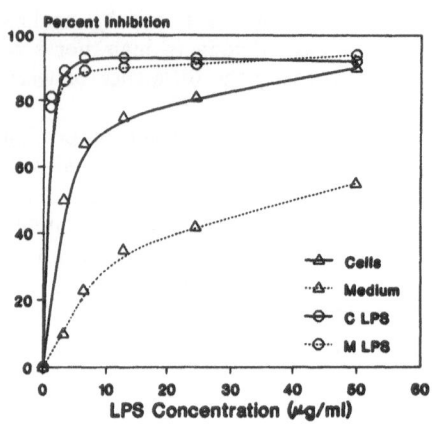

Figure 3. Effects of *A. actinomycetemcomitans* LPS on IgG binding. ELISA inhibition curves plotting percentage inhibition versus inhibitor concentration. Plates were coated with serotype b whole-cell sonicate and developed using a serum immunodominant for serotype b with and without inhibitors at various concentrations. Inhibitors used were ethanol precipitate of the culture medium (Medium), whole-cell sonicate (Cells), and LPS isolated from the culture supernatant (M LPS) or from the cells (C LPS). Endpoint titers were calculated at an absorbance cutoff of 0.350 optical density units and were used to find the PI due to each inhibitor at each concentration as follows: PI = 100 x [1 - (titer with inhibitor / titer without inhibitor)] (from Sims et al., 1991, ref. 4, with permission).

respectively, suggesting crossreactivity between serotypes c and a. There was a positive correlation between the titer ratio and the homologous PI ($r = 0.90$, $P < 0.01$). In contrast, there was no correlation between titer ratio and heterologous PI, and all PI values corresponding to ratios of <10 were below 6.2%.

To obtain additional information as to the relative capacity of various components of Aa sonicates to bind antibody from high-titer sera, we prepared protein, LPS, and polysaccharide fractions from whole-cell sonicates and LPS and polysaccharide from spent culture medium of serotype b. Polysaccharide fractions were prepared both by phenol method[4] and by the method described by Zambon et al.,[7] and in some cases further purified by polymyxin-B Sepharose chromatography to remove contaminating LPS. In addition, we obtained the serotype-specific antigen extracted from Aa Y4 by Amano et al.,[9] designated as "Koga antigen." The composition of all fractions was assessed using GLC (data not shown; see ref. 5). The antibody binding capacity of all of the fractions was compared using a dotblot assay. As shown in Figure 4, row A, intense immunostaining was still apparent in the whole-cell sonic extract, even at dilution step 12 where the dot contained only 1.95 ng (dry weight) of the sample. Staining was also intense in phenol-water-extracted LPS through dilution step 7 and faintly positive at step 12 (row B). In contrast, the phenol-water-prepared polysaccharide from spent culture medium and not further purified was immunopositive, but staining was not visible past step 7, where the dot represented 63 ng of the sample (row C). The polysaccharide fraction, after further purification with polymyxin-B agarose affinity chromatography to remove contaminating LPS, was only faintly immunopositive even at step 1 (row D). The polysaccharide fraction obtained from the spent medium by the method of Zambon et al.[7] was immunopositive at the highest concentrations, but no staining was visible past dot 4 (250 ng) (row E). The rhamnose-fucose serotype antigen (Koga antigen)[9] was intensely immunopositive at step 1, but staining could not be seen past step 7 (63 ng) where staining was comparable to that of LPS at step 12 (row F). Purified protein was intensely positive at step 1, but staining was not visible beyond step 8 (32 ng of antigen) (row G). Patterns obtained with an individual high-titer juvenile periodontitis (JP) serum did not differ significantly from those developed with pooled high-titer patient sera (data not shown).

To characterize this fraction further, we separated LPS purified by the method of Westphal and Jann[6] on the basis of molecular mass by chromatography on Sephacryl S-300, and monitored the sugar content of the effluent by the method of Dubois et al.[13] A typical chromatograph is shown in Figure 5. A small peak was observed at the void volume of the column, followed by a plateau and then a large peak. Selected fractions were subjected to SDS-PAGE and stained with silver. As shown in Figure 6, a faint, relatively wide band of silver-staining material can be observed beginning with fraction 42 and extending with increasing molecular mass through fraction 56. Intensely staining low molecular mass material was observed from fractions 64 through 76; the high molecular mass material was not visible in these fractions. Western blots were prepared with fractions 35 through 74 and developed with pooled high-titer JP sera. As shown in Figure 7, the high molecular mass material in fractions 42 through 63 stained rather intensely. Faintly staining material of intermediate molecular mass was observed in fractions 64 through 68, and faintly staining low molecular mass material was observed in fractions 69 through 74.

Pooled S-300 fractions were dialyzed, lyophylized, and their immunoreactivity assessed in an ELISA inhibition assay using plates coated with LPS or with whole Aa

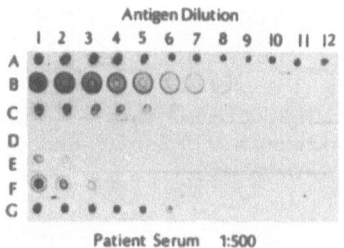

Patient Serum   1:500

Figure 4. Dotblot assays developed with a 1:500 dilution of pooled high-titer JP patient sera. The beginning antigen concentration was 1 mg/ml; 1:1 dilutions were made at each of the 12 steps, and 4 μl was applied at each step. All rows contained material derived from serotype b organisms. Rows: A, whole-cell sonic extract; B, LPS prepared from cells by the phenol-water method of Westphal and Jann (ref. 6); C, unpurified polysaccharide prepared by the phenol method from spent culture medium; D, phenol polysaccharide from spent culture medium purified by polymyxin B-agarose chromatography; E, medium-derived polysaccharide prepared by the method of Zambon et al. (ref. 7); F, rhamnose-fructose serotype antigen prepared by Amano et al. (ref. 9); G, purified protein fraction (from Page et al., 1991, ref. 5, with permission).

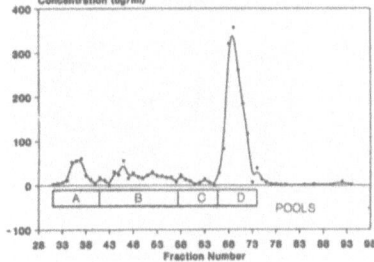

Figure 5. Representative chromatogram of the fractionation of LPS purified by the method of Westphal and Jann (ref. 6) from whole cells on Sephacryl S-300. The effluent was monitored by the sugar content measured by the method of Dubois et al. (ref. 13). Runs were combined as designated by the bar labeled A through D (from Page et al., 1991, ref. 5 with permission).

sonic extract (Fig. 8). In this particular assay, the inhibition by purified LPS using plates coated with LPS as the antigen was set at 100 activity units. Notably, fraction B was highly inhibitory, with values of greater than 90 activity units for plates coated with LPS and about 130 activity units for those coated with whole-cell sonic extract. Fraction C was weakly inhibitory. In marked contrast, inhibition was negligible from fractions A and D, as well as for the Koga antigen preparation,[9] and for polysaccharide prepared by the method of Zambon et al.[7] Inhibition by our protein fraction and by an unrelated *Salmonella* LPS was also negligible.

Composition of Fraction B was assessed by GLC. Almost 50 mol% of the recovered material was accounted for by rhamnose and roughly 20 mol% was accounted for by fucose. Glucose and galactose together accounted for about 15 mol%, and glucosamine was present. Heptulose was not detected, but the fatty acids, tetradecanoic acid (14:0) and [3]OH-tetradecanoic acid (3:OH,14:0), accounted for about 3 mol% and 5 mol%, respectively. Based on the sugar and fatty acid composition, we concluded

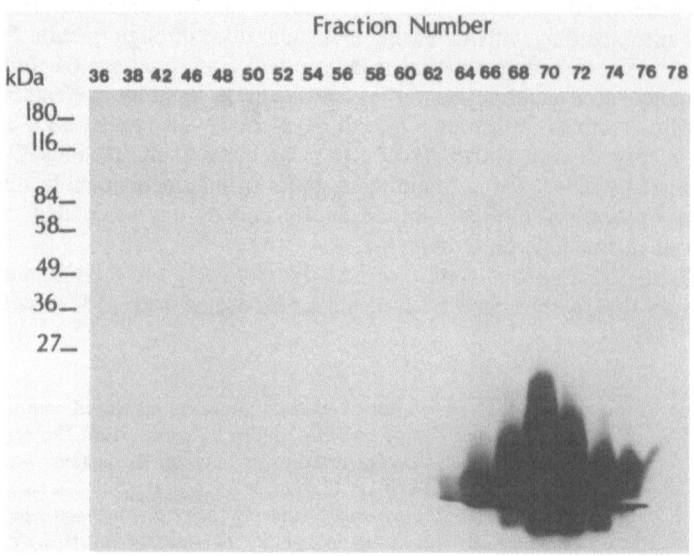

Figure 6. SDS-PAGE analysis of fractions obtained from a Sephacryl S-300 column run of LPS purified from whole cells by the method of Westphal and Jann (ref. 6) and stained with silver by the method of Hitchcock and Brown (ref. 11) (from Page et al., 1991, ref. 5, with permission).

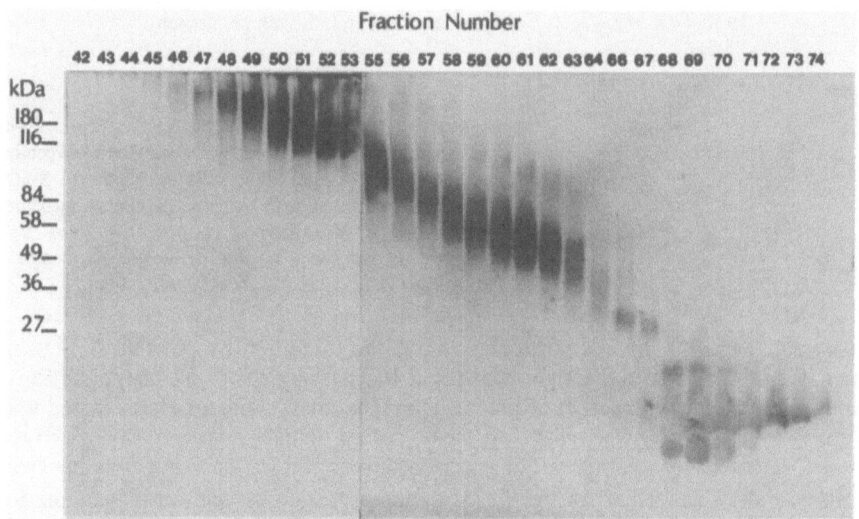

Figure 7. Western blot and SDS-PAGE pattern of fractions obtained from a Sephacryl S-300 column run of LPS developed with pooled high-titer JP patient serum (from Page et al., 1991, ref. 5, with permission).

that the dominant serotype-specific antigen of Aa recognized by patients during the course of their infection was located on the 0 sidechains of LPS.

In our most recent experiments, we have hydrolyzed purified LPS in 1% acetic acid at 100°C for 1.5 hr and removed lipid A by centrifugation. Residual material has

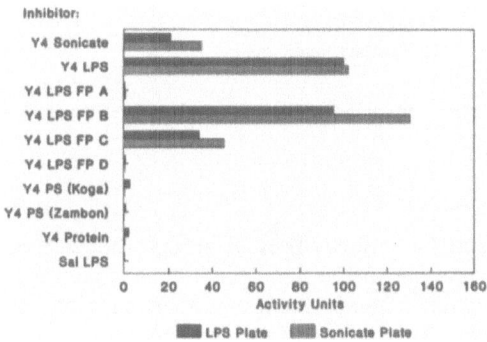

Figure 8. Inhibition of binding of IgG from a pool of high-titer JP patient sera to antigens of A. actinomycetemcomitans measuredd by ELISA inhibition and reported as units of inhibitor activity. Plate antigens were purified LPS and the whole-cell sonic extract. Inhibitors were Y4 whole-cell sonic extract, purified Y4 LPS, column fractions A to D (Fig. 4)(FPA to FPD), Y4 polysaccharide (Y4 PS Koga), Y4 polysaccharide (Y4 PS Zambon), Y4 protein, and <u>Salmonella enteritidis</u> LPS (Sal LPS) (Sigma; catalog no. L-6011) (from Page et al., 1991, ref. 5).

been separated by HPLC using a Bio-Sil TSK-250 column eluted with 0.25 M ammonium acetate buffer, pH 7.0, and the effluent monitored for sugar content. A typical chromatogram is shown in Figure 9 along with dotblots assessing immunoreactivity of the eluted material. A small peak following the void volume, followed by a low plateau and a large peak, were noted. Fractions eluting just before the large sugar peak were richest in immunoreactive material. The composition of this material as determined by GLC is shown in Table 3. Material prepared in this manner

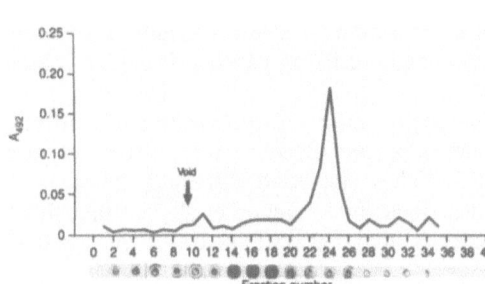

Figure 9. Representative chromatogram of the fraction of material remaining following hydrolysis of LPS purified by the method of Westphal and Jann (ref. 6) in 1% acetic acid at 100°C for 1.5 hr and removal of lipid A by centrifugation on a Bio-Sil TSK-250 column (300 x 7.5 mm) eluted with 0.25 M ammonium acetate, pH 7.0), with effluent monitored for carbohydrate by phenol-sulfuric acid assay. Every second fraction was subjected to dotblot assay developed using pooled patient serum serodominant for serotype b.

is quite similar in both immunoreactivity and sugar and fatty acid composition to the high molecular mass material obtained using the S-300 column. Efforts are currently underway to further purify and characterize this material.

Table 3. Carbohydrate and fatty acid composition of pooled fraction #13-19 (Fig. 9).

Component	Percent Mol	Component	Percent Mol
Rhamnose	39.3	L-glycero-D-mannoheptose	4.3
Fucose	27.7	Glucosamine/Galactosamine	14.0
Galactose	0.95	Tetradecanoic acid	1.5
Glucose	8.8	[3]OH-Tetradecanoic acid	2.1
D-glycero-D-mannoheptose	1.5		

DISCUSSION

Our data show that the bulk of the anti-Aa antibody present in high-titer sera of patients with early-onset periodontitis is serotype-specific and binds to a nonprotein component located on the cell surface, while a much smaller proportion binds to proteins and is crossreactive among serotypes. The serotype-specific material is present in the high molecular mass fraction of purified LPS, which is rich in sugars including glucose, rhamnose, and fucose, characteristic of the 0 sidechains. Immunoreactive material of very similar composition was obtained by HPLC using hydrolyzed material from which lipid A had been removed. The major components of this material are the sugars characteristic of O sidechains with very small amounts of the fatty acids and sugars typical of lipid A and the inner core.

Serotype-specific antigens have been isolated and partially characterized from various serotypes of Aa previously by Zambon et al.,[7] Califano et al.[14] and Koga and coworkers.[9] Serotype-specific antigen from Aa serotype a, b, and c were purified by Zambon et al.[2,7] by DEAE-Sepharose and gel filtration chromatography of ethanol precipitates from spent culture medium, and shown by GLC to consist predominantly of mannose. The serotype b-specific antigen was too large to enter 5% SDS-PAGE gels, and it was predominantly mannose. The polysaccharide fraction we obtained from spent culture medium using the method of Zambon et al.[7] was similar in that it contained 48% mannose and was too large to enter SDS-PAGE gels. Our preparation, however, contained relatively little antigen as assessed by dotblots, and antibody binding to it could be abrogated by removal of contaminating LPS by passage through a column of polymyxin-B Sepharose.

Califano et al.[14] prepared a serotype-specific antigen from serotype b by affinity chromatography of the ethanol precipitate of a cold phenol-water extract of cells previously treated with lysozyme and SDS. They suggested that their material was comparable to the mannan polymer described by Zambon et al.,[7] although its composition was not reported and the material migrated on SDS-PAGE in a manner identical to that we describe here.

A serotype-specific antigen consisting of repeating disaccharide units of L-rhamnose and D-fucose, with a small amount of fatty acid and peptide and not containing mannose, was extracted from Aa serotype b by autoclaving and then chromatographing on DEAE-Sephadex A-25 and Sephacryl S-300 by Amano et al.[9] This preparation, kindly provided to us by Dr. Koga, manifested more antibody binding activity in the dotblot assay than did the polysaccharide we prepared by the method of Zambon et al.,[7] but a great deal less than the fraction we purified from LPS. Furthermore, the Koga antigen did not significantly inhibit binding of antibody from high-titer sera to plates coated with either serotype b sonic extract or purified LPS. Although they used a somewhat different approach, Wilson and Schifferle[15] arrived at the same conclusions as we, (e.g., that the polysaccharide moiety of LPS contains the serotype b antigenic determinants of Aa).

It has not yet been shown that the humoral response to the immunodominant antigen(s) of Aa is important in host defense against this organism, although our lab and others are presently examining this question using opsonization-neutrophil killing assays. Assuming that this antigen does prove to be functionally important in host defense, it would be a prime candidate for use in a vaccine.

REFERENCES

1. J.J. Zambon, *Actinobacillus actinomycetemcomitans* in human periodontal disease, *J. Clin. Periodontol.* 12:1 (1985).

2. J.J. Zambon, T. Umemoto, E. DeNardin, F. Nakazawa, L.A. Christersson, and R.J. Genco, *Actinobacillus actinomycetemcomitans* in the pathogenesis of human periodontal disease, *Adv. Dent. Res.* 2:269 (1988).

3. T.J. Sims, B.J. Moncla, and R.C. Page, Serum antibody response to antigens of oral gram-negative bacteria by cats with plasma cell gingivitis-pharyngitis, *J. Dent. Res.* 69:877 (1990).

4. T.J. Sims, B.J. Moncla, R.P. Darveau, and R.C. Page, Antigens of *Actinobacillus actinomycetemcomitans* recognized by patients with juvenile periodontitis and periodontally normal subjects, *Infect. Immun.* 59:913 (1991).

5. R.C. Page, T.J. Sims, L.D. Engel, B.J. Moncla, B. Bainbridge, J. Stray, and R.P. Darveau, The immunodominant outer membrane antigen of *Actinobacillus actinomycetemcomitans* is located in the serotype-specific high molecular mass carbohydrate moiety of lipopolysaccharide, *Infect. Immun.* 59:3451 (1991).

6. O. Westphal and K. Jann, Bacterial lipopolysaccharides: Extraction with phenol-water and further applications of the procedure, *in*: "Methods in Carbohydrate Chemistry," Vol. 5, R.L. Whistler, ed., Academic Press, New York (1965).

7. J.J. Zambon, J. Slots, K. Miyasaki, R. Linzer, R. Cohen, M. Levine, and R. Genco, Purification and characterization of serotype c antigen from *Actinobacillus actinomycetemcomitans*, *Infect. Immun.* 44:22 (1984).

8. A. Issekutz, Removal of gram-negative endotoxin from solutions by affinity chromatography, *J. Immunol. Meth.* 61:275 (1983).

9. K. Amano, T. Nichihara, N. Shibuya, T. Noguchi, and T. Koga, Immunochemical and structural characterization of a serotype-specific polysaccharide antigen from *Actinobacillus actinomycetemcomitans* Y4 (serotype b), *Infect. Immun.* 57:2942 (1989).

10. K. Bryn and E. Jantzen, Analysis of lipopolysaccharides by methanolysis, trifluoroacetylation, and gas chromatography on a fused-silica capillary column, *J. Chromatogr.* 240:405 (1982).

11. P. Hitchcock and T. Brown, Morphological heterogeneity among *Salmonella* lipopolysaccharide chemotypes in silver-stained polyacrylamide gel, *J. Bacteriol.* 154:269 (1983).

12. H. Towbin, T. Staehelin, and J. Gordon, Electrophoretic transfer of proteins from polyacrylamide gels to nitrocellulose sheets: Procedure and some applications, *Proc. Natl. Acad. Sci. USA* 76:4350 (1979).

13. M. Dubois, K.A. Gilles, J.K. Hamilton, P.A. Rebers, and F. Smith, Colorimetric method for determination of sugars and related substances, *Analyt. Chem.* 28:350 (1956).

14. J.V. Califano, H.A. Schenkein, and J.G. Tew, Immunodominant antigen of *Actinobacillus actinomycetemcomitans* resides in the polysaccharide moiety of lipopolysaccharide, *Infect. Immun.* 59:1544 (1991).

15. M.E. Wilson and R.E. Schifferle, Evidence that the serotype b antigenic determinant of *Actinobacillus actinomycetemcomitans* Y4 resides in the polysaccharide moiety of lipopolysaccharide, *Infect. Immun.* 59:1544 (1991).

IMMUNIZATION WITH FIMBRIAL PROTEIN AND PEPTIDE PROTECTS AGAINST *PORPHYROMONAS GINGIVALIS*-INDUCED PERIODONTAL TISSUE DESTRUCTION

Richard T. Evans,[1] Bjarne Klausen,[2] and Robert J. Genco[1]

[1]School of Dental Medicine
Department of Oral Biology
State University of New York at Buffalo
Buffalo, NY 14214

[2]School of Dentistry
Department of Periodontology
University of Copenhagen
DK-2200 Copenhagen N
Denmark

INTRODUCTION

Porphyromonas gingivalis is frequently described as associated with actively progressing deep periodontal pockets in humans and is considered to be a major pathogenic component of the Gram-negative subgingival flora.[1,2] Development of a vaccine to control periodontal disease caused by this organism would be of benefit to those individuals who are at risk of disease induced by this widely distributed oral pathogen. Identification of target or "critical" antigens for the development of vaccines which may provide protection against *P. gingivalis*-induced periodontal tissue destruction requires the identification and isolation of such antigens and a suitable test animal in which to measure the effects of immunization[3].

In this series of studies we have tested a variety of cell surface antigens of *P. gingivalis* for their protective effect. These include heat-killed intact cells, a partially purified cell surface extract, purified 43 kDa and 75 kDa cell surface components, and a synthetic 20 amino acid peptide based on the known structure of the 43 kDa fimbrillin protein. The animal system we use to test the effects of immunization with these antigens is the gnotobiotic rat. Measures of periodontal tissue destruction include radiographic alveolar bone loss, histometric loss of alveolar crestal height, and increases in enzyme activity of several host-derived proteolytic enzymes in gingival tissue, which make the results obtained from this animal model highly reproducible and reliable.

METHODS AND MATERIALS

Age-matched three-week-old male gnotobiotic Sprague-Dawley rats were obtained from Taconic Farms (Germantown, NY) and maintained in plastic film isolators as previously described.[4,5] Immunizations were carried out subcutaneously (2x) 1 week apart in Freund's incomplete adjuvant as described in detail by Klausen et al.[4] and Evans et al.[6] Animals were infected on 3 alternate days with viable cells of *P. gingivalis* strain 381 in 1 ml of 5% low viscosity carboxymethylcellulose 1 week following immunization. All animals were sacrificed 42 days following the initial infection, a time at which near maximum alveolar bone loss is seen in infected animals and with little or no loss seen in germfree (GF) animals.

Antigens were prepared from *P. gingivalis* strains 381 and the genetically identical strain 2561 (ATCC strain 33277). *P. gingivalis* whole-cell antigen was made by heating strain 381 at 60°C for 15 min on 2 successive days to kill the cells. Highly purified 43 and 75 kDa proteins used as antigens were obtained from *P. gingivalis* strain 2561 as previously described[7,8]. Briefly, these cell surface antigens were sheared from the cells by mild sonication, and the proteins precipated at 40% ammonium sulfate saturation, and then each were differentially precipitated as follows. The 43 kDa subunit was further purified with 1% SDS and 0.2M Mg Cl$_2$ at pH 6.5 using repetitive precipitation steps.[7] Purification of the 75 kDa membrane protein was achieved by precipitation at pH 5.0 and resolubilization at pH 7.4. By repeated cycles of isolectric precipitation at pH 5.0, a homogeneous preparation of the 75 kDa protein was obtained[8]. The synthetic peptide used in this study was prepared using a solid-phase F moc peptide synthesis procedure on an Applied Biosystems model 431A peptide synthesizer as described in detail by Lee et al.[9]

Periodontal tissue destruction was estimated by measuring alveolar bone loss morphometically and radiographically.[4,5,10] Gingival tissue was extracted and analyzed for collagenase, cathepsins B and L and gelatinase activities as previously described.[4,6,11]

Antibody measurements were made using the particle concentration immunofluorescence assay (PCFIA) described by Jolley et al.[12] The procedure measures antibody using a sequential two-step immunoassay. In the first step antigen coated on polystyrene beads (Pandex Company, Mundelein, IL) was reacted with sera (1:200 dilution) or saliva (1:10 dilution) containing antibody from the immunized and/or infected rats. The plates were washed (3x) to remove unreacted rat antibody and then incubated with fluorescein-conjugated goat anti-rat Ig. Measurement of the bound second antibody was made in a PCFIA fluorometer (Pandex Corporation, Mundelein, IL). The intensity of the fluorescence is directly proportional to the amount of conjugated second antibody bound to the antigen-antibody complex and is expressed as relative fluorescence units (RFU).

RESULTS

Figure 1 (upper) shows horizontal bone-level changes seen in immunized and infected animals. GF rats which were both sham-immunized and sham-infected served as a baseline control. SHAM-immunized animals (IF) infected with *P. gingivalis* strain 381 served as positive controls showing significant periodontal disease. Animals immunized with heat-killed intact whole cells (WC), purified 43 kDa fimbrial component (43) and a French press extract of cell components (CF) all significantly ($P < 0.01$) exhibited less periodontal bone loss than the sham-immunized animals. Comparable results were seen when bone support was measured using the radiographic technique (Figure 1, lower). The latter method measures vertical intrabony bone lesions in the mandible and is expressed as percent bone support. The fimbrillin (43) and whole cell immunized (WC) had bone support comparable to GF animals while 75 and CF showed less protection.

Examination of gingival tissue for proteolytic enzyme activity (Figure 2) in the same animals reveals highly elevated levels of collagenase, cathepsin B and L, and gelatinase in the strain 381-infected rats (IF). In each case immunization with WC, 43, CF and, to a lesser extent, 75 reduced the levels of proteolytic activity.

Immunization with WC, 43, 75 and CF increased both serum (Figure 3) and salivary (Figure 4) antibody titers when tested against either the 43 kDa fimbrial component (upper graph) or the 75 kDa antigen (lower graph). Antibody titers in the gnotobiotic animals rose against both test antigens following infection (IF) but were greatest in the immunized animals (Figures 3 and 4, WC, 43, 75 and CF).

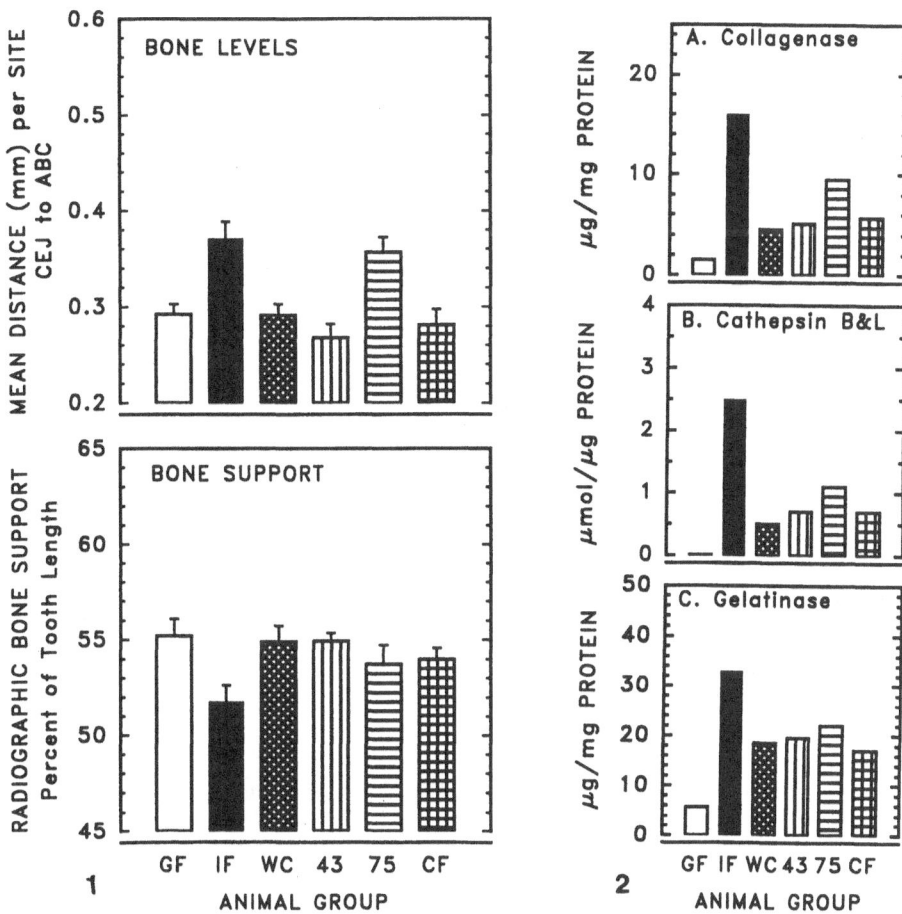

Figure 1 (left). Bone loss (upper) or support (lower) in gnotobiotic rats infected and/or immunized with cell components. Abbreviations: GF, nonimmunized, uninfected; IF, nonimmunized, infected with *P. gingivalis* strain 381, WC, immunized with whole cells, infected; 43, immunized with purified 43 kDa fimbrilin, infected; 75, immunized with 75 kDa cell surface component, infected; CF, immunized with partially purified cell extract, infected.

Figure 2 (right) Results of gingival tissue proteolytic enzyme activity. Results in both figures are given as mean values (N=8).

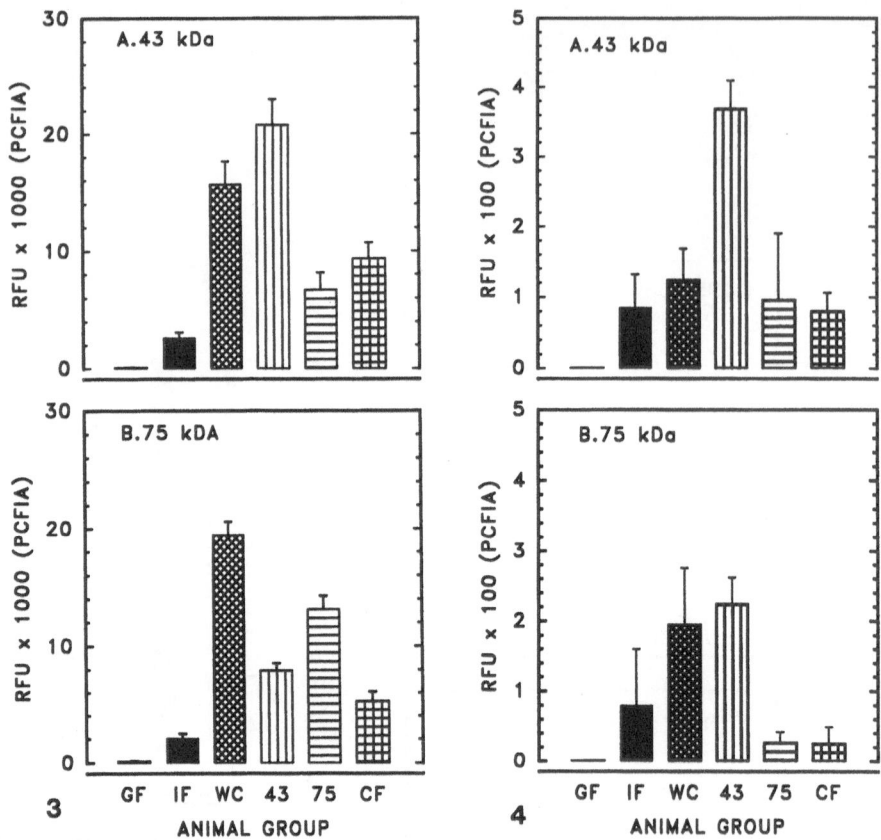

Figure 3 (left). Antibody measurement in serum to the purified 43 kDa protein (upper) and purified 75 kDa component (lower).

Figure 4 (right). Antibody in saliva to the same antigens as figure 3. Abbreviations as in figure 1. X ± SEM (N=8) for both figures.

The effects (Figure 5) of immunizing with a 20 mer peptide representing an exposed epitope of the native fimbrillin molecule were next assessed. This peptide was examined since it represented one of several peptides that inhibited binding *P. gingivalis* cells to sHAP.[9] Peptide plus carrier (P+C) immunized animals were totally protected against horizontal bone loss (Figure 5, upper). Peptide (P) used as an antigen without carrier also protected. Carrier (C) alone (thyroglobulin) exerted no protective effect. Interestingly, peptide plus carrier (P+C) or peptide (P) (Figure 5, lower) did not protect against intrabony (vertical) lesions in the mandible indicating a possible separation of protective abilities between the intact fimbrillin molecule and a subunit component of the fimbrillin molecule.

Patterns of gingival enzyme activities in animals immunized with peptide and infected (Figure 6) are similar to those seen in animals immunized with intact cells or cell-surface components. Animals immunized with peptide together with or without carrier showed a dramatic decline in activities when compared to carrier immunized (C) or infected (IF) animals.

Antibody titers to fimbriae or the 75 kDa antigens reveal an anomalous pattern (Figure 7). Contrary to expectations, P+C-immunized animals did not have elevated antibody titers. In each instance (i.e., with either fimbriae or 75 kDa antigen), the titer was lower than that seen using peptide (P) alone as the immunogen. The animals were competent to produce antibody as shown by the carrier (C)-immunized animal group using thyroglobulin as antigen. As in the previously described experiments, the sera from these animals had antibody likely resulting from infection with *P. gingivalis* strain 381.

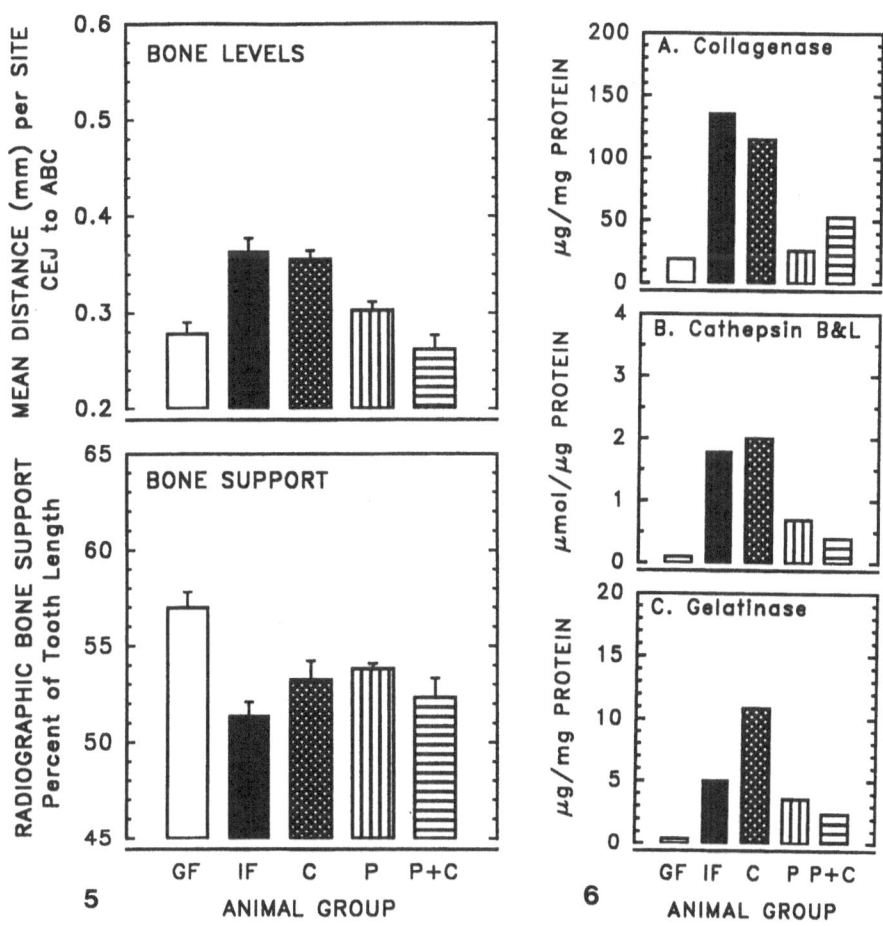

Figure 5 (left). Bone loss (upper) or support (lower) in gnotobiotic rats immunized with peptide. Abbreviations: GF and IF as figure 1; C, carrier immunized, infected; P, peptide immunized, infected; P+C, peptide-carrier conjugate immunized, infected.

Figure 6 (right). Gingival proteolytic enzyme activities. Results in both figures are given as mean values (N=8).

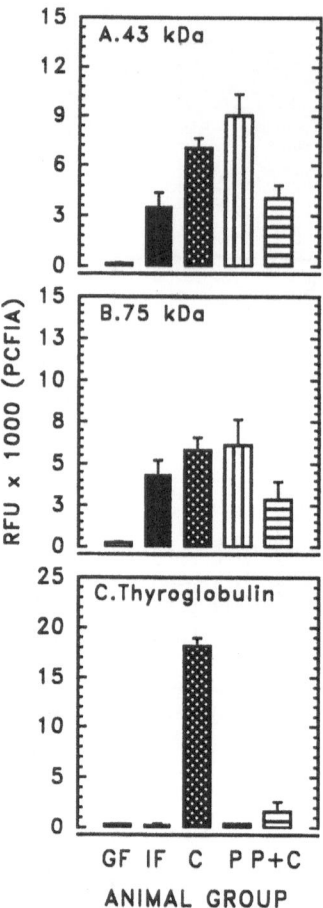

Figure 7. Serum antibody measurements to 43 kDa fimbrillin and 75 kDa component in peptide-immunized animals. Abbreviations as given in figure 5.

DISCUSSION

Development of immune protection against periodontal disease destruction requires the identification of critical or target antigens. By using the gnotobiotic rat, investigations into the protective nature of these critical antigens can be made uncomplicated by exogenous factors such as might be found in the flora of conventional animals. As candidate antigens are identified they can be systemically investigated in the axenic animal, (i.e., an animal mono-infected with the pathogen under study) and then introduced into other animal models such as conventional rats or primates. Periodontal disease induced by *P. gingivalis* has been the major focus for studies on this project.[4,5,6,10] The disease process theoretically should be ameliorated by the host responses since the oral structures involved are all subject to the humoral and cell-mediated immune systems. The nature of the immune system and immune response that accounts for the protection can only be speculated at the present time. For example, it may be that protection is afforded by antibodies which bind to the fimbriae and prevent fimbrial-mediated adherence important for colonization. Other explanations of the protective effect against fimbriae are possible including inhibition of fimbrial-mediated stimulation of host-destructive factors.

In this study development of antibody was used as a measure of immunization effectiveness. Animals were immunized with the appropriate antigen, the response allowed to develop and the animals then infected. Infection alone also serves as a source of antigenic stimulation (Figures 3,4 and 7); however, deliberate vaccination produced a greater response, and one which provided protection. Development of antibody to synthetic vaccines coupled to carriers can frequently be suppressed when a priming dose of carrier is given followed by a second immunization of hapten plus carrier.[13,14] This phenomenon is termed epitope suppression. In the experiment described here hapten plus carrier is used for both primary and secondary immunization to avoid carrier effects. However, a low response was noted (Figure 7) when the fimbrial peptide linked to thyroglobulin was used as immunogen. Speculation is therefore raised that antigen from subsequent infection may exert a form of epitope suppression. Further experimentation is required to clarify this point. In any event, immunization with peptide plus carrier (P+C) gave excellent protection against bone loss (Figure 5) and markedly reduced gingival proteolytic enzyme activity (Figure 6).

Enzyme activities in the gingivae induced by infection were suppressed by immunizing the animals (Figures 2,6). The proteolytic enzyme activities in the animals are measured 42 days following initial infection. In previous experiments it was shown that collagenase and gelatinase activities in strain 381-infected animals remain well above baseline values for up to 84 days. Proteolytic enzyme inhibitor studies also show that collagenase and gelatinase activity are of mammalian and not bacterial origin.[4,6] These results, together with the cathepsin B and L data, raise speculation that gingival tissue destruction, while initiated by the *P. gingivalis* infection, is continued by host mechanisms.

In this study fimbrillin and in particular, a peptide derived from the fimbriae subunit is identified as protective. Fimbriae of *P. gingivalis* were shown to mediate adherence of *P. gingivalis* of this organism to cells to salivary control hydroxyapatite *in vitro* studies.[9] This initial adherence is likely important in colonization of salivary-coated tooth surface as seen in the germfree rat. Colonization is likely for disease initiation[9]. There are at least three mechanisms by which immunization may disrupt the disease process. First, antibodies in the saliva may inhibit *P. gingivalis* cells from adhering to salivary-coated oral surfaces and hence prevent colonization. Second, antibody in the sera might inhibit the downgrowth of the bacteria subgingivally. And finally, the antibody induced might affect other functions of the *P. gingivalis* fimbriae such as stimulation of macrophages. In addition to these antibody-mediated effects, cell-mediated immune responses to this Gram-negative infection must also be considered in evaluating the basis for immune protection. It is clear, however, that *P. gingivalis* fimbriae are key virulence factors in the pathogenesis of periodontal destruction in this model. The mechanisms by which this virulence is expressed is yet to be determined.

SUMMARY

In these studies we have attempted to show that cell surface structures are critical antigens for protection against *P. gingivalis*-induced periodontal destruction. Fimbrillin, and in particular a synthetic 20-amino-acid fimbrillin peptide, exerts a protective effect in gnotobiotic rats, thus identifying them as potentially useful in the development of a vaccine.

ACKNOWLEDGMENTS

The authors wish to acknowledge Drs. N.S. Ramamurthy and L. Golub, State University of New York at Stony Brook, who performed and analyzed the proteolytic enzyme assays. We are grateful to J-Y. Lee and to Drs. G.S. Bedi and H.T. Sojar for preparations of purified fimbrial and cell membrane components and for the synthetic peptides used in this study. We also acknowledge the technical assistance of Cornelia Sfintescu and Alice Wendt. Supported in part by grants DE 04898, DE 07034, and DE 08240 from the National Institutes of Health, U.S. Public Health Service, The Procter and Gamble Co., and Gangsted-fonden. Figures 1, 2, 3, and 4 are reprinted from Evans et al., Immunization with *Porphyromonas (Bacteroides) gingivalis* fimbriae protects against periodontal destruction, *Infect. Immun.*, 60:7 (1992), with permission.

REFERENCES

1. J.J. Zambon, Microbiology of periodontal disease, *in* "Contemporary Periodontics," R.J. Genco, H.M. Goldman and D.W. Cohen, eds., The C.V. Mosby Co., St. Louis (1990).
2. T.J.M. van Steenbergen, A.J. van Winkelhoff, and J. de Graaff, Black-pigmented oral anaerobic rods: classification and role in periodontal disease, *in* "Periodontal Disease Pathogens & Host Immune Responses," S. Hamada, S.C. Holt and J.R. McGhee, eds., Quintessence Publishing Co., Ltd., Tokyo (1991).
3. B. Klausen, Microbiological and immunological aspects of experimental periodontal disease in rats, *J. Periodont.* 62:59 (1991).
4. B. Klausen, R.T. Evans, N.S. Ramamurthy, L.M. Golub, C. Sfintescu, J.-Y. Lee, G. Bedi, J.J. Zambon, and R.J. Genco, Periodontal bone level and gingival proteinase activity in gnotobiotic rats immunized with *Bacteroides gingivalis*, *Oral Microbiol. Immunol.* 6:193 (1991).
5. B. Klausen, Cornelia Sfintescu, and R.T. Evans, Asymmetry in periodontal bone loss of gnotobiotic Sprague-Dawley rats, *Arch. Oral Biol.* 36:685 (1991).
6. R.T. Evans, B. Klausen, H.T. Sojar, G.S. Bedi, C. Sfintescu, N.S. Ramamurthy, L.M. Golub, and R.J. Genco, Immunization with *Porphyromonas (Bacteroides) gingivalis* fimbriae protects against periodontal destruction, *Infect. Immun.* 60:7 (1992).
7. H.T. Sojar, J-Y. Lee, G.S. Bedi, M.-I. Cho, and R.J. Genco, Purification, characterization and immunolocalization of fimbrial protein from *Porphyromonas (Bacteroides) gingivalis*, *Biochem. Biophys. Res. Comm.* 175:713 (1991).
8. H.T. Sojar, J-Y. Lee, G.S. Bedi, M.-I. Cho, and R.J. Genco, Purification, characterization and localization of a major envelope protein antigen from *Porphyromonas (Bacteroides) gingivalis*, *Biochem. Int.* 25:437 (1991).
9. J-Y. Lee, H.T. Sojar, G.S. Bedi, and R.J. Genco, Peptide analogs to fimbrillin inhibit binding of *Porphyromonas gingivalis*, *Infect. Immun.* (In press 1992).
10. B. Klausen, R.T. Evans, and C. Sfintescu, Two complementary methods of assessing periodontal bone level in rats, *Scand. J. Dent.* 97:494 (1989).
11. R.T. Evans, B. Klausen, N.S. Ramamurthy, L.M. Golub, Cornelia Sfintescu, and R.J. Genco, Periodontopathic potential of two strains of *Porphyromonas gingivalis* in gnotobiotic rats, *Arch. Oral Biol.* (Submitted 1992).
12. M.T. Jolley, C.-H. Wang, S.J. Ekinberg, M.S. Yuelke, and D.M. Kelso, Particle concentration fluorescence immunoassay (PCFIA): a new rapid immunoassay technique with high sensitivity, *J. Immunol. Methods.* 67:21 (1984).
13. M-P. Schutze, C. LeClarc, M. Jolivent, F. Audibert, and L. Chedid, Carrier-induced epitopic suppression, a major issue for future synthetic vaccines, *J. Immunol.* 135:2319 (1985).
14. M-P. Schutze, E. Deriand, G. Prezewlocki, and C. LeClarc, Carrier-induced epitopic suppression is initiated through clonal dominance, *J. Immunol.* 142:2635 (1989).

VACCINE DEVELOPMENT: PROGRESSION FROM TARGET
ANTIGEN TO PRODUCT

Ronald W. Ellis

Cellular and Molecular Biology
Merck Sharp & Dohme Research Laboratories
West Point, PA 19486

INTRODUCTION

The research and development of vaccines entail certain common steps. The goal of the research phase is the specification of the design of the vaccine, either live, inactivated or subunit; this includes the definition of the target antigen for a subunit vaccine. The development phase encompasses multiple overlapping elements, which range from those related to the production process through human clinical trials and ultimately licensure. This chapter will review all the pertinent elements in the development of vaccines starting with the target antigen and through to licensure. The development of the *Hemophilus influenzae b* conjugate vaccine will be reviewed as a model case for other vaccines, given that this group of subunit vaccines has been licensed recently by the United States Food and Drug Administration (F.D.A.).

EPIDEMIOLOGY

· *Hemophilus influenzae* type b (Hib) is an encapsulated rod-like Gram-negative bacterium which causes ca. 12,000 cases per year of meningitis in the United States,[1] with all but a small percentage in children 2 months to 5 years of age. The mortality rate from Hib meningitis is up to 5%, while ca. 35% of survivors of Hib meningitis develop permanent neurological sequelae. An equal number of other Hib invasive diseases, such as pneumonia, epiglottitis, septicemia, cellulitis, osteomyelitis, pericarditis and septic arthritis also exact a severe toll in infants and children. In most populations, ca. 75% of cases of invasive diseases occurs in children under 2 years of age, a critical issue with respect to vaccine design as discussed below. Furthermore, half of the cases of Hib disease in countries such as the United States occurs in infants less than 12 months old, and the peak incidence of disease in developing areas such as tropical Africa and in Eskimos and Native

Americans occurs as young as 3-9 months of age.[2,3]   Thus, the epidemiology of the disease dictates that, for maximal health benefit, Hib vaccines should be immunogenic in infants as young as 2 months old in order to assure protective immunity in the period of peak disease incidence.

TARGET ANTIGEN

The natural focus for vaccine research for most pathogens is the surface macromolecule(s), whose antigenic structures are expected to elicit neutralizing antibodies. The surface of Hib organisms is covered by the capsular polysaccharide polyribosyl ribitol phosphate (PRP).   Therefore, PRP was tested as the first-generation Hib vaccine. Antibodies to PRP (anti-PRP) activate complement for bactericidal and opsonic activity.[4,5]   Levels of anti-PRP 1.0 μg/mL [defined by radioimmunoassay (RIA)] were correlated with protection against Hib disease following the immunization of children with the PRP vaccine.[6]   A further study identifying anti-PRP as directly mediating protection against Hib disease was the demonstration by Santosham et al.[7] that passive administration to infants of hyperimmune globulin (IgG) enriched for anti-PRP mediates a significant reduction in the incidence of Hib disease at the minimal level of 0.05-0.15 μg anti-PRP/mL. Since the structure of PRP is common to all strains of Hib, PRP has been considered the (optimal) target antigen for Hib vaccines.

Being a T cell-independent antigen as are polysaccharides in general, PRP vaccine is poorly immunogenic and non-protective in children under 18-24 months of age.[8-10] It is not highly efficacious even in children 2 years of age and older,[11] in whom it induces an IgM anti-PRP response lacking memory T-cell development. Given that the prime target for Hib vaccines is children under 2 years of age, it became necessary to present PRP in an immunogenic form in which it could elicit anti-PRP in infants.   Conjugation technology has provided this type of antigenic presentation.

CARRIERS

The concept of conjugating a carbohydrate or polysaccharide antigen to a protein carrier to enhance its immunogenicity was conceived by Avery and Goebel.[12] This has become understood immunologically as coupling the T-independent antigen to a T-dependent carrier, thereby inducing a T-dependent anti-polysaccharide immune response. Since infants can respond with antibodies to some T-dependent antigens, this concept was extended to PRP, hence creating a class of vaccines known as Hib conjugate vaccines. Potentiation of the anti-PRP response in children under 2 years of age has been achieved by four groups which have conjugated PRP to different carriers.[13]   The carriers and their respective vaccines are tetanus toxoid (PRP-T, Merieux Laboratories), diphtheria toxoid (ProHIBIT®, Connaught Laboratories), mutant non-toxic diphtheria toxin (HibTITER®, Praxis Laboratories), and meningococcal outer membrane protein complex (PedvaxHIB®, Merck Sharp & Dohme). Each of these vaccines utilizes PRP molecules of different chain lengths and distinct conjugation chemistries to their respective carriers to achieve different patterns of immune responses in infants and young children. The specification of the design of Hib conjugate vaccines, based on immunological and epidemiological rationales, concluded the research phase for these vaccines. These serve as models for other bacterial and viral antigens, whether polysaccharide or peptide in nature, which might be conjugated to carrier proteins to enhance immune responses. The rest of this chapter will be devoted to one of these conjugates as a model for the development of vaccines in general.

PROCESS

PedvaxHIB®, a conjugate of PRP to the outer membrane protein complex (OMPC) from Neisseria meningitidis group B (NMB), is the Hib conjugate vaccine which elicits the highest level of anti-PRP in a single dose in 2-month-old infants.[13] Its biochemical constituents are produced in microbial cultures grown in fermentors. PRP is shed from Hib cells, and OMPC is derived by deoxycholate buffer extraction from NMB cells. The process for the PRP-OMPC conjugate vaccine[14] is depicted in Fig. 1. The OMPC carrier is a spherical particle ca. 0.1-0.2 μm in diameter which contains three of the NMB proteins as well as lipids, including LPS. Antigenic presentation on the surface of a relatively large carrier (e.g., OMPC) should provide excellent immunological presentation for eliciting anti-PRP antibodies. Following conjugation, chemicals and unconjugated PRP are removed from the PRP-OMPC product. The bulk conjugate can be stored in aqueous solution or lyophilized in individual vials prior to formulation on aluminum hydroxide. There are several important aspects of the process which can be generalized to all vaccines. The process should be scalable readily to a pilot plant, then an ultimate manufacturing facility, without a significant change in the quality of the product. The process must generate a consistent product, since only a very small number of vaccine vials can be quality-control tested and only a limited number of vaccine lots can be tested clinically prior to licensure. The process is validated for its consistency in removing unwanted constituents by spiking experiments and extensive in-process monitoring.

FORMULATION

The PRP-OMPC conjugate vaccine is formulated by mixture with aluminum hydroxide, to which it adsorbs through ionic interaction. Immunogenicity studies of the conjugate vaccine have been performed in the infant Rhesus monkey, which has been developed as a model for the immunogenicity of Hib conjugate vaccines in human infants.[15] The aluminum hydroxide-adsorbed PRP-OMPC conjugate vaccine was shown to elicit ca. 10-fold higher levels of anti-PRP in this model relative to unadjuvanted aqueous conjugate vaccine. Moreover, in very early human clinical trials, the tolerability of aluminum hydroxide-adsorbed vaccine was superior to that of the aqueous vaccine.[16] As a consequence, all subsequent trials utilized the adjuvanted vaccine, which ultimately became licensed.

It should be noted that the current product consists of two vials stored at 4°C, one containing lyophilized PRP-OMPC conjugate and the other containing aluminum hydroxide in saline which is used to rehydrate the lyophilized conjugate on the day of administration. An alternative form of storage currently under development (Fig. 1) is a single vial containing the conjugate already adsorbed to aluminum hydroxide. Ongoing studies of stability and immunogenicity would validate the equivalence of this alternate form of storage to the current lyophilized conjugate. This process represents a stage in the evolution of the vaccine which will facilitate its use in the field. Any such change for another vaccine product should be supported by appropriate studies of stability and immunogenicity.

PedvaxHIB - PROCESS

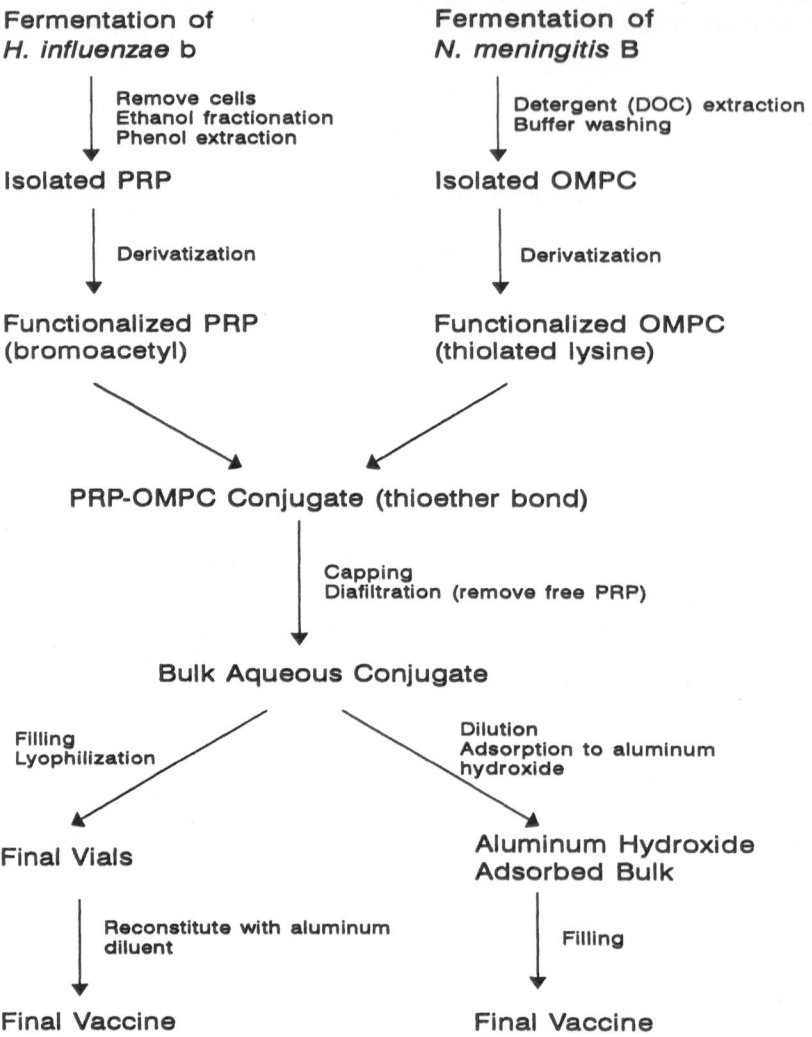

Figure 1. Process for the PRP-OMPC conjugate vaccine.

QUALITY CONTROL

Extensive testing is performed upon samples at various stages in the production of the conjugate vaccine to assure consistency and high quality from lot to lot. Some of the stages at which testing is performed are purified PRP, purified OMPC, aqueous PRP-OMPC conjugate, and final vaccine vial (lyophilized, rehydrated, or preadsorbed). The quality control assays, too numerous to list here, afford a thorough biochemical characterization of the vaccine and its intermediates. These assays test for product attributes such as: 1) purity--including freedom from undesirable or unnecessary Hib/NMB microbial macromolecules and chemicals; 2) identity--a serological or chemical proof that the product is the one intended for manufacture; 3) potency--*in vitro* test(s) that the product is composed of the necessary quantity of antigen (PRP) and/or an *in vivo* test in an experimental species showing that the final vaccine elicits a sufficient level of specific anti-PRP; and 4) safety--*in vitro* and *in vivo* tests which can detect a substance that might be deleterious to a vaccinee. Specifications are set for each of these tests based upon the range of test results for lots used during clinical trials, where the lots have proven to be relatively safe and efficacious in these trials. Following licensure of the vaccine, an unexplained failure of a test result to fall within the range of specifications may disqualify that lot for release to marketing.

Each vaccine lot is labelled for expiration dating, based on an interval during which the vaccine is expected to remain relatively safe and efficacious. To establish the dating period, certain quality control tests are selected as being predictive of the stability of the product. These stability tests are performed on lots of vaccine produced at manufacturing scale, by the process which ultimately would be licensed, and stored at the temperature at which the vaccine would be stored in the field. A dating period of 2 years at 4°C is typical of a vaccine.

Extensive preclinical immunogenicity tests in mice and infant monkeys were performed on the PRP-OMPC conjugate vaccine in the course of its development.[15] These tests showed that the covalent linkage of PRP to OMPC is required for an immune response in infant monkeys, insofar as a conjugate in which covalent linkage of PRP to OMPC could not be shown[17] was not immunogenic in either infant monkeys or infant humans. The conjugation chemistry enables proof of the covalent linkage of PRP to OMPC through chromatographic analysis of an acid hydrolysate of the conjugate for the presence of a novel amino acid that can be formed solely as a result of covalent conjugation. This test was implemented as a quality control assay for the conjugate vaccine. While many types of quality control tests are general to all vaccines, this test for covalent linkage is an excellent example of an assay customized to an important attribute of a particular vaccine.

CLINICAL SAFETY

Upon completion of the initial phases of product development which include definition of the product, of the initial process and of quality control assays and their tentative specifications, an initial vaccine lot for clinical evaluation is produced. An Investigational New Drug (I.N.D.) application is filed with the F.D.A., which requests permission to evaluate the new product in human volunteers under Informed Consent. If evaluation of the I.N.D. application does not suggest the risk of untoward reactions or compromise to the health of the volunteers, then clinical evaluation begins. The evaluation is divided into three phases which encompass (I) safety, (II) immunogenicity, and (III) efficacy.

The Phase I evaluation for the PRP-OMPC conjugate vaccine involved the reporting of temperature and local and systemic reactions post-vaccination at 6 hours, 24 hours, 48 hours and daily up to 5 days and of serious events for up to 14 days. Such studies began in adult volunteers and then older children before proceeding with infants. The vaccine was found to be generally very well tolerated, with a low rate of clinical complaints which did not increase with time following vaccination or dose number.[18] This type of safety evaluation continued throughout the clinical trials for the vaccine and is typical of that for other subunit or inactivated vaccines.

CLINICAL IMMUNOGENICITY

One of the keys to Phase II evaluation of a vaccine is the identification of the most useful serological assay for following the immune response after vaccination. Ideally, the assay would be a correlate of protective immunity (i.e., measure antibodies which mediate protection). In this sense, the anti-PRP assay measures antibodies which prevent Hib infection, as discussed in the Target Antigen Section above. Since the RIAs which measure anti-PRP have shown significant interlaboratory variation,[19] an attempt has been made to standardize the RIA based upon the adoption of a common protocol and the same serological standard. Nevertheless, direct comparisons of RIA data for different vaccines can be made only when sera for different Hib vaccines have been assayed in the same laboratory.

Some of the key clinical questions asked during Phase II trials of a vaccine relate to age of immunization, dosage regimen and dosage level of the vaccine. As described above in the Epidemiology Section, it is important to begin the immunization regimen at 2 months of age, since the incidence of Hib meningitis rises beginning at that age and peaks in many populations during the first year of life. Early studies showed that administration of the second dose of the PRP-OMPC vaccine at 4 months of age elicited higher titers of anti-PRP than a second dose at 3 months of age.[20] Moreover, the regimen of 2 and 4 months was favored since DTP and polio vaccines also are administered by pediatricians in the United States at those ages. A third dose at 6 months of age does not boost anti-PRP levels significantly higher than post-dose 2 levels. However, a third dose at 12-15 months of age was shown to boost anti-PRP levels to significantly higher levels than post-dose 2 levels. Therefore, the vaccine has become indicated for a three-dose regimen at 2, 4 and 12-15 months of age. Initial dose-ranging studies supported the use of a 15-μg (PRP content) dose of PRP-OMPC conjugate vaccine. This dosage level was used in most of the Phase II studies and in the Phase III efficacy study described below (Clinical Efficacy Section) and has become the dosage level for the licensed vaccine.

A single 15-μg dose of the PRP-OMPC conjugate vaccine elicits a geometric mean titer (GMT) of 2.4 μg anti-PRP/mL in 2- to 6-month-old infants, with 78% achieving titers $\geq 1.0 \mu$g anti-PRP/mL and 97% at $\geq 0.5 \mu$g/mL.[18] The corresponding GMT and percent $\geq 1.0 \mu$g and percent $\geq 0.15 \mu$g/mL are 4.6 μg/mL, 88% and 99% following a second dose 2 months later. Most of the anti-PRP response is IgG, essentially all of which is IgG1 in infants. The vaccine can prime infants to produce IgG1 and IgG2 anti-PRP in response to a third dose of PRP-OMPC given in the second year of life.[21] The anti-PRP elicited at all ages in infants is active at sufficient levels in three Hib functional assays, opsonophagocytic with intracellular killing,[22] bactericidal and passive protection of infant rats from Hib bacteremia.[18] Therefore, a very high proportion of 2-month-old infants attain a level of anti-PRP that had been associated with protection against clinical Hib infection in the passive immunization studies described above (Target Antigen Section) and which is biologically active in 3 Hib functional assays. Studies of this type described for the PRP-OMPC conjugate are representative of those performed in Phase II studies for other vaccines and lay the groundwork for a protective efficacy trial.

CLINICAL EFFICACY

Phase III clinical studies test the protective efficacy of the vaccine. An ideal design is a double-blind, placebo-controlled study, which eliminates vaccinee (parental) and physician bias, either of which might influence clinical observations. For the PRP-OMPC vaccine, there were two additional stringencies placed upon the design. It was considered worthwhile to evaluate efficacy in a study population which is hyporesponsive to Hib conjugate vaccines and which has a peak incidence of Hib disease early during the first year of life. If efficacy were demonstrated in such a population, then efficacy would be assured in other hyporesponsive and normally responsive populations regardless of the time of peak incidence of disease. The Navajo Indians in North America fulfill these two criteria, with a peak incidence of disease at 3-9 months of age and GMT anti-PRP of 0.9 and 1.3 μg/mL post-doses 1 and 2[23] (compared to 2.4 and 4.6 μg/mL for the general population).

When the study was carried out in this population, the rates of efficacy were 100% after a single dose of vaccine and 93% after 2 doses of vaccine.[23] This demonstration of efficacy following a single dose in 2-month-old infants is consistent with the immunogenicity profile of the PRP-OMPC conjugate vaccine. While this study validated the use of the anti-PRP RIA as a qualitative correlate of protection, it has not been possible from this or other protective efficacy studies to assign an absolute quantitative level of anti-PRP which correlates with protection against Hib clinical disease following vaccination of infants with conjugate vaccines. In employing the additional criteria beyond the classic double-blind placebo-controlled study design, the PRP-OMPC efficacy trial is a useful model for other vaccines where an appropriate study population may be utilized which tests the performance of the vaccine in a more stringent fashion.

MANUFACTURING PROCESS AND FACILITY

While Phase I and II clinical studies are underway, plans are made for the ultimate manufacture of the vaccine for licensure and marketing. An estimate is made of the annual market size for the vaccine which, adjusted for dosage level, translates into the annual bulk vaccine requirement. The initial process is developed in a piloting facility and scaled up to support annual dosage requirements, to be as consistent as possible, to minimize cost and manipulations, and to minimize capital requirements. Once the process is developed, the final production facility is constructed. Typically, while the Phase III study is underway, three to five lots are produced in the final manufacturing facility by manufacturing personnel using Standard Operating Procedures for the production process to be licensed. Following release by validated quality control assays, three or more of these manufacturing consistency lots are tested clinically for safety and immunogenicity, to be at least comparable to those of lots used in the Phase III study. These lots also are placed on long-term stability studies, with periodic testing (e.g., 0, 1, 2, 6, 12, 18, 24, 30, 36 months) in the designated stability-indicating assays, in order to establish a dating period for the vaccine. All of the preclinical, clinical, process, and stability data are assembled into a Product License Application (P.L.A.). At the same time, a detailed description of the manufacturing facility is assembled into an Establishment License Application (E.L.A.). Following submission of the P.L.A. and E.L.A., discussions are held with the F.D.A., typically including session(s) before an External Advisory Committee made up of academic experts in the field. Following the granting of licensure by the F.D.A., marketing of the product commences.

REFERENCES

1. C.V. Broome, Epidemiology of Haemophilus influenzae type b infections in the United States, Pediatr. Infect. Dis. J. 6:779 (1987).

2. H.A. Bijlmer, L. van Alphen, B.M. Greenwood, J. Brown, G. Schneider, A. Hughes, A. Menon, H.C. Zanen, and H.A. Valkenburg, The epidemiology of Haemophilus influenzae meningitis in children under five years of age in The Gambia, West Africa, J.Infect. Dis. 161:1210 (1990).

3. J.I. Ward, Invasive Haemophilus influenzae type b disease in Alaska: Background epidemiology for a vaccine efficacy trial, J. Infect. Dis. 153:17 (1986).

4. D. Ambrosino, J.R. Schreiber, R.S. Daum, and G.R. Siber, Efficacy of human hyperimmune globulin in prevention of Haemophilus influenzae type b disease in infant rats, Infect. Immun. 39:709 (1983).

5. K.L. Cates, H.K. Marsh, and D.M. Granoff, Serum opsonic activity after immunization of adults with Haemophilus influenzae type b PRP-D conjugate vaccine, Infect. Immun. 48:183 (1985).

6. H. Kayhty, H. Peltola, V. Karanko, and P.H. Makela, The protective level of serum antibodies to the capsular polysaccharide of Haemophilus influenzae type b, J. Infect. Dis. 147:1100 (1983).

7. M. Santosham, R. Reid, D.M. Ambrosino, M.C. Wolff, J. Almeido-Hill, C. Priehs, K.M. Aspery, S. Garrett, L. Croll, S. Foster, G. Burge, P. Page, B. Zacher, R. Moxon, and G.R. Siber, Prevention of Haemophilus influenzae type b infections in high-risk infants treated with bacterial polysaccharide immune globulin, N. Engl. J. Med. 317:923 (1987).

8. H. Peltola, H. Kayhty, A. Sivonen, and P.H. Makela, Haemophilus influenzae type b capsular polysaccharide vaccine in children: a double-blind field study of 100,000 vaccinees 3 months to 5 years of age in Finland, Pediatr. 60:730 (1977).

9. S.B. Black, H.R. Shinefield, R.A. Hiatt, B.H. Fireman, and The Kaiser Permanente Pediatric Vaccine Study Group, Efficacy of Haemophilus influenzae type b capsular polysaccharide vaccine, Pediatr. Infect. Dis. J. 7:149 (1988).

10. M.T. Osterholm, J.H. Rambeck, K.E. White, J.L. Jacobs, L.M. Pierson, J.D. Neaton,C.W. Hedberg, K.L. MacDonald, and D.M. Granoff, Lack of efficacy of Haemophilus b polysaccharide vaccine in Minnesota, J. Am. Med. Assoc. 260:1423 (1988).

11. D.M. Granoff and M.T. Osterholm, Safety and efficacy of Haemophilus influenzae type b polysaccharide vaccine, Pediatr. 80:590 (1987).

12. O.T. Avery and W.F. Goebel, Chemo-immunological studies on conjugated carbohydrate proteins. II. Immunological specificity of synthetic sugar-protein antigens. J. Exp. Med. 50:533 (1929).

13. P.P. Vella and R.W. Ellis. "Vaccines - New Approaches to Immunological Problems," R.W. Ellis (ed.), pp. 2-20, Butterworth-Heinemann, Boston (1992).

14. S. Marburg, D. Jorn, R.L. Tolman, B. Arison, J. McCauley, P.J. Kniskern, A. Hagopian, and P.P. Vella, Bimolecular chemistry of macromolecules: Synthesis of bacterial polysaccharide conjugates with Neisseria meningitidis membrane protein, J. Am. Chem. Soc. 108:5282 (1986).

15. P.P. Vella, J.M. Staub, and J. Armstrong, Immunogenicity of a New Haemophilus influenzae type b conjugate vaccine (meningococcal protein conjugate) PedvaxHIB,) Pediatr. 85 (Supplement):668 (1990).

16. M.S. Einhorn, G.A. Weinberg, E.L. Anderson, P.D. Granoff, and D.M. Granoff, Immunogenicity in infants of Haemophilus influenzae type b polysaccharide in a conjugate vaccine with Neisseria meningitidis outer-membrane protein, Lancet 2:299 (1986).

17. J.Y. Tai, P.P. Vella, A.A. McLean, A.F. Woodhour, W.J. McAleer, A. Sha, C. Dennis-Sykes, and M.R. Hilleman, Haemophilus influenzae type b polysaccharide-protein conjugate vaccine, Proc. Soc. Exp. Biol. Med. 184:154 (1987).

18. V.I. Ahonkhai, L.J. Lukacs, L.C. Jonas, H. Matthews, P.P. Vella, R.W. Ellis, J.M. Staub, K.T. Dolan, C.M. Rusk, G.B. Calandra and R.J. Gerety, Haemophilus influenzae type b conjugate vaccine (meningococcal protein conjugate) (PedvaxHIB,): clinical evaluation, Pediatr. 85 (Supplement):676 (1990).

19. J.I. Ward, D.P. Greenberg, P.W. Anderson, K.S. Burkart, P.D. Christenson, L.K. Gordon, H. Kayhty, J.S.C. Kuo, and P. Vella, Variable quantitation of Haemophilus influenzae type b anticapsular antibody by radioantigen binding assay, J. Clin. Microbiol. 26:72 (1988).

20. A.A. Lenoir, P. Granoff, and D.M. Granoff, Immunogenicity of Haemophilus influenzae type b polysaccharide-Neisseria meningitidis outer membrane protein conjugate vaccine in 2- to 6-month-old infants, Pediatr. 80:283 (1987).

21. D.M. Granoff, A. Chacko, K. Lottenbach, and K. Sheetz, Antibodyresponses of vaccine failure (VF) patients reimmunized wit Haemophilus influenzae type b polysaccharide (HIB PS) or HIB-PS meningococcal outer membrane protein (OMP) conjugate vaccine, Pediatr. Res. 23:916A (1988).

22. B.M. Gray, Opsonophagocidal activity in sera from infants and children immunized with Haemophilus influenzae type b conjugate vaccine (meningococcal protein conjugate), Pediatr. 85 (Supplement):694 (1990).

23. M. Santosham, M. Wolff, R. Reid, M. Hohenboken, M. Bateman, J. Goepp, M. Cortese, D. Sack, J. Hill, W. Newcomer, L. Capriotti, J. Smith, Marjorie Owen, S. Gahagan, D. Hu, R. Kling, L. Lukacs, R.W. Ellis, P.P. Vella, G. Calandra, H. Matthews, and V. Ahonkhai, The efficacy in Navajo infants of a conjugate vaccine consisting of Haemophilus influenzae type b polysaccharide and Neisseria meningitidis outer-membrane protein complex, N. Engl. J. Med. 324:1767 (1991).

SIGNIFICANCE OF IMMUNE RESPONSES TO ORAL ANTIGENS

IN DENTAL DISEASES

Martin A. Taubman and Daniel J. Smith

Department of Immunology
Forsyth Dental Center
140 Fenway
Boston, MA 02115

INTRODUCTION

Immunological Correlates of Protection: Role of Antibodies in Dental Caries

Since dental caries is an infectious disease, one might expect that some of the general rules of infections might pertain. However, mutans streptococcal infections of the teeth are sequestered from many components of the systemic immune system. Therefore, local secretory immunity may play a major role in oral host defenses against these infections. After dental caries occurs, convalescence might be expected to be followed by a rise in antibody level. Frequently such antibody may exist (in serum) for months or years.[1]

RELATIONSHIP BETWEEN ANTIBODY AND DISEASE

Serum Antibody

Correlations have been sought between dental caries experience and the levels of serum or salivary antibody to oral streptococcal antigens. The purpose has been to ascertain whether there is protection afforded by antibody by seeking associations with either serum or salivary antibodies. These types of studies have only been vaguely informative because of the numerous subtle differences in experimental details and the difficulty in interpretation of retrospective measures of disease and immunity. Despite this, several investigations have reported that high levels of serum IgG antibody to mutans streptococcal cells or antigens were associated with low levels of caries in young adults (see Table 1). Serum antibody from caries-free young adults inhibited glucosyltransferase enzyme (GTF) from mutans streptococci to a greater extent than serum from caries-prone individuals.[2] Others suggested that low numbers of mutans streptococci in dental plaque were associated with relative elevations in serum IgG antibody in children approximately 2 to 5 years old.[3] It has been suggested that the ability to sustain critical antibody levels during initial infectious challenge with mutans streptococci may be a key element. Similar findings of raised serum antibody to antigen I/II in a low caries population have been described.[4] In general, these cross-sectional observations in young adults favor an interpretation that serum IgG antibody

Table 1. Associations of serum antibody to mutans streptococci and dental caries: studies in humans.

STUDY	SUBJECTS	RESULTS
Challacombe, 1980[10]; Challacombe et al., 1984[4]	Young adults	High serum IgG antibody correlated with low dental caries
Gregory et al., 1986[7]	Adults	As above
Block et al., 1979[2]	Young adults	Serum IgG antibody from caries free inhibits GTF more than IgG from caries prone
Aaltonen, et al., 1985[3]	Children (> 2.6 years)	High serum IgG antibody associated with low numbers of mutans streptococci
Lehner et al., 1978[67]	Children (> 2.5 years)	High serum IgG antibody associated with low dmfs
Kent et al., 1991[40]	Older adults	Elevated serum IgG antibody associated with more caries experience

to mutans streptococcal antigens can be associated with lower caries experience. Of greater interest is the finding that young individuals with active caries have an elevated specific serum antibody level to mutans streptococci. In contrast, in older adults serum IgG antibody levels to mutans streptococci are significantly and positively associated with clinical variables which emphasize measures of cumulative caries experience. Thus, some aspects of the serum IgG antibody response to mutans streptococci seem to be directly related to the antigenic challenge.

Salivary Antibody

Cross-sectional studies relating dental caries experience with levels of salivary IgA, or IgA antibody to mutans streptococci, or their antigens, have been confusing and subject to conflicting interpretations. Initial investigations attempted to relate the concentrations of IgA in whole or glandular saliva with the level of dental caries in the subjects.[5] With respect to antibody levels, some studies found elevated levels in subjects with little or no caries[6-8], while other studies found elevated levels of IgA antibody only in subjects with active lesions.[1,7,9,10] Thus, correlations based on adult salivary IgA antibody concentrations have been inconclusive (see Table 2). However, some of these findings may support a more general concept that an increased antigenic load (from bacteria shed from lesions) can give rise to elevated antibody. More importantly, the most critical aspect of the IgA antibody level may be the coordinated presence of elevations in individual IgA antibody levels to critical streptococcal antigens at phases of infection or disease which are vulnerable to such intervention.

There have been several confounding variables in these types of studies, both with respect to clinical features and to the measurement of antibody. Specifically, these include: 1) Salivary antibody levels are more variable than serum antibody levels; 2) the IgA levels in saliva are markedly dependent on the rate of salivary flow; and 3) IgA1-specific proteases[11,12] produced by microorganisms and present in the oral cavity including *Streptococcus mitis* and *S. sanguis* which are among the first colonizers of the tooth surface. Each of these factors may be significant in regulating the IgA antibody concentration and repertoire of specificities in the oral cavity.

Table 2. Associations of salivary antibody to dental caries: studies in humans.

STUDY	SUBJECTS	RESULTS
Brandtzaeg, 1983[5]	Adults	Salivary IgA levels not correlated with dental caries
Lehtonen et al., 1984[6]	Adults	High IgA antibody to mutans streptococci associated with low or no dental caries experience
Gregory et al., 1986[7]	Adults	As above
Bolton & Hlava, 1982[8]	Children (3-14 years)	As above
Huis in't Veld et al., 1978[9]	Young adults	High IgA antibody only associated with active lesions
Challacombe, 1980[10]	Adults	As above

DENTAL CARIES AND IMMUNOLOGICAL DEFICIENCIES

In a group of studies, selective IgA deficient subjects were subdivided on the basis of the presence of IgM compensation. Patients without salivary antibody showed markedly elevated levels of dental caries when compared to matched controls that had antibody in their saliva.[13,14] On the other hand, IgA-deficient subjects with IgM compensatory antibodies, had significantly less caries than IgA- and IgM-deficient subjects. In fact, those with this form of compensation may have actually had less caries than matched normal controls. Little investigation of dental caries in other human deficiency states have been performed.

At present little or no information is available as to the relationship of HIV infection and subsequent dental caries experience. Therefore, this population can be useful to dissect the effects of immune components on dental caries infection.

Studies in rats deficient in T cells (e.g., congenitally athymic) indicate more dental caries after mutans streptococcal infection than in euthymic counterparts.[15] T cell function is critical in the generation of normal secretory IgA and other antibody isotypes. Athymic rats produced much less antibody with greater infection, suggesting that the immunologic defect may lead to a greater infection level because of the loss of the regulatory aspect of the host immune response on the oral flora. In general, the results of these animal studies support the findings in human immunodeficient subjects and reinforce the concept that antibody affects oral cariogenic microorganisms.

PREVENTIVE ASPECTS OF VACCINES IN DENTAL CARIES

Immunological Protection and Dental Caries

The discovery of a separate and unique mucosal immune system in 1965[16] suggested that it might be possible to interfere with caries by stimulating salivary antibodies to appropriate bacterial antigens.[17-19] A variety of routes of antigen administration have resulted in protection from dental caries in animals. Most significant were the initial studies designed to stimulate directly the gut-associated lymphoreticular tissues (GALT) by ingestion of killed mutans streptococci.[20,21] These studies demonstrated that consistent induction of secretory IgA antibody to a dental pathogen resulted in reductions in both disease and cariogenic infection. Importantly, the previous notion that salivary IgA antibody was a protective principle was reinforced by these investigations since virtually no serum (IgG) antibody could be detected.

Immunization with intact streptococcal cells would elicit immune responses to an array of cell-bound components likely to be involved in several of the phases of colonization, accumulation, or acid production. Antibody to the serotype carbohydrate is uniformly distributed on the mutans cell surface.[22] Conceivably, antibody to this or any antigen at or near the cell surface could interfere with important bacterial pathogenic features or result in clearance by virtue of stearic hindrance. However, undesirable antibody responses may follow. For example, membrane-bound components can induce antibody that cross reacts with host tissue.[23]

Table 3. Clinical trials of response to glucosyltransferase or tetanus toxoid vaccine.[31,35]

TRIAL	N	ANTIGEN	ROUTE	PAROTID AB	LABIAL AB	SERUM AB	MUTANS* ACCUMULATION
I	25	GTF	Oral capsules	+ +	±	-	Reduced
IIA	23	GTF	Topical	+	±	-	Reduced
IIB	23	Tetanus toxoid	Topical	+	Insufficient fluid	-	Not tested
IIC	13	Tetanus toxoid	Topical	+	Insufficient fluid	-	Not tested
III	27	GTF	Oral/topical	+	±	-	Reduced

* Significant difference in reaccumulation of indigenous mutans streptococci in immunized groups after a dental prophylaxis.

Vaccine Approach Related to Mutans Streptococci

Several steps in the colonization, accumulation, or acid production of the mutans streptococci may be vulnerable to immunological attack.[24] The initial interaction in colonization takes place with the pellicle-coated tooth surface. The components on the mutans streptococci that participate in pellicle binding appear to be protein.[25] Following the initial attachment phase, mutans streptococci accumulate. These streptococci produce GTFs, which then synthesize water-soluble and water-insoluble glucans from sucrose. Mutans streptococci and certain other oral streptococci express cell surface proteins which serve as binding sites for these extracellular glucans. The interactions between GTF, glucans and cell-bound glucan-binding proteins appear to be fundamental to the accumulation process. This mechanism results in an increase in numbers of potentially cariogenic bacteria in plaque and can be inhibited *in vitro* by specific antibody to GTF.[24,26]

There has been a focus on identifying safe and effective antigens from the mutans streptococci for use in experimental immunization. Examples of such antigens are the cell wall antigen I/II[27] also known as SpaA, SpaP, B or P1, which may be involved in the initial attachment phase, and components of the GTF enzyme system that participate in the accumulation phase.[24,28,29] Present strategies promise to provide safe and effective vaccines. Most of the strategies involve active immunization.

We have utilized vaccines as probes for associating immunological and microbiological correlates of dental diseases (studies summarized in Table 3). A vaccine containing GTF prepared from *S. sobrinus*, suspended in aluminum phosphate (AP), was orally administered in gelatin capsules to 25 adult males aged 18 to 36 years.[30] The subjects were assigned to a vaccine or placebo group. Saliva samples were taken during an initial screening and antibody levels were compared with those in the parotid saliva of these same individuals taken prior to the immunization procedure. Parotid saliva S-IgA antibody rose after oral antigen ingestion. A unique feature of this study was the use of a defined soluble antigen combined with a suspension of aluminum phosphate. We also found that GTF administration resulted in retardation of reaccumulation of indigenous mutans streptococci after a dental prophylaxis.[30] Presumably these effects were related to the salivary IgA antibody to *S. sobrinus* GTF which inhibits GTF of other serotypes. Antibody to water-insoluble glucan synthesizing GTF can markedly inhibit glucan synthesis and subsequent accumulation of mutans streptococci *in vitro*[26] and *in vivo*. This effect may result from direct inhibition of GTF-mediated glucan synthesis giving rise to alteration in the nature of dental plaque and restrictions in availability of the glucan to binding sites for accumulation of mutans streptococci.

INDUCTIVE CAPABILITIES OF SALIVARY GLANDS

An approach to determine if inductive elements for a mucosal response are present in salivary glands has involved more direct antigen placement on or in oral mucosa or salivary glands to stimulate local or distant GALT elements. Direct instillation of mutans streptococci into parotid ducts resulted in elevated antibody production and interference with bacterial colonization.[31] Intramucosal injection of monkeys with *S. mutans* antigen resulted in serum, but no detectable salivary antibody. Topical application of a small *S. mutans* antigen (3.8 kDa) and dimethyl sulfoxide to monkey gingival tissues[32] resulted in crevicular IgG, no serum IgG antibody, and whole saliva IgA antibody, and in reduced caries and proportions of plaque *S. mutans*.[33] The mechanism of this IgA induction remains unclear although the antibody could have been derived from crevicular fluid. Other studies of topical mucosal application in rodents with whole *S. mutans* did not result in detectable antibody but did appear to reduce *S. mutans* colonization.[34] Questions of duration remain.

Local application of GTF (combined with AP) onto lower labial surfaces was investigated in 23 subjects.[35] GTF (or placebo) was administered to the lower lip daily for five days (Table 3). The delay in reaccumulation of indigenous mutans streptococci was significantly associated ($p < 0.025$) with elevations in parotid IgA antibody levels. These results suggest that labial application of GTF-AP can modify the salivary immune environment to retard the reaccumulation of indigenous mutans streptococci.

Salivary tissues contain the principal cellular components required for immune response to antigen.[36] Deposited antigen could be captured by resident mononuclear phagocytes after endocytosis by ductal epithelium[37] or other specialized cells.[38] Antigen could be presented directly to immunologically competent cells in the labial gland, or be transported into local lymphatics. In either case, the recirculating IgA precursor cell population could be expanded and migrate to the parotid gland. To investigate this further, tetanus toxoid was also applied topically to the minor salivary gland region of 23 subjects[35] in two experiments (see Table 3). In these experiments parotid salivary IgA antibody levels were elevated. In a third clinical trial we studied the ability to drive precommitted IgA to differentiate as these cells traversed the minor salivary glands[39] in 27 patients (Table 3). An oral dose of antigen (GTF) was followed 8 and 9 days later by doses of labial applied antigen. Some elevation of parotid salivary IgA antibody to GTF was observed, but the most striking finding again was the very significant retardation of indigenous mutans streptococcal reaccumulation in the immunized group.

Table 4. Regulatory role of antibody: studies in humans.

STUDY	SUBJECTS	RESULTS
SALIVARY IgA ANTIBODY		
Gahnberg & Krasse, 1983[28]	Adults	Increased clearance of labelled *S. mutans* challenge in subjects with high salivary IgA
Gregory et al., 1985[41]	Adults	Preexisting salivary IgA to one serotype resulted in rapid elimination of challenge with the same serotype
Camling & Kohler, 1987[42]	Children	Increased IgA antibody in children with no caries vs. children with greater than two df surfaces
SERUM IgG ANTIBODY		
Challacombe, 1980[10] Challacombe et al., 1984[4] Gregory et al., 1986[7]	Young adults	High levels of serum IgG antibody associated with low caries experience
GINGIVAL CREVICE FLUID (GCF) ANTIBODY		
Krasse et al., 1987[68]	Adults	Inverse relationship between implanted *S. mutans* and antibody to *S. mutans* in GCF
Camling et al., 1991[43]	Adults	Higher recovery of implanted *S. mutans* from surfaces with low GCF IgG antibody
Smith et al., 1991[44]	Older adults	No detectable GCF antibody to mutans streptococci associated with increased colonization

ANTIBODY REGULATION OF ORAL MUTANS STREPTOCOCCAL INFECTIONS

Congenitally athymic rats develop more dental caries and are more extensively colonized with mutans streptococci than control euthymic rats.[15] These findings, along with the human experiments in which immunization was accompanied by a reduction in indigenous cariogenic bacteria[30,35] were interpreted to indicate that antibody could be viewed as regulating the colonization or accumulation of the mutans streptococci. While this concept has not been tested directly, there are several compelling experiments and data which suggest, indirectly, that such may be the case. Data are presented in Table 4 for both root and coronal surfaces, and for IgA and IgG antibodies.

Serum IgG Antibody. We[40] found that only serum IgG antibody to *S. mutans* showed a significant positive correlation with subject decayed missing filled (DMF) scores. No strong associations were found with serum antibody to other supragingival bacteria. The results support the involvement of mutans streptococci in the root caries processes and suggest that substantial antibody has been formed in response to infection. We consider that the presence of such antibody could potentially exert a regulatory effect on the cariogenic bacteria.

Salivary IgA Antibody. Studies[41] have suggested that the presence of preexisting indigenous antibody to a particular serotype can result in rapid elimination of that serotype, but not of serotypes to which little or no salivary IgA antibody is present. An increased clearance of labelled *S. mutans* challenge has been observed in subjects with high salivary IgA antibody to *S. mutans* antigens.[28] Also, IgA antibody to *S. mutans* was significantly higher in children with no caries experience (n=22) when compared to children with more than two DF surfaces(n=17).[42] Conceivably, salivary IgA can interfere with colonization or accumulation of mutans streptococci on more coronal

Table 5. Mutans streptococcal recolonization of 46 cleaned root surfaces.[44]

IgG Antibody to Mutans Streptococci in Adjacent Gingival Crevicular Fluids	Sites with Mutans Increase after 24 Hours		
	< 2 Fold	≤ 10 Fold	> 10 Fold
Antibody Present	50%	74%	26%
Antibody Absent	29%	36%	64%

* p < 0.05, Chi Square analysis.

surface aspects, whereas serum or gingival crevice IgG may influence the areas most approximating the gingivae.

Gingival Crevicular Fluid (GCF) IgG Antibody. Camling et al.[43] have demonstrated a tendency to higher recovery of mouth rinse implanted *S. mutans* from surfaces with low GCF IgG antibody than from surfaces with high GCF antibody activity. We have investigated[44] the effect of preexistent levels of GCF antibody to mutans streptococci on recolonization and accumulation of indigenous microbiota after cleaning of exposed root surfaces. The increase in indigenous mutans streptococci was monitored 24 hours after the cleaning. Sites were ranked with respect to the increase in the respective organism. Gingival crevice fluids containing antibody were significantly associated with root surfaces having low increases in mutans streptococcal colony forming units (CFU). In a total of 46 sites examined, 74 percent of the sites with antibody to mutans streptococci showed an increase of 10-fold or less, while 64 percent of the sites with no GCF antibody showed increases greater than 10-fold (Table 5). Also, 50 percent of the sites with antibody showed an increase of mutans streptococci equal to, or less than 2-fold, whereas only 29 percent of sites with no detectable antibody showed such a low increase. These findings provide preliminary evidence that early recolonization of exposed root surfaces can be regulated, to some extent, by the level of GCF IgG antibody to mutans streptococci on a site-specific basis. Some indication as to the mechanism(s) of this regulatory effect of antibody is suggested by the demonstration[45] that levels of salivary

IgA and serum IgG antibodies to *S. mutans* antigens were significantly higher in caries-resistant (CR) subjects than in caries-susceptible (CS) subjects. Salivas and sera from CR subjects significantly inhibited *S. mutans* growth and adherence, acid production and GTF and glucose-phosphotransferase activities more than salivas or sera from CS. Inhibition of one or more of these pathogenic processes *in vivo* could result in stasis of *S. mutans* infection.

SYNTHETIC PEPTIDES AS POTENTIAL VACCINES

The concept of synthetic vaccines for infectious bacterial or viral diseases emanates from the demonstration that chemically synthesized peptides can elicit antibodies which react with the original protein.[46] Two characteristics of critical importance in investigations of synthetic peptide vaccines are immunogenicity and T cell recognition. Immunogenicity is dependent on the quality of Th cell function, immunization protocol, nature of T cell - B cell interactions and the genetics of the immunized animal. Dual recognition of antigen by T cells and MHC molecules is determined by an individual's phenotype. High percentages of human T lymphocytes proliferate to single peptides.[47] Several factors can be managed in synthetic vaccine design which could enhance its use as antigen instead of the parent protein. The feasibility of this approach as been emphasized by Lehner[48] who used synthetic peptides to elicit secretory antibody and to confer immunity to humans. Studies have deduced the B and T cell regions of the amino terminal epitope of a 3,000 dalton streptococcal peptide. The minimal peptide containing 17 amino acids as a dimer will elicit both B cell and T cell immunity. Most individuals appear to be capable of responding to this material with serum and salivary antibody and also with CD4+ T cell proliferative responses.[49] The 17 amino acid peptide (as a dimer) in dimethyl sulfoxide was applied directly to the gingivae of monkeys and appeared to initiate both gingival IgG and salivary IgA antibodies.[32] Other studies[50] employed peptides from the *S. mutans* PAc antigen coupled to cholera toxin B subunit for intranasal immunization. The immunization suppressed colonization of mouse teeth after a challenge of mutans streptococci. Synthetic peptide technology and various passive immunization stratagems are mechanisms that could also avert any potential heart reactive antibody formation [23]. We[69] have recently synthesized a 22 residue peptide whose sequence surrounds a GTF catalytic region, DSIRVDAVD, described by Mooser and coworkers.[51] This peptide contains nearly complete homology between GTFs from *S. sobrinus* and *S. mutans*. Furthermore, extensive sequence homology exists in this region between GTFs which synthesize insoluble and soluble glucan. Thus, epitopes contained on this peptide are likely to elicit cross-reactive immune responses among GTFs from the most common mutans streptococci that infect man. We have shown this peptide (CAT) to be antigenic in humans and immunogenic in rodents. We have also synthesized another peptide (GLUa), derived from a putative glucan-binding region of GTF. Twenty-nine human sera and salivas were screened for IgG or IgA antibody activity, to this peptide. Five sera and four salivas contained significant reactivity. The GTF-inhibitory activity of four of the five sera could be significantly diminished by preincubation with the GLUa. Furthermore, antibody to intact GTFs from both *S. sobrinus* and *S. mutans* react with both the GLUa and CAT peptides, indicating shared epitopes between the synthetic peptides and native GTF and that antibody to epitopes on these peptides may affect GTF-mediated activities *in vivo*. The immunogenicity of the CAT and GLUa synthetic peptides was measured after injection of Sprague-Dawley rats. These peptides elicited vigorous serum IgG antibody responses. Sera from rats immunized with CAT or GLUa reacted with intact GTF from both *S. mutans* and *S. sobrinus* in ELISA and Western blot. Also, incubation of splenocytes or lymph node lymphocytes from *S. sobrinus* GTF-injected rats with the GLUa induced significant proliferative responses. These combined observations support the presence of T cell epitopes and the existence of shared epitopes between CAT or GLUa peptides and intact mutans streptococcal GTF.

TARGET POPULATIONS FOR DEVELOPMENT OF A DENTAL CARIES VACCINE

The target population for the caries vaccine might very well be a young age group. The most appropriate age for immunization remains a major question that must take into account the ontogeny of the secretory immune system and the host age of initial colonization with mutans streptococci. The secretory immune system is not fully developed at birth.[52] At this time, salivary IgA concentrations are very low but increase dramatically during the first year of life.[53] Secretory IgA levels are somewhat lower in children less than two years of age than adult levels. However, these children demonstrate secretory antibody responses to viral antigens that are quite similar to those of older children[54] and also show immunologic memory.[55] Furthermore, most infants less than three months old have measurable levels of salivary IgA antibody to antigens of bacteria that colonize the oral cavity after birth.[56] Therefore, infants can demonstrate secretory immune responses to antigens that are actually encountered.

The colonization of the oral cavity with many oral streptococci, including those that are cariogenic, requires the presence of teeth. *Streptococcus sanguis*, a prominent component of dental plaque, can be found in nearly all children by the end of the first year of life.[54] Salivary antibody to *S. sanguis* GTF is present in many 3- to 5-year old children.[56] Mutans streptococci are not consistently found in the oral cavity until the third year of life or later.[53,54] Salivary antibody to *S. mutans* GTF is not detectable in the great majority of children 4 years of age, although most adults demonstrate detectable IgA antibody to this antigen. Therefore, antibody to *S. mutans* GTF is not present when children are colonized with mutans streptococci.

Theoretically, immunization with GTF or other virulence-associated antigens at approximately 12 months of age would encounter a competent immune system. Salivary IgA antibody to these antigens would then be present to interfere with subsequent colonization and accumulation of mutans streptococci in children. Clinical control of the level of mutans streptococcal infection in mothers[55] has been shown to limit or virtually eliminate the infectious challenge. Children with no detectable mutans streptococci do not have dental caries.[56] Reduction of infection coupled with the presence of appropriate salivary antibody before permanent colonization by mutans streptococci should ultimately result in caries prevention. Development and testing of further strategies utilizing information derived from the molecular pathogenesis of dental infections and the ontogeny and capabilities of the host immune response promise to result ultimately in the complete prevention of dental infections.

PERIODONTAL DISEASES

Immunological Correlates

In individuals with various types of periodontal diseases there are clear-cut systemic immune responses to specific oral bacteria.[57] Some of these systemic responses in various diseases are illustrated in Table 6. Thus, individuals with adult periodontitis show elevated antibody to *Porphyromonas gingivalis*, patients with localized juvenile periodontitis (JP) show antibody to *Actinobacillus actinomycetemcomitans* serotype b, patients with generalized JP show elevated antibody to *A. actinomycetemcomitans* serotype a and patients with rapidly progressive periodontitis show elevated antibody to *A. actinomycetemcomitans* serotype b.[58] Antibody levels to other species may also vary, but these are not necessarily significantly elevated above antibody levels in "normal" subjects with no periodontal disease. Despite the consistency of these antibody elevations and the association with various disease states, the significance of these elevated responses has not been completely delineated.[59]

Antibody in Periodontal Disease Accompanies Infection

Often the specific bacteria to which there are elevated serum responses can be isolated from the active lesions of patients, supporting the concept of a distinct infection leading to immune response.[60] We have confirmed this observation in a separate study in which a different method of bacterial identification was employed.[61]

Role of Antibody in Periodontal Diseases

Human studies have not permitted the evaluation of a protective or destructive role of antibody in periodontal diseases. In human periodontal disease there is little indication that antibody is a correlate of protection. However, studies in animal systems may shed some light on this problem. The earliest immunization studies indicated that such antibody might be protective.[62] In studies in animals utilizing adoptive transfer of cloned antigen-specific T helper cells, subsequent abundant antibody formation and reduced disease suggested that antibody may be protective.[63] However, on the basis of theoretical considerations and several previous demonstrations[64] there is also some evidence that the immune response can contribute to periodontal diseases.

Table 6. Serum antibody levels (ELISA units) of 51 subjects grouped by clinical characteristics.[58]

	CLINICAL DIAGNOSIS			
ANTIBODY TO:	AP[a]	LJP[b]	GJP[c]	RPP[d]
Actinobacillus actinomycetemcomitans 29523 (a)	77±19	208±118	425±51	98±32
A. actinomycetemcomitans Y4 (b)	53± 9	269± 73	96±40	180±75
Porphyromonas gingivalis	25± 5	35± 20	10± 2	20± 8
Prevotella intermedia	86±14	61± 9	100±35	83±22
Fusobacterium nucleatum	53±16	22± 9	36±16	22± 8

[a] Adult periodontitis   [b] Localized juvenile periodontitis
[c] Generalized juvenile periodontitis   [d] Rapidly progressive periodontitis

Studies of the cell composition of human periodontal tissues suggest that there may be an abundance of suppressor activity associated with disease.[65] Furthermore, the majority of $CD4^+$ T cells in periodontal disease might be memory cells presumably being reactivated in the tissues, tentatively lending further support to a concept of local, and possibly antigen-specific, immunoregulatory circuits in periodontal diseases.[66]

It is clear that the use of vaccines in periodontal diseases has not been explored to the extent that it has in dental caries. This has been related to the lack of identification of causative agents in periodontal diseases and to the uncertainty of the role of the immune system in the genesis and progression of periodontal diseases. Currently, studies of the microflora of periodontal diseases have identified at least 2 presumptive pathogens and some evidence in animals points to a potential protective role for immunity in periodontal diseases. The time would appear to be opportune to clearly define the role of the host immune systems in periodontal diseases since the potential of vaccines in periodontal diseases has not been fully explored.

ACKNOWLEDGEMENTS

The research reported in this manuscript was supported by grants DE 04733, DE 03420, DE 07009, DE 04881 and DE 06153 from the National Institute of Dental Research.

REFERENCES

1. S.J. Challacombe, L.A. Bergmeier, C. Czerkinsky, and A.S. Rees, Natural antibody to *Streptococcus mutans*: Specificity and quantitation, *Immunology* 52: 143 (1984).
2. M.S. Block, D.J. Smith, J.L. Ebersole, and M.A. Taubman, Effects of saliva and sera from caries-free caries-prone subjects on glucosyltransferase (GTF), *J Dent Res* 58: 145 (1979).
3. A.S. Aaltonen, J.O.-P. Tenovuo, R. Saksala, and O. Meurman, Serum antibodies against oral *Streptococcus mutans* in young children in relation to dental caries and maternal close-contacts, *Archs Oral Biol* 30: 33 (1985).
4. S.J. Challacombe, L.A. Bergmeier, and A.S. Rees, Natural antibodies in man to a protein antigen from the bacterium *S. mutans* related to dental caries experience, *Archs Oral Biol* 29: 179 (1984).
5. P. Brandtzaeg, The oral secretory immune system with special emphasis on its relation to dental caries, *Proceedings of the Finnish Dental Society* 79: 71 (1983).
6. O.-P.J. Lehtonen, E.M. Grahn, T.H. Stahlberg, and L.A. Laitinen, Amount and avidity of salivary and serum antibodies against *Streptococcus mutans* in two groups of human subjects with different dental caries susceptibility, *Infect Immun* 43: 308 (1984).
7. R.L. Gregory, S.J. Filler, S.M. Michalek, and J.R. McGhee, Salivary immunoglobulin and serum antibodies to *S. mutans* ribosomal preparations in dental caries-free and caries-susceptible human subjects, *Infect Immun* 51: 348 (1986).
8. R.W. Bolton and G.L. Hlava, Evaluation of salivary IgA antibodies to cariogenic microorganisms in children. Correlation with dental caries activity, *J Dent Res* 61: 1225 (1982).
9. J. Huis In't Veld, D. Bannet, W. van Palenstein-Helderman, P. Sampaio Camargo, and O. Backer-Dirks, Antibodies against *Streptococcus mutans* and glucosyltransferases in caries-free and caries-active military recruits, *Adv Exp Med Biol* 107: 369 (1978).
10. S.J. Challacombe, Serum and salivary antibodies to *Streptococcus mutans* in relation to the development and treatment of human dental caries, *Archs Oral Biol* 25: 495 (1980).
11. A.G. Plaut, The IgA1 proteases of pathogenic bacteria, *Annu Rev Microbiol* 37: 603 (1983).
12. M. Kilian, J. Mestecky, and M.W. Russell, Defense mechanisms involving Fc-dependent functions of IgA and their subversion by bacterial immunoglobulin A proteases, *Microbiol Rev* 52: 296 (1988).
13. R.R. Arnold, M.F. Cole, S. Prince, and J.R. McGhee, Secretory IgM antibodies to *Streptococcus mutans* in subjects with selective IgA deficiency, *Clin Immunol Immunopathol* 8: 475 (1977).
14. D.W. Legler, Immunodeficiency disease and dental caries in man, *Archs Oral Biol* 26: 905 (1981).
15. W.E. Stack, M.A. Taubman, T. Tsukuda, D.J. Smith, J.L. Ebersole, and R. Kent Caries in congenitally athymic rat, *Oral Microbiol Immunol* 5: 309 (1989).
16. T.B. Tomasi, Jr, E.M. Tan, A. Solomon, and R.A. Prendergast, Characteristics of an immune system common to external secretions. *J Exp Med* 121: 10 (1965).
17. M.A. Taubman, Role of immunization in dental disease. In "Comparative Immunology of the Oral Cavity"(Mergenhagen, S.E and Scherp, H.W. Eds) U.S. Gov. Print. Off., pp. 138 (1973).
18. M.A. Taubman and D.J. Smith, Effects of local immunization with *Streptococcus mutans* on induction of salivary immunoglobulin A antibody and experimental dental caries in rats, *Infect Immun* 9: 1079 (1974).
19. J.R. McGhee, S.M. Michalek, J. Webb, J.M. Navia, A.F.R. Rahman, and D.W. Legler, Effective

immunity to dental caries: Protection of gnotobiotic rats by local immunization with *S. mutans*, *J Immunol* 16: 300 (1975).

20. S.M. Michalek, J.R. McGhee, R.R. Arnold, J. Mestecky and L. Bozzo, Ingestion of *Streptococcus mutans* secretory IgA and caries immunity, *Science* 192: 1238 (1976).

21. D.J. Smith, M.A. Taubman, and J.L. Ebersole, Local and systemic antibody response to administration of glucosyltransferase antigen complex, *Infect Immun* 28: 441 (1980).

22. V.J. Iacono, M.A. Taubman, D.J. Smith, P.R. Garrant, and J.J. Pollack, Structure and function of the type specific polysaccharide of *Streptococcus mutans* 6715, *Immunol Abst Spec Suppl*:75 (1976).

23. M.W. Russell and H. Wu, *Streptococcus mutans* and the problem of heart cross-reactivity, *Crit Rev Oral Biol Med* 1: 191 (1990).

24. M.A. Taubman and D.J. Smith, Oral immunization for the prevention of dental diseases, In "Current Topics In Microbiology and Immunology", Vol 146, J Mestecky and JR McGhee (eds), Berlin, Springer Verlag, p 187 (1989).

25. R.H. Staat, S.D. Langley, and R.J. Doyle, *S. mutans* adherence: presumptive evidence for protein-mediated attachment followed by glucan-dependent cellular accumulation, *Infect Immun* 27: 675 (1980).

26. M.A. Taubman, D.J. Smith, J.L. Ebersole, and J.D. Hillman, Protective aspects of immune response to glucosyltransferase in relation to dental caries, In: R.J. Doyle, J.E. Ciardi, (eds) Glucosyltransferase, Glucans, Sucrose and Dental Caries, Spec Supplement, Chem Senses, Washington, DC, p. 249 (1983).

27. M.W. Russell and T. Lehner, Characterization of antigens extracted from cells and culture fluids of *Streptococcus mutans* serotype c, *Archs Oral Biol* 23: 7 (1978).

28. L. Gahnberg and B. Krasse, Salivary immunoglobulin A antibodies and recovery from challenge of *Streptococcus mutans* after oral administration of *Streptococcus mutans* vaccine in humans. *Infect Immun* 39: 514 (1983).

29. S.M. Michalek and N.K. Childers, Development and outlook for a caries vaccine, *Crit Rev Oral Biol Med* 1: 37 (1990).

30. D.J. Smith and M.A. Taubman, Oral immunization of humans with *Streptococcus sobrinus* glucosyltransferase, *Infect Immun* 55: 2562 (1987).

31. F.G. Emmings, R.T. Evans, and R.J. Genco, Antibody response in the parotid fluid and serum of Irus monkeys (*Macaca fascicularis*) after local immunization with *S. mutans*, *Infect Immun* 12: 281 (1975).

32. T. Lehner, P. Walker, L.A. Bergmeier & J.A. Haron, Immunogenicity of synthetic peptides derived from the sequences of a *S. mutans* cell surface antigen in nonhuman primates, *J Immunol* 143: 2699 (1989).

33. T. Lehner, A. Mehlert, and J. Caldwell, Local active gingival immunization by a 3,800 molecular-weight streptococcal antigen in protection against dental caries, *Infect Immun* 52: 682 (1986).

34. B. Krasse and H.V. Jordan, Effect of orally applied vaccines on oral colonization by *Streptococcus mutans* in rodents, *Arch Oral Biol* 22: 479 (1977).

35. D.J. Smith and M.A. Taubman, Effect of deposition of antigen on salivary immune responses and reaccumulation of mutans streptococci, *J Clin Immunol* 10: 273 (1990).

36. J. Pappo, J.L. Ebersole, and M.A. Taubman, Resident salivary gland macrophages function as accessory cells in antigen-dependent T cell proliferation, *Immunology* 63: 99 (1988).

37. L. Mayer and R. Shlien, Evidence for function of Ia molecules on gut epithelial cells in man, *J. Exp. Med.* 166: 1471 (1987).

38. J.L. Wolf, D.H. Rubin, R. Finberg, R.S. Kauffman, A.H. Sharpe, J.S. Trier, and B.N. Fields, Intestinal M cells: A pathway of entry for Reovirus into the host, *Science* 212: 471 (1982).

39. C. Czerkinsky, M.W. Russell, N. Lycke, M. Lindblad, and J. Holmgren, Oral administration of a streptococcal antigen coupled to cholera toxin B subunit evokes strong antibody responses in salivary glands and extramucosal tissues. *Infect Immun* 57: 1072 (1989).

40. R. Kent, D.J. Smith, K. Joshipura, and M.A. Taubman, Associations among serum IgG antibody specificities and clinical variables in a population at risk for root surface caries, *J Dent Res* 70: 317 (1991).

41. R.L. Gregory, S.M. Michalek, S.J. Filler, J. Mestecky, and J.R. McGhee, Prevention of *Streptococcus mutans* colonization by salivary IgA antibodies, *J. Clin Immunol* 5: 55 (1985).

42. E. Camling and G. Kohler, Infection with the bacterium *Streptococcus mutans* and salivary IgA antibodies in mothers and their children, *Infect Immun* 32: 817 (1987).

43. E. Camling, L. Gahnberg, C.G. Emilson, and B. Lindquist, Crevicular IgG antibodies and recovery of

S. mutans implanted by mouth rinsing, Submitted. Also E. Camling Thesis Goteborg. (1991).

44. D.J. Smith, J. Van Houte, P. Ali-Salaam, W.F. King, and M.A. Taubman, Effect of crevicular fluid antibody on colonization of exposed root surfaces by mutans streptococci and *A. viscosus*, *J Dent Res* 70: 317 (1991).

45. R.L. Gregory, J.C. Kindle, L.C. Hobbs, S.J. Filler, and H.S. Malmstrom, Function of Anti-*Streptococcus mutans* antibodies: Inhibition of virulence factors and enzyme neutralization, *Oral Microbiol Immunology* 5: 181 (1990).

46. M.W. Steward and C.R. Howard. Synthetic peptides: a next generation of vaccines, *Immunol Today* 8: 51 (1987).

47. J.A. Berzofsky, A. Benusassan, B. Cease, J.F. Bourge, R. Cheynier, Z. Lurhuma, J.J. Salaund, R.E. Gallo, and D. Zagury, Antigenic peptides recognized by T lymphocytes from AIDS viral envelope-immune humans, *Nature* 334: 706 (1988).

48. T. Lehner, J. Haron, L.A. Bergmeier, A. Mehlert, R. Beard, M. Dodd, B. Mielnik, and S. Moore, Local oral immunization with synthetic peptides induces a dual mucosal IgG and salivary IgA antibody responses and prevents colonization of *S. mutans*, *Immunol* 67: 419 (1989).

49. A. Childerstone, J. Haron, and T. Lehner, The reactivity of naturally sensitized human CD4 cells and IgG antibodies to synthetic peptides derived from the amino terminal sequences of a 3800 MW *S. mutans* antigen, *Immunology* 69: 177 (1990).

50. I. Takahashi, N. Okahashi, K. Matsushita, M. Tokuda, T. Kanamoto, E. Munekata, M.W. Russell, and T. Koga, Immunogenicity and protective effect against oral colonization by *Streptococcus mutans* of synthetic peptides of a streptococcal surface protein antigen, *J Immunol* 146: 332 (1991).

51. G. Mooser, R.J. Paxton, S.A. Hefta, J.E. Shively, and T. Lee, Sequence of a catalytically significant active site peptide from two *S. sobrinus* glucosyltransferases, *J Dent Res* 69: 325 (1990).

52. M.D. Cooper, P.W. Kincade, D.E. Bockman, and A.R. Lawton, Origin, distribution and differentiation of IgA-producing cells, *Adv Exp Med Biol* 45: 13 (1974).

53. L. Gahnberg, D.J. Smith, M.A. Taubman, and J.L. Ebersole, Salivary IgA antibody to glucosyltransferase of oral microbial origin in children, *Archs Oral Biol* 30: 555 (1985).

54. J. Carlsson, H. Grahnen, and G. Jonsson, Lactobacilli and streptococci in the mouths of children, *Caries Res* 9: 333 (1975).

55. B. Kohler, D. Bratthall, and B. Krasse, Preventive measures in mothers influence the establishment of the bacterium *Streptococcus mutans* in their infants, *Archs Oral Biol* 28: 225 (1983).

56. B. Kohler, I. Andreen, and B. Jonsson, The earlier the colonization by mutans streptococci, the higher the caries prevalence at 4 years of age, *Oral Microbiol Immunol* 3: 14 (1988).

57. M.A. Taubman, J.L. Ebersole, and D.J. Smith, Association between systemic and local antibody and periodontal diseases, In "Host-Parasite Interactions in Periodontal Diseases" Genco RJ; Mergenhagen SE Eds, ASM, Washington DC, p. 283 (1982).

58. M.A. Taubman, A.D. Haffajee, S.S. Socransky, D.J. Smith, and J.L. Ebersole, Longitudinal monitoring of humoral antibody in subjects with destructive periodontal diseases, *J. Periodontal Res.* In press (1992).

59. J.L. Ebersole, Systemic humoral immune responses in periodontal disease, *Crit Rev Oral Biol* 1: 283 (1990).

60. J.L. Ebersole, M.A. Taubman, D.J. Smith, D.E. Frey, A.D. Haffajee, and S.S. Socransky, Human serum antibody responses to oral microorganisms, IV Correlations with homologous infection, *Oral Microbiol Immunol* 12:53 (1987).

61. M.A. Taubman, J.L. Ebersole, J. Sioson, D.J. Smith, A.D. Haffajee, R. Kent, and S.S. Socransky, Serum antibody and homologous infection in subjects with destructive periodontal diseases, *J Dent Res* 67: 330 (1989).

62. J.M. Crawford, M.A. Taubman, and D.J. Smith, The effects of local immunization with periodontopathic microorganisms on periodontal bone loss in gnotobiotic rats, *J Periodontal Res* 13: 445 (1978).

63. K. Yamashita, J.W. Eastcott, M.A. Taubman, D.J. Smith, and D.S. Cox. Effect of adoptive transfer of cloned *Actinobacillus actinomycetemcomitans*-specific T helper cells on periodontal disease, *Infect Immun* 59: 1529 (1991).

64. M.A. Taubman, H. Yoshie, J.R. Wetherell, Jr, J.L. Ebersole and D.J. Smith, Immune response and periodontal bone loss in germ-free rats immunized and infected with *Actinobacillus actinomycetemcomitans*, *J Periodontal Res* 18: 71 (1983).

65. E.D. Stoufi, M.A. Taubman, J.L. Ebersole, D.J. Smith and P.P. Stashenko, Phenotypic analyses of mononuclear cells from healthy and diseased periodontal tissues, *J Clin Immunol* 7: 235 (1987).

66. G.J. Seymour, M.A. Taubman, J.W. Eastcott, E. Gemmell, and D.J. Smith, Characterization of activated

T cells from human periodontal disease sites: Coexpression of CD45RA and CD29 on CD4 positive cells, Submitted (1992).

67. T. Lehner, J.J. Murray, G.B. Winter, and J. Caldwell, Antibodies to *Streptococcus mutans* and immunoglobulin levels in children with dental caries, *Archs Oral Biol* 23: 1061 (1978).

68. B. Krasse, L. Gahnberg, and C.G. Emilson, Effect of antibodies on colonization of gingival tooth surfaces by *Streptococcus mutans*, In Borderland Between Caries and Periodontal Disease, G Cimasoni, T Lehner eds Geneva, Medicine and Hygiene, p. 339 (1987).

69. D.J. Smith, M.A. Taubman, W.F. King, and P. Ali-Salaam, Immunological characteristics of a synthetic peptide derived from a putative glucan-binding domain of GTF. *J Dent Res* 70: 461 (1991).

LABORATORY CORRELATES OF PROTECTION AND

PROTECTIVE IMMUNITY TO *BORDETELLA PERTUSSIS*

Roberta D. Shahin[1], ChrisAnna Mink[1],
Bernhard L. Wiedermann[2], and Bruce D. Meade[1]

[1]Division of Bacterial Products
Center for Biologics Evaluation and Research
Bethesda, MD 20892

[2]Department of Infectious Diseases
Children's National Medical Center
Washington, D.C. 20010

INTRODUCTION

Pertussis is a highly contagious respiratory disease that is typified by episodes of paroxysmal coughing. The disease commences as a mild upper respiratory infection, usually in the absence of fever, and this initial period of non-specific symptoms lasts approximately 7-10 days.[1] The disease usually progresses to episodes of paroxysmal coughing, lasting from 1-4 weeks, which may be accompanied by an inspiratory whoop in young children and may also be followed by vomiting. Convalescence, characterized by a decreasing severity of cough, can last from weeks to months. One of the difficulties in early diagnosis of pertussis is that the organism can be most successfully cultured from the nasopharynx during the early stage of the disease when symptoms are not characteristic; upon onset of paroxysmal coughing, *Bordetella pertussis* is difficult to isolate.

The population at greatest risk of pertussis-associated complications (i.e., pneumonia, seizures, encephalopathy and death) is children younger than 12 months. The World Health Organization estimates that greater than half a million deaths per year worldwide are due to pertussis, most of these occurring in unvaccinated infants.[2] In contrast, in adults and older children, pertussis is often manifested as a milder disease, which has a low mortality and is, at times, clinically non-characteristic.

Widespread use of a parenterally administered whole-cell vaccine, composed of whole killed *B. pertussis*, has been effective in decreasing the incidence of pertussis in the United States[3] and abroad. Clinical protection due to vaccination with whole-cell pertussis vaccine has been correlated in large-scale efficacy trials with the ability of a particular whole-cell vaccine preparation to protect mice from an intracerebral challenge with virulent *B. pertussis*.[4] Thus, the mouse intracerebral challenge test has been the laboratory correlate of protection used to assess the whole-cell pertussis vaccine.

A laboratory correlate of protection may be defined as a standardized and reproducible laboratory procedure that can predict the clinical efficacy of a vaccine. Laboratory correlates of protection are important in the production and control testing of vaccines in order to ascertain vaccine potency and consistency. Such correlates are also used to demonstrate the equivalence of different vaccine formulations in a single population, as well as to predict the efficacy of the same vaccine in different populations. Several examples of laboratory correlates of vaccine efficacy are listed in Table 1.

Genetically Engineered Vaccines, Edited by
J.E. Ciardi *et al.*, Plenum Press, New York, 1992

Although parenteral whole-cell vaccine has been effective against pertussis, its use has been associated with significant local reactions and fever[5]. Endotoxin has been implicated in the etiology of these reactions. Rarely, serious adverse reactions, including brain injury and death, have been observed following whole-cell vaccine, but it is unclear if this association is causal in relationship[1]. With safety issues as an impetus, pertussis vaccine programs have focused on evaluating purified components of *B. pertussis* as protective vaccine candidates in the laboratory and in the clinic, and on understanding the mechanism or mechanisms of vaccine-induced clinical protection.[5]

Table 1. Laboratory correlates of vaccine-mediated protection.

Vaccine	Component	Correlate of protection
Hib	ribosyl-ribitol phosphate (PRP)	anti-PRP serum antibody ≥ 1 µg/ml*
DTP	diphtheria	serum diphtheria toxin-neutralizing antibody ≥ 0.01 U/ml*
DTP	tetanus	serum tetanus toxin-neutralizing antibody ≥ 0.01 U/ml*
DTP	pertussis	protection of mice against intracerebral challenge with virulent *B. pertussis*

*in humans

Antigens currently available in highly purified form that are candidates for inclusion in acellular pertussis vaccine preparations include chemically or genetically inactivated pertussis toxin; filamentous hemagglutinin (FHA), a major adhesin of *B. pertussis*; pertactin, a 69-kDa outer membrane protein; and two serologically distinct types of fimbriae.

A placebo-controlled efficacy trial of two acellular pertussis vaccines, one composed of chemically inactivated purified pertussis toxin, and the second composed of inactivated pertussis toxin combined with FHA, was conducted in Sweden between February 1986 and October 1987.[7] These vaccines were administered parenterally to children 5-11 months of age, with a second dose administered 2 months after the first vaccination. After 16 months of blinded followup, point estimates of vaccine efficacy for cultured-confirmed disease with cough of any duration were 69% (95% confidence interval: 47%-82%) for the two-component vaccine and 54% (95% confidence interval: 26%-72%) for the one-component vaccine. Vaccine efficacy increased with increased severity of case definition, such that both vaccines had estimates of efficacy of approximately 80% for culture-confirmed pertussis with greater than 28 days of cough.[8] However, vaccine-mediated protection did not correlate with post-vaccination serum antibody responses to purified *B. pertussis* antigens in this trial.[7]

A number of possible explanations exist for the lack of correlation between serum antibody and clinical protection in the Swedish trial. The measurement of serum antibodies to pertussis toxin and FHA might include, but not discriminate among, subsets of protective antibodies directed to specific epitopes or antibodies that have particular effector functions. Another possible explanation is that immune responses other than serum antibody may provide a laboratory correlate of protective immunity to pertussis. *B. pertussis* is a pathogen that is tropic for the cilia of the respiratory epithelium, and it does not disseminate from the lungs to cause bacteremia or meningitis.[1] Since pertussis infection remains localized to the respiratory tract, the amount of specific antibody present in the respiratory tract, due either to local synthesis or transudation from the circulation to the lungs, may reflect clinical protection against pertussis.

MUCOSAL ANTIBODY IN CLINICAL AND SUBCLINICAL INFECTION

Our laboratory has been interested in investigating the role of immunity at the respiratory mucosa in protecting against clinical pertussis and *B. pertussis* infection. In a study of families with at least one clinically diagnosed index case of pertussis, serum and saliva were collected from index cases as well as from household contacts that remained well or had no symptoms typical of classic clinical pertussis. Analysis of sera and saliva from household contacts by an enzyme-linked immunosorbant assay demonstrated serum and salivary IgA antibodies to purified antigens of *B. pertussis* (data not shown), even though these individuals had no symptoms of classic clinical disease. In contrast, analysis of post-vaccination sera from healthy children demonstrated serum IgG antibodies reactive to *B. pertussis* antigens in all specimens with infrequent detection of specific IgA antibodies. These results suggest that serum and salivary IgA antibodies to *B. pertussis* components may be induced by subclinical infection, but are infrequently induced by parenteral pertussis vaccination. However, these data do not establish if the immune response elicited by subclinical *B. pertussis* infection protects against typical disease.

MUCOSAL IMMUNIZATION DECREASES EXPERIMENTAL INFECTION

Long-lasting immunity to pertussis occurs after natural infection, while protection afforded by whole-cell pertussis vaccine is of limited duration[9]. The long-lasting immunity conferred by natural infection may be a reflection of immunologic memory in the respiratory tract, which can be recalled at the respiratory mucosa upon subsequent infection. To test this hypothesis, we immunized mice via the respiratory route with FHA. Kimura et al. have shown that a strain of *B. pertussis* containing a deletion in the FHA gene was impaired in the ability to colonize the tracheas of experimentally infected mice.[10] Intranasal immunization of adult female mice with FHA prior to aerosol challenge with wild-type *B. pertussis* resulted in a 2-3 log reduction ($P < 0.01$) of bacteria recovered from the lungs as well as from the tracheas of immunized mice compared to unimmunized controls (Fig. 1).

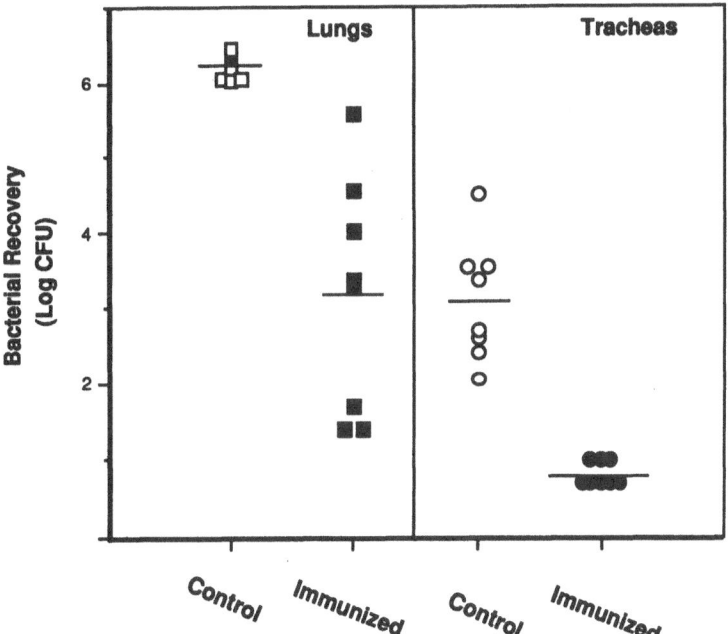

Figure 1. Bacterial recovery from intranasally immunized mice after *B. pertussis* infection.

IgG antibodies to FHA were detected in both the serum and bronchoalveolar lavage fluids of intranasally immunized mice. IgA antibodies to FHA were detected in the bronchoalveolar lavage fluids, but not in the serum, of these mice. (Fig. 2). These same sera and bronchoalveolar lavage fluids were immunoblotted against electrophoretically resolved whole *B. petussis* extracts and only recognized bands corresponding to FHA. Therefore, the decrease in bacterial recovery after infection conferred by intranasal immunization was due to an immune response specifically elicited by FHA, rather than an immunogenic contaminant in the FHA preparations. These experiments, however, do not establish the mechanism or mechanisms of protection due to mucosal immunization with FHA; these mechanisms may include antigen-specific antibodies of several isotypes with different effector functions, as well as antigen-specific T cells.

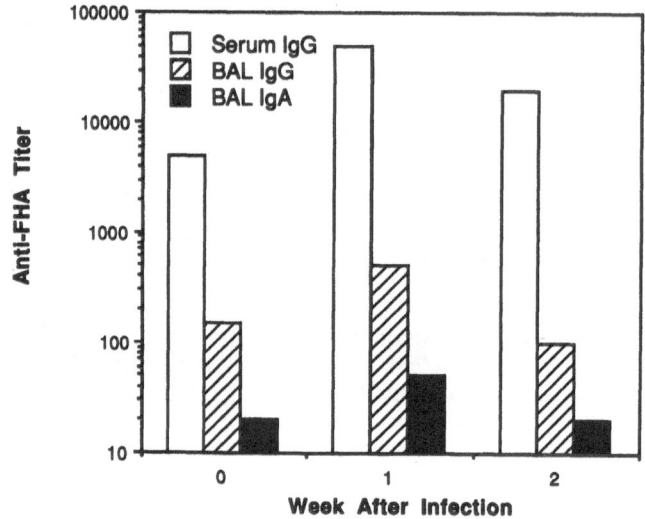

Figure 2. Anti-FHA antibody in serum and BAL of intranasally immunized mice.

Lymphocytes isolated from the lungs and spleens of intranasally immunized mice were analyzed in limiting dilution cultures for the production of anti-FHA antibody. Limiting dilution analysis revealed a 60-fold increase in frequency of FHA-specific B cells isolated from the lungs of intranasally immunized mice in comparison to unimmunized control mice. Thus, respiratory immunization with a major adhesin of *B. pertussis* stimulates a specific local immune response in the respiratory tract and protects against subsequent respiratory infection.

Since antigenic stimulation of the gut-associated lymphoid tissue results in a mucosal immune response in respiratory tissues[11], oral immunization might also protect against respiratory pertussis infection. In addition to providing protective immunity at the site of infection, an oral pertussis vaccine would provide ease of administration, thereby facilitating improved vaccine coverage of at-risk populations.

Investigators in Austria have reported that infants receiving an oral pertussis vaccine composed of killed whole *B. pertussis* demonstrated a rise in salivary, but not serum, pertussis agglutinins.[12] In contrast, children immunized parenterally with whole-cell pertussis vaccine demonstrated pertussis agglutinins in their serum but not their saliva. Although specific antibody responses to individual antigens of *B. pertussis* were not reported, these results suggests that oral administration of pertussis antigens elicits specific antibodies at distal mucosal tissues in infants.

In order to determine if gut administration of a purified pertussis antigen would disseminate protective immunity to the respiratory tract, mice were immunized intraduodenally with FHA. Gut administration of two 100 µg doses of FHA prior to aerosol infection with *B. pertussis* resulted in a 2 log decrease (p< 0.05) in recovery of bacteria from the tracheas and lungs of immunized mice in comparison to unimmunized controls (Fig. 3). A single 100 µg dose of FHA resulted in approximately a 0.5 log decrease, which was not statistically significant. Because FHA is easily proteolyzed, formulation of FHA and other labile antigens, to protect them from rapid degradation in the gut, may stabilize these species as protective oral vaccine candidates.

FHA is a 220-kDa filamentous protein that has been sequenced and found to contain an integrin binding site[13], as well as a region of homology to the hemolysins of *Serratia marcescens* and *Proteus mirabilis*.[14] These particular regions of the FHA molecule may represent the functional binding domains of this adhesin and may be responsible for the capacity of this antigen to bind to the follicle-associated epithelium and interact with antigen-presenting cells to elicit a protective immune response to *B. pertussis* infection.

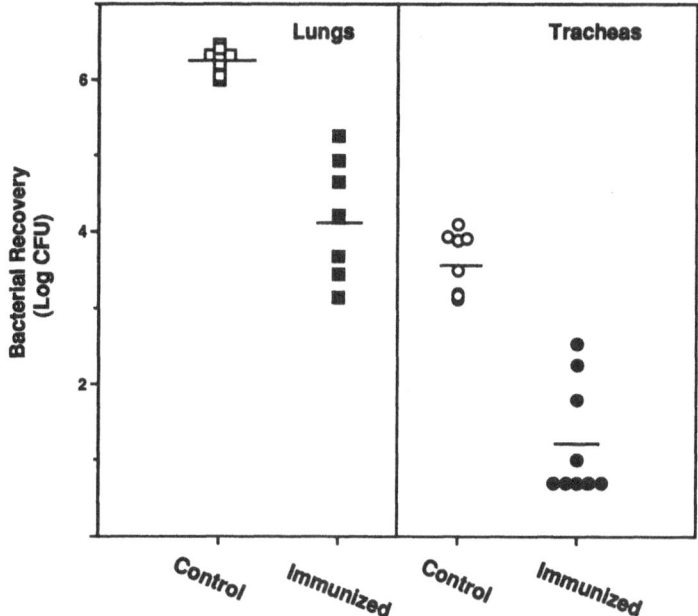

Figure 3. Bacterial recovery from intraduodenally immunized mice after *B. pertussis* infection.

Our data suggest therefore, that the generation of a specific immune response to pertussis antigens at the site of infection, the respiratory mucosa, may protect against *B. pertussis* infection and that specific antibodies to *B. pertussis* components in respiratory secretions may correlate with protective immunity to pertussis. This, in turn, suggests that mucosally administered pertussis vaccines can be designed that would prevent clinical pertussis infection. Such vaccines would be desirable for their safety, ease of administration, and potential to increase vaccine coverage.

References

1. J.D. Cherry, P. A. Brunell, G. S. Golden, and D. T. Karzon, Report of the task force on pertussis and pertussis immunization - 1988. *Pediatrics* 81:939 (1988).
2. A.S. Muller, J. Leeuwenburg, and D.S. Pratt, Pertussis: Epidemiology and control, *Bulletin of the World Health Organization.* 64:321 (1986).
3. Centers for Disease Control: Annual Summary 1981: Reported morbidity and mortality in the United States, *Mortality Morbidity Weekly Rev.* 30:65 (1982).
4. P. Armitage, W. C. Cockburn, D. G. Evans, J. O. Irwin, J. Knowelden, and A. F. B. Standfast, Vaccination against whooping cough. Relation between protection in children and results of laboratory tests, *Br. Med. J.* 2:454 (1956).
5. C.L. Cody, L.J. Baraff, J.D. Cherry, S.M. Marcy and C.R. Manclark, Nature and rates of adverse reactions associated with DTP and DT immunizations in infants and children, *Pediatrics.* 68: 650 (1981).

6. M.J. Brennan, D.L. Burns, B.D. Meade, R.D. Shahin and C.R. Manclark, Recent advances in the development of pertussis vaccines. *in*: "Vaccines," R. Ellis, ed. Butterworth Publishers, Shoreham, MA (1991).
7. Ad Hoc Group for the Study of Pertussis Vaccines. Placebo-controlled trial of two acellular pertussis vaccines in Sweden - Protective efficacy and adverse events, *Lancet* 1: 955 (1988).
8. J. Storsaeter, H. Hallander, C.P. Farrington, P. Olin, R. Mollby and E. Miller, Secondary analyses of the efficacy of two acellular pertussis vaccines evaluated in a Swedish phase III trial, *Vaccine* 8: 457 (1990).
9. H.J. Lambert, Epidemiology of a small pertussis outbreak in Kent County, Michigan, *Public Health Rep.* 80:365 (1965).
10. A. Kimura, K. T. Mountzouros, D. A. Relman, S. Falkow and J. L. Cowell, *Bordetella pertussis* filamentous hemagglutinin: Evaluation as a protective antigen and colonization factor in a mouse respiratory infection model, *Infect. Immun.* 58:7 (1990).
11. P. Weisz-Carrington, S. R. Grimes Jr. and M. E. Lamm, Gut-associated lymphoid tissue as source of an IgA immune response in respiratory tissues after oral immunization and intrabronchial challenge, *Cell. Immunol.* 106(1):132 (1987).
12. E. Baumann, B. R. Binder, W. Falk, E. G. Huber, R. Kurz, and K. Rosanelli, Development and clinical use of an oral heat-inactivated whole cell pertussis vaccine, *Dev. Biol. Stand.* 61: 511 (1985).
13. D.A. Relman, M. Domenighini, E. Tuomanen, R. Rappuoli, and S. Falkow, Filamentous hemagglutinin of *Bordetella pertussis*: Nucleotide sequence and crucial role in adherence, *Proc. Natl. Acad. Sci., USA* 86:2637 (1989).
14. A. Delisse-Gathoye, C. Locht, F. Jacob, M. Raaschou-Nielsen, I. Heron, J. Ruelle, M. De Wilde and T. Cabezon, Cloning, partial sequence, expression, and antigenic analysis of the filamentous hemagglutinin gene of *Bordetella pertussis*, *Infect. Immun.* 58:2895 (1990).

CHALLENGES AND OPPORTUNITIES IN VACCINE RESEARCH

John R. La Montagne and Regina Rabinovich

Division of Microbiology and Infectious Diseases
National Institute of Allergy and Infectious Diseases
Bethesda, MD   20852

INTRODUCTION

In the past century, vaccines have been developed against infectious diseases that previously killed and disabled millions of children and adults worldwide, such as smallpox and polio. In the U.S., use of vaccines has virtually eliminated diseases such as diphtheria, tetanus, mumps and poliomyelitis. Although these might seem like a distant memory, review of the changing epidemiology of diseases such as polio and measles is instructive. We have progressed from summer epidemics of polio that caused 18,000 cases of paralytic disease in the U.S. every year, to the virtual eradication of wild polio in the Americas due to active eradication programs with vaccine. Similarly, it is estimated that the application of measles vaccine to 15-month-olds in the U.S. has prevented over 52 million cases, 5,200 deaths and 17,000 cases of mental retardation since it was introduced into use in the early 1960s.[1]

However, at the present time, there are many serious and life-threatening infectious diseases for which no effective vaccines have been developed, such as certain pediatric respiratory and diarrheal diseases and sexually transmitted diseases (STDs). In addition, there are new and re-emerging diseases, such as Lyme disease, measles, HIV and tuberculosis. Even though by previous standards the current measles epidemic (27,000 cases in 1990) is minor, control measures have included the addition of a second dose of measles vaccine to the immunization schedule, at significant cost to the nation.

Advances in molecular biology and immunology have created opportunities for improving the safety, efficacy and efficiency of existing vaccines for diseases such as pertussis and measles and for enhancing the development of new vaccines. In 1981, National Institute of Allergy and Infectious Diseases founded the Program for the Accelerated Development of Vaccines to focus and enhance research activities leading to new vaccines for important diseases. This workshop is evidence that an important new era in vaccine research and development continues; the new tools in microbiology are beginning to be applied to vaccine problems with the hope that vaccine solutions will emerge.

Genetically Engineered Vaccines, Edited by
J.E. Ciardi *et al.*, Plenum Press, New York, 1992

The investment in the scientific base which underlies vaccine research and development is capable of yielding impressive results. Since 1980, 10 new or improved vaccines have become available in the U.S.[†] Among other significant contributions, National Institutes of Health-supported researchers have been instrumental in the development of first-generation vaccines for *Haemophilus influenzae* type b (Hib) and hepatitis B, and in the rapid development of improved second-generation vaccines for both of these diseases (plasma-derived hepatitis B to a recombinant hepatitis B; PRP polysaccharide to the Hib PRP conjugate vaccines). Both were recommended for inclusion into childhood immunization programs in the past year. The use of conjugate Hib vaccines has already had profound impact on the incidence of invasive disease in infants in this country.

While vaccines have clearly made a profound impact on the health and well-being of the world's population, an added benefit that derives from their use is a decrease in health care costs. The cost effectiveness of vaccination has been demonstrated with smallpox, measles, polio, and most recently, conjugate Hib vaccines. Use of this vaccine will save nearly $400 million for every annual cohort of children that is vaccinated. Moreover, the loss of life to a person cannot be measured--and thousands of lives in the U.S. are saved each year by the use of vaccines. While treatment of patients who are ill with infectious diseases remains an important objective, prevention is a far more cost effective method.

The Program for the Accelerated Development of Vaccines was an attempt to indicate that the scientific base for the development of vaccines had matured to the point that they could be applied to the accelerated development of all vaccines. Therefore, the scope of the Program was quite broad. The list of vaccines targeted by this Program were marked as priorities for vaccine research as a result of a study by the Institute of Medicine (IOM), published in 1985 (Table 1). Some diseases were not considered by the IOM, such as tuberculosis.

Table 1. IOM priorities for vaccine development (1985).

United States	Developing Countries
Hepatitus B Virus	*Streptococcus pneumoniae*
Respiratory Synctial Virus	Rotavirus
Haemophilus influenzae type b	Plasmodium (malaria)
Influenza	Salmonella (typhoid)
Varicella	Shigella

The success of the Program for Accelerated Development of Vaccines has been based on the support for the fundamental cornerstones of vaccine research and development: basic research, including a clear understanding of the pathogenesis of each organism and of the immune response to infection; clinical research for the early evaluation of safety and immunogenicity of candidate vaccines in small groups of adults and children; larger clinical trials for the evaluation of candidate vaccines, to determine

---

[†] Hepatitis b (plasma derived), hepatitis b (recombinant), Haemophilus influenzae type b (polysaccharide) and Haemophilus influenzae type b (conjugates), influenza, pneumococcus (polysaccharide), rabies, typhoid, enhanced potency poliovirus, acellular pertussis booster.

the safety and efficacy of the candidate vaccines; and development of close partnerships with industry.

There are diseases on the IOM list of priorities that cause significant morbidity and mortality which still do not enter into the lexicon of "vaccine preventable"--such as respiratory syncytial virus (RSV) infection and malaria. These unmet challenges should not, to quote Dr. Bernard Fields of Harvard, "be considered as just engineering problems." This implies that the basic information to answer the most vexing problems is not available and waiting to be applied in an ingenious way.

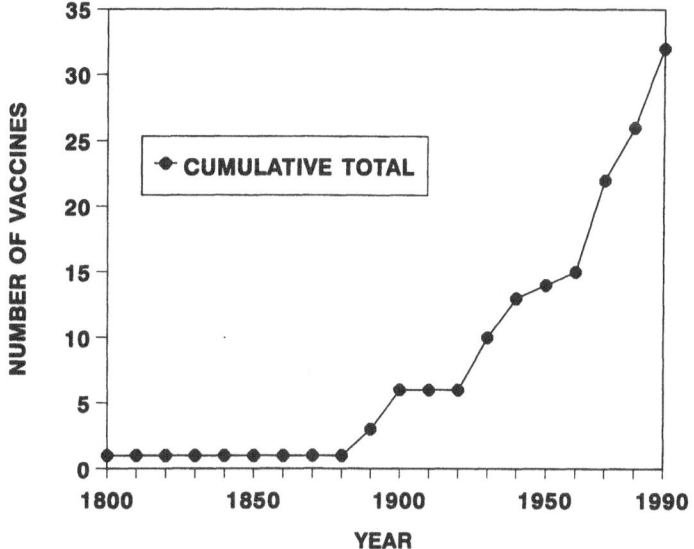

Figure 1. Introduction of new vaccines, Jenner to the present day.

A historical review of the rate at which vaccines have been added to the armamentarium of physicians, from Jenner to the present day, is instructive (Figure 1). Each of the changes noted on this curve is associated with a development that occurred before it, such as the identification of toxins and the methods for creating toxoids. Similarly, the explosion of information about microbial causes of infection which occurred at the end of the 19th century and the development of tissue culture techniques in the 1950s also resulted in the introduction of new vaccines. Most recently, we are experiencing an accelerated growth due to the application of a number of biotechnologic innovations in vaccine research, such as recombinant DNA techniques and monoclonal antibodies.

The timetable for successful development is uncertain and can be lengthy. For example, it took 30 to 40 years from identification of the causative organism to the successful licensure of vaccines for measles, polio and rubella. The scientific community rapidly discovered the "simple vaccines," for which immunity depends on a humoral antibody response, and human antibodies to the natural disease are subsequently protective. The remaining diseases pose special challenges by their complicated pathogenesis, a lack of natural immune response, or route of infection through mucosa, of which there are significant gaps in our understanding of mucosal immunity. For example, the lack of a protective immune response in people infected with gonorrhea raises the question as to whether it is possible to stimulate an artificial immune response that could be better than a "natural" immune response. Many of the infections for which we lack effective vaccines, such as rotavirus, sexually transmitted diseases, and dental vaccines, are going to require effective mucosal immunity. Finally, development of live attenuated derivatives are not always possible when organisms cannot be cultured or there are problems with reliable attenuation. For these diseases, subunit vaccines may provide a solution to isolating epitopes that stimulate immunopathogenic responses from those that stimulate protective responses.

It is clear that the search for new and improved vaccines requires long-term commitment to additional basic, clinical and applied research. The new scientific base may accelerate this rate of development, but there are very real steps that must be passed to assure the safety and efficacy of vaccines.

More than ever before, the issue of safety has taken on a role that it did not have before. The change in the risk/benefit relationship, in polio for example in polio, places emphasis on improving vaccines or immunization strategies so that serious adverse events do not occur. The overwhelming concern with vaccine safety has been a driving force behind significant changes in immunization policy in the U.S. In the 1980s, there was much public concern regarding the safety of the required childhood vaccines, with a resultant crisis in liability costs and the potential for decreased vaccine availability in the U.S.. The escalating liability costs for vaccine manufacturers in the early 1980s, brought about mainly by concerns over the safety of pertussis vaccine, resulted in a crisis for U.S. immunization programs. In order to maintain the availability of vaccines, the U.S. Congress passed the National Childhood Vaccine Injury Act in 1986, followed by the Vaccine Compensation Amendments of 1987. Briefly, these two laws established a no-fault compensation program to compensate injuries alleged to have occurred due to a mandated childhood immunization. Prospective and retrospective claims were included in the program. The payment of prospective claims was to be supported by an excise tax charged on each dose of the mandated childhood vaccines sold in the U.S. Retrospective claims were to be paid from appropriated funds ($80 million in Fiscal Year 1989).

Although projected costs to satisfy the retrospective claims were estimated to be approximately $80 million/year, a large number of such claims have been presented to the U.S. Court of Claims. Estimates vary regarding the cost of settling these claims, but these costs have been projected in the range of $850 million to $2.8 billion. And yet, without adequate funding for this or a similar program, liability actions directly to manufacturers would surely increase and force further spiraling costs of vaccines, in addition to potential concerns over vaccine availability.

The long history of the development of an acellular pertussis vaccine represents both the difficulties and the challenges ahead. The presently available whole-cell vaccine was developed in the 1940s and has remained essentially unchanged. Presently, we still lack a serologic correlate of immunity. However, the high rate of relatively minor reactions such as fever and pain, and the well-publicized charges that this vaccine may cause severe adverse events, has had significant impact on public immunization programs and vaccination policy in the U.S. There are just now several ongoing clinical trials of acellular pertussis vaccines, including a Phase III NIAID trial in Sweden. We believe that the successful development of a less reactogenic acellular pertussis vaccine is an essential component of the success of the Children's Vaccine Initiative (CVI).

The CVI is a visionary goal of universal childhood vaccines that would decrease childhood mortality from infectious diseases when given early in infancy. In addition, the CVI is a global organizational effort to bring together the diverse talents in industry, academia and government to achieve this goal. Launched in September 1990 at the World Summit for Children in New York, the CVI aims for the development of safe, heat-stable, affordable childhood vaccines that are available as combination products, can protect against many diseases and be administered orally soon after birth, and require few doses to maintain protection.

The development of a single "Children's Vaccine" that will fulfill these criteria is a worthy endpoint unlikely to be reached within a single decade. There is consensus that the vaccine research and development community does have within its reach the development of safe and effective vaccines that will simplify immunization schedules, expand the number of childhood diseases for which vaccines are available, and assure the continuing safety of childhood vaccines.

The CVI requires multiagency input to achieve its objectives and is truly a global initiative. Vaccine research, an NIAID mandate, is a cornerstone for long-term progress within the CVI. To achieve the kinds of vaccines envisioned by the CVI, significant scientific advances in both basic science and vaccine research will be necessary. These advances will complement efforts such as the combination of existing antigens currently undertaken by industry. A diverse group of agencies and institutions, including the CVI international sponsors and committees (WHO, Rockefeller, UNDP, UNICEF, and the World Bank), the WHO Expanded Programme on Immunization and the WHO Vaccine Development Program, USAID, FDA, CDC, and other national vaccine research programs and industry, can be expected to support other elements essential to the progress of the CVI, including regulatory issues, legal issues, application of results of research to production and delivery in the developing world, incorporation of new vaccines into U.S. and global immunization strategies, and funding for immunization programs.

The CVI is not just a program for the developing world. The potential applications of simplifying the immunization schedule, increasing the number of vaccine preventable diseases, and improving the safety of vaccines begin at home. The impact of requiring fewer visits for protection leads to decreasing the costs of administration of vaccines.

The goals of the CVI provide an additional focus for vaccine research and development. This focus combines the goals of prevention, clearly a long-term goal of the NIAID, with goals that incorporate issues relevant to vaccine delivery. The CVI aims to develop vaccines that are more effective and efficient in preventing infectious diseases, both in the U.S. and abroad. Immunization is a principal feature of health care of infants and young children throughout the world. Universal immunization programs that deliver vaccines to all children are responsible for much of the increase in life expectancy in the last half-century. The initiative aims to develop vaccines which provide a long-term, cost-effective solution to the problem of access to immunizations. Although most infants receive their first dose of vaccine at an appropriate age, immunization with subsequent doses of DPT is significantly delayed for inner-city and minority children in the U.S., who are therefore needlessly at risk of disease.[5] Vaccines which require fewer doses and could be administered orally would facilitate the immunization of all infants. The impact on the existing programmatic and financial problems in vaccinating children is potentially very large.

In the past year, the outbreaks of measles in the U.S. led to the recommendation for a second dose of measles vaccine in massive immunization efforts, targeting specially the high-risk areas in the inner cities, but making this a part of the routine immunization schedule. These efforts, while laudable and currently necessary, may be made increasingly cost-effective by the successful investment in the CVI.

The goals of the CVI are congruent with the significant NIAID investment in infectious diseases with a high priority for the prevention of childhood morbidity and mortality, and with the potential of universal applicability within the pediatric immunization schedule. This spans a broad range of diseases and includes both vaccine-related basic and clinical research. Potential target diseases for the CVI include the following:

- RESPIRATORY INFECTIONS
  Pertussis, Pneumococcus, Respiratory Syncytial Virus,
  Influenza, Parainfluenza, Tuberculosis, Varicella
- ENTERIC INFECTIONS
  Rotavirus, Cholera, Shigella, Typhoid,
  Enterotoxigenic E. Coli
- HEPATITIS
  Hepatitis A and C
- IMPROVING EXISTING VACCINES
  Measles, Mumps, Tetanus, Polio, Diphtheria, Rubella,
  Haemophilus influenzae type B, Hepatitis B
- OTHER
  Group B Streptococcus, Meningococcus A, B and C
- PARASITIC/TROPICAL DISEASES
  Malaria, Dengue, Filariasis, Trypanosomiasis,
  Leishmaniasis, Schistosomiasis

A number of areas of research are necessary for progress in the CVI. We need to explore alternative ways to enhance the immune response, including the development of better adjuvants, and the multiple applications of glycoconjugate technologies, and increase our understanding of vaccine immunology, including mucosal immunity and neonatal immunology. We need to consider alternative approaches to protecting infants, such as maternal immunization. There is much work that needs to be done on the combination of antigens. Innovative vaccine delivery technologies applicable to oral vaccine delivery and delivery of combinations of antigens, such as microencapsulation technologies, ISCOMs, and vaccine vectors, must be pursued.

Finally, we want to emphasize that, in the process of advancing the CVI, we can and must improve and assure the safety of vaccines. Aside from routine regulatory guidelines on safety, and the fundamental premise that arises from the Hippocratic oath ("Above all, do no harm"), assurance that vaccines are as safe as they can be remains an important issue, not only for Congress, but also for the parents of the babies to be immunized.

Given the imperative for improved health for the nation and, indeed, for the world at an affordable cost, enhanced development of vaccines should be a priority for biomedical research. In 1991, the National Institutes of Health (NIH) developed a strategic plan for vaccine development which recognizes the capacity of diverse components of the biomedical research community to address urgent issues in the vaccine arena. It provides the framework under which progress can occur in developing safe and effective vaccines, both for traditional infectious diseases and new targets, including emerging microbes, specific cancers, and immunologic diseases. The plan for vaccine development is organized into three areas:

- Expansion of the basic scientific foundations for vaccine development;

♦ Enhancement of efforts focused on vaccine development for common serious diseases such as HIV, other sexually transmitted diseases, and pediatric diseases targeted by the CVI;

♦ Development of vaccines for the prevention of diseases not normally thought to be preventable through vaccination, such as cancer and autoimmune diseases.

Finally, the plan recognizes that these strategic directions cross-cut and complement individual Institute priorities. The strategic directions also fit national priorities detailed in the Secretary of Health and Human Service's Initiatives and Program Directions, and congressional priorities.

In light of the liability concerns facing private vaccine manufacturers in the U.S., and the multiple barriers to vaccine development and testing in the developing world, it is imperative that the NIH continue to provide the global leadership to reach the potential offered by the impressive advances in biomedical research.

REFERENCES

1. Centers for Disease Control, "Center for Infectious Diseases Annual Report," (1992).
2. E.D. Shapiro, E.R. Wald, A.G. Margolis, and M.E. Ortenzo, The decreasing incidence of *Haemophilus influenzae* type b (Hib) disease in both Connecticut and greater Pittsburgh, *P.A. Abstract*, Society for Pedatric Research, (1992).
3. R.M. Davis, L.E. Markowitz, S.R. Preblud, W.A. Orenstein, and A.R. Hinman, A cost effectiveness analysis of measles outbreak control strategies, *American Journal of Epidemiology*. 126(3):450 (1987).
4. National Institute of Allergy and Infectious Diseases, *Potential cost savings from biomedical research–two examples from the NIAID*. Administrative Report, Policy and Analyis Branch, NIAID, (March 1991).
5. Centers for Disease Control, Retrospective assessment of vaccination coverage among school-aged children-selected U.S. cities, 1991, *Morbidity and Mortality Weekly Report*. 41(6):103 (1992).

SUMMARY AND RECOMMENDATIONS FOR FUTURE RESEARCH

Joseph E. Ciardi, Jerry R. McGhee and Jerry M. Keith, Editors

Several general findings were gleaned from the workshop discussions:

- There is strong evidence of a common mucosal immune system in humans that is shared by all mucosal sites, including the oral cavity. This mucosal immune system develops early in infancy and is separate from the body's systemic immune system. The mucosal and systemic immune systems appear to interact and possibly could be exploited together in developing optimal preventive measures for infectious disease.

- A better understanding of the common mucosal immune system is needed to appreciate all of the dimensions of oral and systemic disease. Continued efforts to develop a vaccine(s) against oral disease not only will have practical benefit, but also will accelerate basic understanding of mucosal immune processes.

- Different infectious diseases and environments will require different strategies and approaches in developing vaccine delivery systems. No one approach will be perfect, and a variety of alternative delivery systems should be explored.

- Several approaches are particularly promising today for developing a vaccine(s) against oral disease. These involve the use of subunit vaccines, recombinant vaccines, synthetic peptides, peptomers (crosslinked peptide polymers), and monoclonal antibodies. All of these approaches require the application of advanced biotechnology and genetic techniques and the combination of skills and expertise from many disciplines.

- An effective vaccine(s) would be useful in combatting dental caries and periodontal diseases in susceptible, high-risk populations. Because some of the oral pathogens causing these diseases are also implicated in other serious and systemic diseases, a preventive oral vaccine could have additional benefit in preventing associated diseases.

- Promising research is already being conducted to develop vaccines against herpes simplex virus, human papillomavirus, and human immunodeficiency virus--diseases with significant oral complications. Some vaccines are already

Genetically Engineered Vaccines, Edited by
J.E. Ciardi *et al.*, Plenum Press, New York, 1992

being tested in clinical trials; others are only in the early stages of development. This research, involving dental scientists, should continue to be supported.

- The successful development of a vaccine(s) against oral, as well as other, diseases requires the concerted effort of industry, government, and academia; well-trained scientists in appropriate fields; and adequate resources to support basic research and clinical testing.

Summarized on the pages that follow is a synthesis of the deliberations and recommendations in the three key areas covered by workshop sessions, panel discussions (chaired by Drs. Michael Lamm, Roy Curtiss, III, and Arnold Bleiweis), and contributions from members of an advisory committee, consisting of Drs. Arnold Bleiweis, Suzanne Michalek, Edward Shillitoe, and Ms. Linda Richardson who worked with us on this document. The rationale for these recommendations is also provided. The recommendations are not listed in any order of priority.

OPTIMIZING THE MUCOSAL IMMUNE RESPONSE

Discussions on optimizing the mucosal immune response addressed four general areas: the common mucosal immune system in relation to vaccines, mucosal memory and relevance for mucosal vaccines, development of the mucosal immune system, and oral tolerance and vaccines. It was noted that four generalizations of the mucosal immune system have been reasonably well established in humans:

1.   The common mucosal immune system is operative.

2.   The mucosal immune system exhibits immunological memory.

3.   The mucosal immune system is functional as early as infancy.

4.   The mucosal immune system does not necessarily follow the rules established for systemic immunity.

The Common Mucosal Immune System In Relation To Vaccines

There is compelling evidence for a common mucosal immune system in experimental animals and humans. However, there are many gaps in our knowledge of this system and, thus, in our ability to use this knowledge to develop mucosal vaccines.

It is becoming clear that the mucosal immune system can be divided broadly into two parts: inductive sites and effector sites. Inductive sites are those in which the initial reaction to a foreign antigen occurs. In the mucosal immune system, these sites include the Peyer's patches in the gastrointestinal tract and lymphoid aggregates in the upper respiratory tract.   These components are often collectively termed gut-associated and bronchus-associated lymphoreticular tissues (GALT and BALT), respectively. The former include tissues in the small and large intestine, as well as the appendix. The latter include adenoids, tonsils, and bronchus-associated nodes. Certain types of immunoglobulin A precursors appear to predominate in GALT, while other types appear to predominate in BALT. The interaction between these tissues and their relative contribution to the mucosal immune responses that take place at effector sites (including the major and minor salivary glands in the oral cavity) need to be better understood.

Recommendation:

> Continue studies to elucidate mucosal immune reactions at inductive and effector sites, with particular attention to the relative contribution of gut-associated and bronchus-associated lymphoreticular tissues to immune responses in the human oral cavity.

The relative importance of the migration of specific immune B cells from GALT to other mucosal sites such as the salivary glands needs to be established, as well as the potential immunoglobulin responses in these glands. The SCID mouse may be a very useful animal model for such studies.

Research also is needed on how to take optimal advantage of the common mucosal immune system for immunization (both priming and boosting), including potential interactions between the peritoneal cavity and mucosal sites.

The interplay between humoral (IgA) and cell-mediated (T cell, NK cell, intraepithelial lymphocytes) immune effector mechanisms in the mucosal immune system deserves more study.

The exact portals of entry, receptors, and mechanisms of infection of mucous membranes by pathogenic microorganisms need to be determined.

Mucosal Memory and Relevance for Mucosal Vaccines

The mucosal immune system responds to foreign antigens by producing primarily immunoglobulin A (IgA) antibodies. This response is observed even after a primary immunization, suggesting that there is mucosal memory and that it favors IgA antibodies. These observations, which have import for the development of mucosal vaccines, need to be confirmed by further studies. Pertinent questions are: What constitutes a primary immune response in mucosal tissue? Does primary oral immunization lead to the induction in Peyer's patches of immune B cells with IgA memory? Does maintenance of this population of cells with IgA memory require additional immunization? How long do these cells last? Are they restricted to the site of induction, or do they recirculate to other mucosal or systemic tissues, or only within the mucosal immune system?

Recommendation:

> Continue studies to determine definitively whether there is mucosal memory and the nature of this memory. Continue to evaluate the role of IgA in mucosal immunity and its importance for the development of oral vaccines.

Studies are needed to determine if primary oral immunization induces IgA responses with a lower affinity for antigens, and if additional immunization selects IgA memory B cells that have undergone changes that would make them less or more effective. Such studies would have profound implications for the development of mucosal vaccines. They could show, for example, that several immunizations would be needed to effect a maximal immune response or that undesirable responses would be elicited to the carriers or vectors used for delivering the vaccine.

Mucosal T cells also are involved in mucosal immunity, and it is well established that the CD4+ T helper cells are required for mucosal IgA responses to most, if not all, vaccines. However, very little is known about the memory associated with these cells and the importance of CD4+ memory T cells in mucosal IgA responses. Similarly, little is known about the induction of CD4+ (or CD8+) T cells that result in cytokine-induced, cell-mediated immune responses. This information will be important when considering the use of recombinant bacteria for vaccine delivery, since *Salmonella*, BCG, and other bacteria induce strong cell-mediated immune responses. It also is known that mucosal immunization with virus vaccines induces CD8+ cytotoxic T lymphocyte responses at mucosal sites. These responses, and the memory that may be associated with them, need to be described better.

Recommendation:

Better characterize T cell responses, including the cytokine profiles that develop in mucosal sites after oral immunization with microbial vaccines.

Development of the Mucosal Immune System

Studies of the early development of mucosal IgA responses suggest that newborns can undergo mucosal immune (IgA) responses to external stimuli, although not at the same levels as seen in adults. Specific responses have been shown to the oral streptococci implicated in dental caries (e.g., *Streptococcus mutans*). Salivary IgA antibodies are increased, but not to the level required to protect against dental caries.

It must also be recognized that infants during the first few months of life are protected by maternal antibodies. Thus, the infant exhibits both serum and mucosal immunity, obtained from the mother (serum) and from its own natural defenses (mucosal), prior to the development of any systemic antibody responses. These findings have broad implications for the development of mucosal vaccines to protect against dental caries or any other disease. The mucosal immune system appears to be accessible at the very earliest ages, even before the development of systemic immunity. Thus, it may be possible to induce mucosal immunity early with oral immunization. Care must be taken, however, to avoid damaging the naturally protective mechanisms already in place.

Recommendation:

Further explore the possibility of developing mucosal vaccines for infants that can be used, prior to systemic immunization, to protect against oral and other diseases. Consider developing mucosal delivery strategies for all childhood vaccines.

Oral Tolerance and Vaccines

Administration of mucosal vaccines could lead to the development of oral tolerance (i.e., systemic unresponsiveness) to orally delivered vaccines. Because of this potential problem, a more fundamental understanding is needed of the immune mechanisms involved in oral tolerance. T cells are strongly suspected of playing a role in the induction of oral tolerance to protein antigens, and they deserve special focus. Also of interest is whether oral tolerance to carbohydrates can be induced in humans, or whether it is only limited to proteins.

Recommendation:

> Conduct fundamental studies of oral tolerance and antibody responses. Determine the role of T cells, including putative T suppressor cells, in the induction of oral tolerance. Establish whether humans are already tolerant to commonly encountered protein antigens in the diet (e.g., ovalbumin, milk, and meat proteins) and whether tolerance can be induced to proteins not normally encountered.

Current knowledge can already be applied to the design and development of mucosal vaccines. Most vaccines and oral or other delivery systems should be designed in such a way to avoid or minimize the development of oral tolerance. In some situations (e.g., for autoimmune diseases), it may be possible to design vaccines that can selectively induce oral tolerance in order to prevent autoimmune responses.

Recommendation:

> Apply existing and developing knowledge on oral tolerance to the design of mucosal vaccines.

Other Areas of Research

Mucosal Adjuvants. To stimulate specific mucosal immune responses, it will be necessary to understand the criteria for optimal mucosal adjuvants as well as the most effective ways of expressing and presenting antigens. At present, only aluminum hydroxide is suitable for boosting immune responses to vaccines in humans. The adjuvant properties of existing vaccines need to be evaluated carefully in terms of their mucosal immune effects. Desirable properties and potential side effects, including inherent toxicity, need to be determined.

Recommendation:

> Determine the characteristics and criteria for optimal mucosal adjuvants.

Antigen Presenting Cells. The role of different antigen presenting cells, including class II MHC positive macrophages, dendritic cells, B cells, and epithelial cells, also needs to be defined at the mucosal inductive and effector sites. Similarly, the most effective cells (e.g., M cells, enterocytes) and routes for introducing antigens into the mucosal immune system (the inductive phase) need to be determined. Studies of the biology of these cells, including their isolation and growth *in vitro*, are needed. The distinctive features for antigen presentation in the mucosal immune system should be compared to those of the systemic immune system.

Recommendation:

> Determine the relative roles for known antigen presenting cells in the induction of mucosal immunity, and define the most effective ways of expressing and presenting antigens for the mucosal immune system.

Passive Immunity. The dental caries model in rodents and primates has been instrumental in demonstrating the usefulness of passive antibodies in protection.

These studies should be extended to humans. The technology developed could also be extended to studies of the major pathogens implicated in periodontal diseases.

Recommendation:

> Give more emphasis to studies of the role of passive immunity, including the use of monoclonal antibodies, in the prevention and treatment of mucosal diseases in humans.

ANTIGEN DELIVERY SYSTEMS

Research needs were analyzed for the development and testing of antigen delivery systems that would be effective in inducing protective immune responses against various infectious diseases. Emphasis during the scientific presentations was especially given to the appropriateness of delivery systems for oral diseases, such as dental caries, periodontal diseases, herpes simplex virus infections, and human immunodeficiency virus infection. The members agreed that the delivery systems must be tailored to specific diseases since, in most cases, it is desirable to induce a mucosal response that would prevent infection by a specific microorganism.

More than 95 percent of all infectious diseases occur on or invade the body through the mucosal surfaces. The mucosal immune system is the major defense against these diseases at the mucosal surfaces. Because the mucosal immune system matures very early in life, the knowledge gained by stimulating mucosal responses with various antigen delivery systems will be beneficial in developing vaccines against infectious diseases occurring in childhood, including dental caries.

Evidence continues to accumulate on the importance of the mucosal immune system, especially secretory IgA antibodies, in protecting against a variety of viral and bacterial infections. However, the mechanisms involved in the induction of mucosal responses are not well understood. Studies have shown that the oral administration of an antigen will result in the induction of IgA antibodies in external secretions. Yet, the antigen, administered orally, must survive in the gastrointestinal tract and must be presented to the appropriate host cells to induce an immune response.

Approaches to Antigen Delivery

Innovative approaches of antigen delivery are being used for introducing microbial antigens to appropriate inductive sites of the mucosal immune system. However, a single "perfect delivery system" has not been developed. Clearly, it is necessary to continue research on a variety of delivery systems to establish the ability of each system to induce the desired response in a host.

Delivery systems that are being studied for use in the development of oral vaccines include live, avirulent recombinant bacterial (*Salmonella*, BCG, etc.) and viral (adenovirus, poliovirus, etc.) vectors, liposomes, and microcapsules; novel methods for delivery of peptides (e.g., peptomers, or crosslinked peptide polymers), conjugates, and antigenic subunits; and monoclonal antibodies for passive immune protection. All show much promise. However, modern, practical approaches involving these delivery systems are being developed and will require in-depth characterization. Further information on the safety and effectiveness of each system is needed, and especially comparative information on the safety and effectiveness of different systems when the same antigen(s) is tested. Immune responses in different animals or with different HLA haplotypes should be emphasized.

Recommendation:

> Continue studies on a variety of antigen delivery systems to establish their ability to induce a desired immune response. Characterize their mechanisms of action and evaluate their safety and effectiveness individually and comparatively.

Additional information on the use of molecular adjuvants as part of a delivery system also needs to be obtained. These adjuvants include conjugates of proteins or peptides with cholera toxin B subunit and muramyl dipeptide derivatives. The safety and effectiveness of each in potentiating specific responses, especially mucosal responses, still need to be established. This testing will be particularly important in light of the recent interest in using the B subunit of cholera toxin as an oral adjuvant.

Recommendation:

> Establish the safety and effectiveness of molecular adjuvants used in delivery systems.

The mechanisms by which the delivery systems and adjuvants can stimulate specific mucosal immune responses need to be clarified. Many approaches are being investigated using potentially protective antigens of *S. mutans*. These hallmark studies have established a solid foundation of valuable information that will be applicable to developing vaccine protection against dental caries as well as other infectious diseases.

Recommendation:

> Clarify the mechanisms by which delivery systems and adjuvants stimulate specific mucosal immune responses.

Understanding the Mucosal Immune System

Recent studies on the mucosal immune response suggest that there may be a type of compartmentalization within the common mucosal immune system. Studies with antigen delivery systems should be encouraged to help define this possibility and the means for effective presentation of the antigen to optimize the induction of an immune response at the site of microbial challenge.

Recommendation:

> Direct more research toward understanding the mucosal immune system and mechanisms by which induction of mucosal immune responses can be regulated with vaccine delivery systems.

Studies of the pathogenesis of a variety of microbial pathogens at early stages of infection are needed to elucidate processes that could be blocked by a secretory immune response.

Recommendation:

> Target studies on the pathogenesis of microbial pathogens to early stages of infection to elucidate processes that could be blocked by stimulating a secretory immune response.

Caries Vaccine

Studies aimed at the development of a caries vaccine have made significant contributions toward novel vaccine delivery systems and understanding the mechanisms important in inducing protective immune responses. This information has been useful for the development of safe and effective vaccines against a variety of infectious diseases affecting the mucosa, as well as dental caries. During the past decade, safe, practical, and economical antigen delivery systems have been developed.

Recommendation:

> Support human clinical trials to evaluate the safety, immunogenicity, and efficacy of the most promising caries vaccine delivery systems. Encourage further research on recombinant bivalent and/or polyvalent vaccines that also protect against life-threatening diseases.

Herpes Simplex Virus Vaccine

A variety of approaches are being pursued in the development of a vaccine for herpes simplex virus (HSV). These include recombinant live virus vectors (e.g., vaccinia, adenovirus, etc.), synthetic peptides, and subunit and recombinant glycoprotein vaccines. Several of these approaches are promising, and some are already being tested in clinical trials. The use of vaccinia virus recombinants that express HSV glycoproteins D or B appears very promising in animal models, and clinical trials are anticipated once data on the safety of other vaccinia recombinants (e.g., a vaccinia/human immunodeficiency virus) in humans become available. Further research on the development of vaccinia/HSV recombinants as vaccines against HSV should focus on improvements in efficacy and safety as a prelude to human studies.

Recommendation:

> Continue basic and clinical research on promising approaches to the development of vaccine delivery systems for herpes simplex virus, including the use of vaccinia/HSV recombinants.

Test Systems

Improved *in vitro* systems are needed to study the mechanisms regulating induction of mucosal immune responses. Standardized techniques are needed to quantitate these responses accurately.

Recommendation:

> Develop *in vitro* assay systems to study more effectively the mechanisms regulating induction of mucosal immune responses.

Recommendation:

Develop standardized techniques for accurately quantifying mucosal immune responses, including evaluation of T and B cell responses at the single-cell level.

Strategies for Enhancing Research

Workshop members agreed that the development of safe and effective vaccines will require substantial coordinated support from academia, industry, and government. Both scientific interchange and adequate funding will be needed to facilitate the development and clinical testing of these vaccines. Specific strategies for enhancing research were suggested.

1. Establish knowledgeable review groups for evaluating research applications and projects aimed at vaccine development. Members of these review groups should be knowledgeable in the pathogenesis of bacterial, viral, fungal, and parasitic diseases; mucosal and systemic immune systems; vaccine formulation and delivery systems; and vaccine testing strategies, including clinical trials.

2. Consider legislation to limit liability with regard to vaccine testing and use. Reconsider the "no fault" principle in order to promote private investment in vaccine development.

3. Enhance support for graduate and postgraduate research training in the life sciences with emphasis on studies of infectious diseases and mucosal immunity.

4. Provide support to international agencies (e.g., the World Health Organization; United Nations Education, Social, and Cultural Organization) which subsidize research and development of vaccines and which produce, distribute, and administer vaccines to control infectious diseases in the developing world.

5. Develop the means for cooperative (government, academia, industry), expedited testing and licensing of vaccines against infectious diseases.

6. Promote public education on the safety and effectiveness of vaccines.

TARGET ANTIGENS AND IMMUNOLOGICAL CORRELATES OF PROTECTION

Possibilities were evaluated for enhancing immunological protection against three major oral diseases: dental caries, periodontal diseases, and herpes simplex viral infection. The possibilities for immunologically protecting against the oral complications of infection with human papillomavirus and human immunodeficiency virus were also considered. During discussions, workshop participants reiterated that the successful development of safe and effective vaccines will require a consortium of industry, government, and academia: industry to offer marketing and production abilities, government to provide research funds, and academia to offer the flexibility and environment in which to nourish research.

Dental Caries

Dental caries is a significant disease in two American populations: children, who are initially infected, and elderly persons, who are more susceptible to caries resulting from recession of the gingiva and exposure of tooth root surfaces. Although not life-threatening, dental caries can have profound effects on one's quality of life and may be associated with other diseases, especially in individuals whose immune systems are compromised. Mostly limited to high-risk groups in the United States and other industrialized countries, dental caries is widespread in many developing countries.

In considering whether a vaccine against dental caries would be accepted in the United States, it was determined that the medical community would be amenable to participation in immunization programs and clinical trials. By combining the caries immunogen with other oral vaccines (e.g., polio), acceptance would be even easier to achieve. Further, it was agreed that recruitment of the mucosal immune response was the more desirable route of administration for a caries vaccine and that systemic administration by injection of a streptococcal vaccine is probably not desirable or achievable at the present time.

Although much has been achieved in the past 15 years of research on a caries vaccine, total mucosal suppression of cariogenic pathogens, such as streptococci, has not been achieved. Several approaches are promising, however, and deserve continued study. One of the most promising strategies is the use of synthetic peptides or recombinant-specific polypeptides based on the glycosyltransferase (GTF) enzyme. GTF is a unique virulence factor found in the mutans group of streptococci. However, because related genes that encode enzymes which are immunologically cross-reactive to GTF are found in certain less-cariogenic streptococci, further studies are needed to determine the effects, if any, of antibodies to GTF on the oral microflora of vaccinated individuals as compared to appropriate control populations.

A second approach is based on the development of synthetic peptides or recombinant-specific polypeptides using the knowledge gained from the cloning and sequencing of *S. mutans* spaP, *S. sobrinus* spaA, and related genes within the entire I/II superfamily of streptococci. For these gene products, the adhesion portion will need to be identified and the polypeptide should be tested for immunogenicity. The potential applicability of this approach is less than that of the GTF system, but both approaches could be combined for an enhanced benefit.

A third, less developed, approach involves the use of serotype-specific wall polysaccharides of *S. mutans*, which are currently considered nonimmunogenic. Attempts to produce immunogenic glycoconjugates based on this antigen should be encouraged.

An alternative method, passive immunization using monoclonal antibodies, may be useful as well and needs to be tested clinically. Passive immunization of monkeys with monoclonal antibodies to *S. mutans* surface antigen I/II has been shown to decrease both the colonization of teeth by *S. mutans* and the development of dental caries. Topical application of these monoclonal antibodies in humans was also found to prevent colonization by implanted exogenous strains, as well as indigenous strains, of *S. mutans*. However, because long-term prevention of colonization of *S. mutans* in humans was not consistent with the short-term application of monoclonal antibody or the rapid loss of the monoclonals from the teeth, further studies are needed to determine the mechanism(s) of action.

Recommendation:

>Conduct further human studies to evaluate the safety and efficacy of passive immunization with monoclonal antibodies to prevent dental caries. Also elucidate the mechanism(s) of action.

Passively administered antibody may also influence the accumulation of microorganisms in elderly individuals who are susceptible to root surface caries.

The potential for interfering with the development of dental caries at very early ages is good. The mucosal immune system appears well developed in very young children, and antibody to GTF or other mutans streptococcal antigens may be easily stimulated at this age. The means and effects of such stimulation must be studied, however. Clinical trials in children before they become infected with mutans streptococci also are needed.

Recommendation:

>Continue research on various approaches for intervening immunologically in the development of dental caries in children, the elderly, and high-risk populations. Evaluate the mechanisms and effects of these approaches.

Antigens (including purified antigens) of mutans streptococci have been shown to be safe in limited clinical trials, and they have demonstrated the ability to induce immune responses. Based on this evidence, support should be extended to clinical studies of such antigens in caries-prone individuals, especially in developing countries where dental caries may be rampant. Clinical trials to test the effectiveness of newly developed vaccines in inducing mucosal immune responses and in protecting against disease should include vaccines against dental caries and other oral diseases, such as periodontal diseases and herpes simplex infections, as well as a variety of other diseases.

Recommendation:

>Extend support to clinical studies of antigens of mutans streptococci in caries-prone individuals and of newly developed vaccines against dental caries and other diseases to evaluate their effect on inducing mucosal immune responses and protection from disease. Support collaborative international efforts aimed at worldwide control of dental caries, including testing of a caries vaccine.

Dental caries and periodontal diseases are infectious diseases caused by particular bacterial pathogens. Basic studies are needed to evaluate the molecular pathogenesis of the early stages of these infections and of the initial immune responses. These studies are the background against which safe and effective interventions can be developed.

Recommendation:

>Support basic research on the molecular pathogenesis of early stages of dental and periodontal infections and of initial immune responses to these infections as a basis for developing optimal interventions.

Periodontal Diseases

The role of the immune system in periodontal diseases is receiving considerable attention. A number of bacterial pathogens have been identified and research is being focused on them. The target organisms include *Actinobacillus actinomycetemcomitans*, *Porphyromonas gingivalis*, and *Bacteroides forsythus*. Target antigens include the lipopolysaccharide (LPS) and leukotoxin of *A. actinomycetemcomitans* and the hemolysins, fimbrillin, and protease of *P. gingivalis*. Identification of a target antigen of *B. forsythus* awaits further studies of this bacterium at the molecular level.

As with dental caries, several approaches to combatting these pathogens appear to be viable. These include the development of immunogenic recombinant-specific polypeptides or synthetic peptides representing the active sites of enzymes or toxins. For this, a multivalent humoral or mucosal vaccine (or both) appears to be desirable. A mucosal vaccine could prevent colonization of gingival tissue in disease-free individuals, and the humoral vaccine could prevent the spread of organisms in the gingival crevices of diseased patients. A well-defined animal model will need to be investigated, however, before any clinical trials in humans are initiated.

Passive immunization using monoclonal antibodies may be valuable in controlling periodontal diseases which could require a unique mix of antibodies for each periodontitis patient.

In all cases, great care must be taken to ensure that the correct genus-species and the correct isolates of the pathogen are chosen when formulating vaccines. One must also recognize that the periodontal diseases of today may change and that new pathogens may emerge, requiring different approaches in the future.

Recommendation:

> Support basic and clinical research on a variety of approaches for intervening immunologically in periodontal diseases. In tandem, pursue basic studies of the role of the mucosal immune system in periodontal disease and the use of this system in prevention of disease.

Important to this research is the development of animal models to evaluate completely the role of immune mechanisms in periodontal diseases. Numerous studies are needed and should be performed in systems that allow manipulation of the host's immune response. Potential animal models include rodents, primates, and dogs. Further evaluation of host immune responses in human periodontal diseases is especially important.

Recommendation:

> Develop well-defined animal models for investigating the role of the immune system in periodontal diseases and for testing vaccines developed against these diseases prior to their clinical use.

Herpes Simplex Virus

Herpes simplex virus (HSV) is a well-defined disease affecting many Americans. It is a health care problem with much visibility in the medical community. The antigens of HSV have been well studied, and the gB, gD, and gH glycoproteins have been cloned and sequenced. gD can be obtained in large quantities. Separate human trials are under way with two immunogenic agents: Chiron, using truncated gD2, and Lederle, using full-length gD1. Whether truncated molecules will be protective is not yet known, and immunodominant regions and other important domains still need to be determined. A vaccine for HSV could be available commercially within 5 years.

Recommendation:

Continue clinical trials of potential vaccines against herpes simplex virus to evaluate their safety and effectiveness.

Papillomaviruses

Human papillomaviruses (HPV) are found in many oral benign and malignant diseases, including oral cancer which kills more than 9,000 Americans each year, hairy leukoplakia, and laryngeal papillomatosis. These viruses can be detected in the oral cavity of up to 50 percent of Americans, and they appear to be a cofactor in the development of cervical cancer. Their specific role in the pathogenesis of oral disease is unclear however. Further research is needed to determine the epidemiology, pathogenesis, and treatment of diseases caused by HPV.

Very little is known about the immune response to papillomaviruses since these viruses cannot be grown in tissue culture systems and antigens cannot be prepared by standard means. Recently, using gene transfer techniques, investigators have been able to inhibit papillomaviruses in oral cancer cells by the expression of antisense RNA to the E6 and E7 genes of papillomaviruses. The optimal means of delivering these constructs to oral epithelial cells needs to be developed.

Vaccine development against HPV is in its infancy, but may prove to be important in the eventual control of HPV infection. Further progress is expected in this area given the recent cloning and expression of the viral genome.

Recommendation:

Support basic research on the role of papillomaviruses in the pathogenesis of oral diseases. Using advanced genetic techniques, pursue the possibility of developing a vaccine in the future for human papillomaviruses.

Human Immunodeficiency Virus

Human immunodeficiency virus (HIV) infection often results in changes in the oral cavity, including an increased incidence of opportunistic infections such as candida, mucositis, and hairy leukoplakia. The continuing spread of AIDS resulting from HIV infection demands the development of better therapies and a vaccine for HIV.

However, appropriate target antigens for anti-HIV immunization were not identified because of the genetic instability of the virus. Further research on the antigenic structure of HIV is required before candidate antigens become obvious. In one study, vaccines using the major virus antigens have shown promising results in slowing the decline in numbers of CD4 cells in patients with HIV infection, and additional follow up studies should be pursued vigorously.

Further research is needed on the pathogenesis of HIV infection in the oral cavity. Specific areas of research that are already under way and should be continued are: investigation of the mucosal immune response to HIV and other viruses to determine if infected patients secrete protective antibodies; examination of the hypothesis that an autoimmune response is responsible for some of the effects of HIV; exploration of the possible prevention of HIV infection by a "salivary inhibitory substance"; and study of the role of opportunistic infections in HIV, such as *Candida albicans*, HSV, HPV, Epstein-Barr virus, and cytomegalovirus.

Recommendation:

Vigorously support research to develop a vaccine against HIV, and continue studies of the pathogenesis of HIV infection in the oral cavity.

SPEAKERS AND MODERATORS

Anthony C. Allison, Ph.D., B.M.
Vice President for Research
Syntex Research S3-5
3401 Hillview Avenue
Palo Alto, California 94304

Arnold S. Bleiweis, Ph.D.
Professor and Chairman
Department of Oral Biology
College of Dentistry
University of Florida
Box J424 JHMHC
Gainesville, Florida 32610-0446

Joseph E. Ciardi, Ph.D.
Director
Caries, Nutrition and Fluoride Program
Extramural Program
National Institute of Dental Research
National Institutes of Health
Westwood Building, Room 509
Bethesda, Maryland 20892

Gary H. Cohen, Ph.D.
Professor and Chairman
Department of Microbiology
School of Dental Medicine
University of Pennsylvania
4001 Spruce Street
Philadelphia, Pennsylvania 19104

Roy Curtiss III, Ph.D.
Professor and Chairman
Department of Biology
College of Arts and Sciences
Washington University
Campus Box 1137
St. Louis, Missouri 63130

Ronald W. Ellis, Ph.D.
Executive Director
Cellular and Molecular Biology
Merck, Sharp and Dohme
    Research Laboratories
West Point, Pennsylvania 19486

Richard T. Evans, Ph.D.
Associate Professor of Oral Biology
School of Dental Medicine and
    Associate Professor of Microbiology
School of Medicine
State University of New York at Buffalo
Buffalo, New York 14214

Robert J. Genco, D.D.S., Ph.D.
Professor of Oral Biology & Periodontics
Department of Oral Biology
School of Dentistry
State University of New York at Buffalo
Foster Hall
Buffalo, New York 14214

Shigeyuki Hamada, D.D.S., Ph.D.
Professor and Chairman
Department of Oral Microbiology
Osaka University Faculty of Dentistry
1-8 Yamadaoka, Suita Osaka 565 Japan

Albert Z. Kapikian, M.D.
Chief, Epidemiology Section
Laboratory of Infectious Diseases
National Institute of Allergy
    and Infectious Diseases
National Institutes of Health
Building 7, Room 103
Bethesda, Maryland 20892

Jerry M. Keith, Ph.D.
Chief, Laboratory of Microbial Ecology
National Institute of Dental Research
National Institutes of Health
Building 30, Room 316
Bethesda, Maryland 20892

Mogens Kilian, D.D.S., Ph.D.
Professor of Medical Microbiology
    and Chair of Bacteriology
Institute of Medical Microbiology
The Bartholin Building
University of Aarhus
DK-8000 Aarhus C, Denmark

Hiroshi Kiyono, D.D.S., Ph.D.
Professor of Dentistry
Department of Oral Biology
    and Preventive Dentistry
School of Dentistry
University of Alabama at Birmingham
Birmingham, Alabama 35294

Michael E. Lamm, M.D.
Professor and Chairman
Department of Pathology
School of Medicine
Case Western Reserve University
2085 Adelbert Road
Cleveland, Ohio 44106

John R. LaMontagne, Ph.D.
Director
Division of Microbiology and
    Infectious Diseases
National Institute of Allergy and
    Infectious Diseases
Solar Building, Room 3A18
Bethesda, Maryland 20892

Thomas Lehner, M.D., B.D.S.
Professor and Chairman
Division of Immunology
United Medical and Dental Schools
Guy's and St. Thomas' Hospital
    and Medical School
London Bridge
London SE1 9RT England

Harald Löe, D.D.S.
Director
National Institute of Dental Research
National Institutes of Health
Building 31, Room 2C39
Bethesda, Maryland 20892

Francis L. Macrina, Ph.D.
Professor and Chairman
Department of Microbiology
    and Immunology
School of Basic Health Sciences
Virginia Commonwealth University
Box 678, MCV Station
Richmond, Virginia 23298-0678

Jerry R. McGhee, Ph.D.
Professor, Department of Microbiology
Director, Immunobiology Vaccine Center
Medical/Dental Basic Health Sciences
University of Alabama at Birmingham
Birmingham, Alabama 35294

John J. Mekalanos, Ph.D.
Professor, Department of Microbiology
    and Molecular Genetics
Harvard Medical School
200 Longwood Ave.
Boston, Massachusetts 02115

Jiri Mestecky, M.D.
Professor of Microbiology, Medicine
    and Oral Biology
Department of Microbiology
School of Medicine
University of Alabama at Birmingham
Birmingham, Alabama 35294

Suzanne M. Michalek, Ph.D.
Professor, Department of Microbiology
School of Dentistry
University of Alabama at Birmingham
Birmingham, Alabama 35294

Brian R. Murphy, M.D.
Chief, Respiratory Viruses Section
Laboratory of Infectious Diseases
National Institute of Allergy
    and Infectious Diseases
National Institutes of Health
Building 7, Room 106
Bethesda, Maryland 20892

Marian R. Neutra, Ph.D.
Professor of Pediatrics
Director, GI Cell Biology Laboratory
Children's Hospital
    and Harvard Medical School
300 Longwood Avenue
Boston, Massachusetts 02115

Pearay L. Ogra, M.D.
Professor and Chairman
Department of Pediatrics
Medical Branch
University of Texas
C-51
Galveston, Texas 77550

Roy C. Page, D.D.S., Ph.D.
Professor and Director
Research Center in Oral Biology
Office of Research SM-42
School of Dentistry
University of Washington
Seattle, Washington 98195

Robert Redfield, M.D.
Chief, Department of Retroviral Research
Walter Reed Army Institute of Research
13 Taft Court
Rockville, Maryland 20850

Frank A. Robey, Ph.D.
Research Chemist
Laboratory of Cellular Development
    and Oncology
National Institute of Dental Research
National Institutes of Health
Building 30, Room B17
Bethesda, Maryland 20892

James F. Rooney, M.D.
Laboratory of Oral Medicine
National Institute of Dental Research
National Institutes of Health
Building 30, Room 230
Bethesda, Maryland 20892

Michael W. Russell, Ph.D.
Research Associate Professor
Department of Microbiology
    and Oral Biology
School of Medicine and Dentistry
University of Alabama at Birmingham
Birmingham, Alabama 35294

Roberta D. Shahin, Ph.D.
Senior Staff Fellow
Division of Bacterial Products
Center for Biologics Evaluation
    and Research
Food and Drug Administration
Building 29, Room 414
Bethesda, Maryland 20892

Edward J. Shillitoe, Ph.D.
Professor and Chairman
Department of Microbiology
University of Texas Dental Branch
P.O. Box 20068
Houston, Texas 77225

Charles Kendall Stover, Ph.D.
Senior Scientist
Molecular Genetics Department
Medimmune Inc.
19 Firstfield Road
Gaithersburg, Maryland 20878

Martin A. Taubman, D.D.S., Ph.D.
Head, Department of Immunology
Forsyth Dental Center
140 Fenway
Boston, Massachusetts 02115

AUTHOR INDEX

SUBJECT INDEX